T0348772

Nano and Giga Challenges
in Microelectronics

Nano and Giga Challenges
in Microelectronics

Edited by:

Jim Greer
NMRC, Cork, Ireland

Anatoli Korkin
Motorola, Inc., Phoenix, Arizona, USA

Jan Labanowski
Ohio Supercomputing Centre, Columbus, Ohio, USA

ELSEVIER

2003

Amsterdam - Boston - London – New York – Oxford – Paris
San Diego – San Francisco – Singapore – Sydney - Tokyo

ELSEVIER B.V.
Radarweg 29
P.O. Box 211, 1000 AE Amsterdam
The Netherlands

First edition 2003

Library of Congress Cataloging in Publication Data
A catalog record from the Library of Congress has been applied for.

British Library Cataloguing in Publication Data
A catalogue record from the British Library has been applied for.

ISBN: 0 444 51494 5

Transferred to digital print 2007

Printed and bound by CPI Antony Rowe, Eastbourne

Preface

Microelectronics technologies have reached a new stage in their development: the ultimate miniaturization of transistors is in sight us gate lengths approach atomic dimensions, the interconnect bottleneck is posed to limit circuit speeds, new material sets are being introduced into microelectronic manufacture at an unprecedented rate, and alternative technologies to mainstream CMOS are seriously being considered in many cases off financial, not performance, grounds.

It is against this backdrop of technology change that the series Nano and Giga Challenges in Microelectronics is being launched. Spawned by a meeting of the same name held in Moscow in 2002, the series is intended to provide tutorial and expository articles on advanced technology problems related to micro- and nano- technology development. The articles capture the flavor and excitement of the Nano and Giga meeting (future meetings are planned, with the next to be held in St. Petersburg in 2004) with the articles solicited from leading researchers in representative subject areas. Each chapter is intended as a self-contained introduction to an advanced research topic ranging from microelectronics materials to molecular electronics, and extending through to nanoelectronic circuit architectures.

The book's intention is to act as an introduction for engineers and researchers wishing to obtain a fundamental knowledge and a snapshot in time of the cutting edge in technology research. As a natural consequence, the Nano and Giga Challenges is also a useful reference also for the "gurus" wishing to keep abreast of the latest directions and challenges in microelectronic technology development and future trends. The combination of viewpoints presented within the book can help to foster further research and cross-disciplinary interaction needed to surmount the barriers facing future generations of technology design. This book also serves as a companion to the special issue of the journal Microelectronics Engineering, which documents the technical papers presented during the Nano and Giga 2002 meeting.

It is a great pleasure for the Editors of Nano and Giga Challenges to present these collected chapters, and we thank each of the authors of this volume for sharing their insights and expertise and the sponsors of NGCM2002 for their gracious support: Digital DNA Lab Motorola, Russian Research Center Kurchatov Institute, US Department of Energy - Nuclear Cities Initiative, Nunn & Turner Foundation - Nuclear Threat Initiative, Moscow State University, International Science & Technology Center, Elsevier Science, European Office of Aerospace Research and Development United States Air Force, Russian Federal Nuclear Center (VNIIEF), Russian Foundation for Basic Research, US Office of Naval Research, KINTECH Kinetic Technologies, and Ohio Supercomputer Center are gratefully acknowledged.

Jim Greer   Anatoli Korkin   Jan Labanowski

NMRC   Motorola   OSC

Cork, Ireland   Phoenix, Arizona   Columbus, Ohio

May, 2003

Contents

Nano and Giga Challenges in Microelectronics
Greer at al (Editors)

1

Integrated Circuit Technologies: From Conventional CMOS To The Nanoscale Era [*]

Peter M. Zeitzoff [a,1], James A. Hutchby [b], Gennadi Bersuker [a], and Howard R. Huff [a].

[a]*International SEMATECH Inc., 2706 Montopolis Drive, Austin, TX 78741.*
[b]*Semiconductor Research Corp., P.O. Box 12053, Durham, NC 27709.*

Abstract

The development of advanced MOSFETs for future integrated circuit technology generations is discussed from the perspective of the 2001 International Technology Roadmap for Semiconductors (ITRS). Starting from overall chip circuit requirements, MOSFET and front-end process integration technology requirements and scaling trends are discussed, as well as some of the key challenges and potential solutions. These include the use of high-k gate dielectrics, metal gate electrodes, and perhaps the use of non-classical devices such as multiple-gate MOSFETs in the later stages of the ITRS.

Key words: MOSFET scaling, high-k gate dielectric, non-classical CMOS, FinFET, strained Si on SiGe: enhanced mobility channel.

1. Introduction

For over thirty-five years, the integrated circuit (IC) industry has rapidly and consistently scaled (reduced) the design rules, increased the chip and wafer size, and improved the design of devices and circuits [1], [2]. In doing so, the industry has been following the well-known Moore's Law [1], [3], [4], which in its simplest form states that the number of functions per chip is doubled while the cost

per function is halved every one-and-a half to two years. As a result of following Moore's Law, chip speed and functional density have increased exponentially with time while average power dissipation per function and cost per function have decreased exponentially with time [5]. In particular, the number of memory bits per chip has quadrupled every three to four years, while the speed of microprocessors has more than doubled every three years, based on the increase from about 2 MHz for the Intel® 8080 in the mid-1970's to well over 1 GHz for current leading-edge chips [6], [7]. The design rules have been scaled from about 8 μm in 1972 to the 130 nm (0.13 μm) dynamic random access memory (DRAM) half pitch of 2001's leading-edge technologies. (This half pitch is defined as half the minimum metal or polysilicon pitch for DRAMs,

---

[*] It is an expanded version of a paper, *MOSFET and Front-End Process Integration: Scaling Trends, Challenges, and Potential Solutions Through The End Of The Roadmap*, by Peter M. Zeitzoff, James A. Hutchby, and Howard R. Huff *International Journal of High-Speed Electronics and Systems*, vol. 12, pp. 267–293 (2002), Word Scientific Publishing Co. Pte Ltd, Singapore.

[1] peter.zeitzoff@sematech.org

2

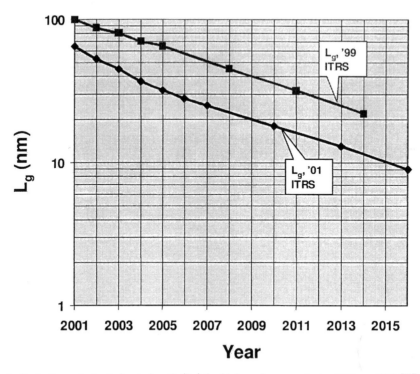

Fig. 1. Projections of physical gate length (Lg) for high-performance logic: 2001 vs. 1999 ITRS[2,9].

and since this is the finest feature size for densely packed structures, it is a key parameter characterizing the technology.) The overall rate of scaling is about 0.87 per year, or $(1/\sqrt{2}) \sim 0.7$ in a time interval of between two and three years, where the $(1/\sqrt{2})$ factor is the traditional reduction in design rules between successive technology generations. Owing to this design rule scaling, the size of logic gates, memory cells, etc., is reduced by a factor of one-half between successive generations, and this reduction is a major component, along with the chip size increase and the circuit design improvements mentioned above, of the Moore's Law doubling in the number of functions per chip every one-and-a half to two years. However, because certain key material, process, and MOSFET limits are being approached, there are increasing difficulties in continuing to scale the design rules at this rate. This paper is focused on the scaling of the design rules, along with the consequences, difficulties, and potential solutions associated with this scaling.

In the International Technology Roadmap for Semiconductors (ITRS) [8], the progression of IC technology is projected over the next 15 years. The goal is to aid the IC industry and its supporting infrastructure (especially the material and equipment suppliers and university and research laboratory researchers) in continuing to follow Moore's Law during that period. Since it presents a consensus view from many experts across the IC industry, the ITRS is influential and has credibility in the industry, as evidenced by its being widely cited. One key change in the most recent ITRS is a major acceleration in the scaling of the physical gate length (L_g) of the MOSFET transistors, where L_g is the final, etched length at the bottom of the gate electrode. The L_g projections for high-performance transistors from both the 1999 [9] and 2001 [2] ITRS are plotted in Figure 1. For any given year, the 2001 projection is reduced from the 1999 projection by an amount ranging from about 35% to 50%. Alternatively, the 2001 projections lead the 1999 projections by four to six years. This accelerated scaling is important because reduction

Table 1
Key parameters for high-performance logic, data from 2001 ITRS[2].

Year		Near-term years							Long-term years		
		2001	2002	2003	2004	2005	2006	2007	2010	2013	2016
DRAM Half Pitch	nm	130	115	100	90	80	70	65	45	32	22
MPU Physical Gate Length, L_g	nm	65	53	45	37	32	28	25	18	13	9
V_{dd}	V	1.2	1.2	1.1	1	0.9	0.9	0.8	0.6	0.5	0.4
Maximum on-chip local clock frequency	GHz	1.7	2.3	3.1	4.0	5.2	5.6	6.7	11.5	19.3	28.8
Allowable maximum power dissipation, with heatsink	W	130	140	150	160	170	180	190	218	215	288
Number of transistors per chip	Millions of transistors	276	348	439	553	697	878	1106	2212	4424	8848

of L_g is a critical factor enabling increased MOS-FET speed. The accelerated L_g scaling is thus an enabler of the IC industry's drive to maximize the chip speed, and it reflects actual trends in the industry since 1999, which are projected to continue.

The overall chip requirements for speed, power dissipation, functional density, cost, etc., drive the ITRS projections for MOSFETs. For logic chips there are two main application categories: high-performance and low-power logic. Since the overall chip requirements for each of these application areas are distinct, the scaling goals are different for each of them. High-performance logic is mainly used for high-end desktop and server applications (such as for microprocessor unit (MPU) chips), where the main need is to maximize chip speed. Consequently, the key MOSFET design goal is to maximize transistor speed, and the tradeoff is relatively high MOSFET leakage current, for reasons to be explained later. In contrast, low-power logic is used mainly in mobile systems, where the over-riding goal is to preserve battery life by minimizing the chip power dissipation, particularly the static power dissipation. Hence, the device design is aimed at minimizing the MOSFET leakage current, and the tradeoff is reduced MOSFET speed compared to high-performance logic MOSFETs, for reasons to be explained later. Owing to the rapid scaling projected in the 2001 ITRS, the IC industry faces a number of difficult MOSFET scaling challenges in the foreseeable future. These include excessive gate leakage current and boron penetration from the P+ polysilicon gate electrode as the gate dielectric thickness is scaled, as well

as an increasing impact of polysilicon depletion in the gate electrode. Also, with MOSFET scaling, it will become increasingly difficult to simultaneously achieve shallow junction depth, low sheet resistance, low contact resistance, and low junction leakage, as is required for the source/drain (S/D) extension and contact regions. In the long term, improved electron and hole mobility in the inversion layer may be required to meet the transistor performance goals. Furthermore, the classical planar bulk MOSFET may eventually be inadequate, especially as a result of excessive short-channel and quantum effects as well as excessive statistical variability in dopant placement for very small transistors [10]. As a result, non-classical MOSFETs, such as ultra-thin body double-gate devices, may be used.

The discussion in this paper is limited to logic technology, since many of the key technology issues are driven by the logic technology. For both high-performance and low-power logic technologies, the 2001 ITRS scaling approach and results are presented, as well as key front-end process challenges and potential solutions. Finally, possible solutions for mobility improvement and the use of non-classical transistors for very small MOSFETs are discussed.

2. Circuit and MOSFET Scaling Approach and Results

Table 1 gives projections from the 2001 ITRS [2] of several key parameters for high-performance logic. The overall chip circuit requirements are in the last three rows. As is typical of the ITRS, the projections are given for each of the near-term years, from 2001 through 2007, while for the long-term years beyond 2007, projections are given only for every third year. As explained in the Introduction, the DRAM half pitch is a key parameter characterizing the leading-edge technology, and this half pitch is projected to be reduced by a factor of $(1/\sqrt{2}) \sim 0.7$ every three years, consistent with the long-term trends noted in the Introduction. As a result, the technology generations are the 130 nm generation in 2001, the 90 nm generation in 2004, the 65 nm generation in 2007, etc. To maximize the MOSFET speed, the L_g is projected to be as small as possible, and in all cases is half or less than half the DRAM half pitch. Defining such small L_g features is difficult, requiring special lithographic and etching techniques [11]. The local clock frequency and the number of transistors per chip are both projected to increase rapidly, becoming almost 29 GHz and 9 billion transistors per chip, respectively, at the end of the roadmap in 2016. In contrast, the maximum allowable power dissipation per chip is projected to increase slowly because it is limited by system level cooling, test constraints, and the ability of the package to dissipate heat.

The chip power dissipation and clock frequency scaling requirements in Table 1 drive the scaling of key MOSFET parameters for high-performance logic. The total chip power dissipation, P, is the sum of the dynamic power dissipation, P_d, and the static power dissipation, P_{st}. The former is:

$$P_d = C_a V_{dd}^2 f_c \qquad (1)$$

where C_a is the total chip active capacitance (i.e., the total capacitance being switched during an average clock cycle), V_{dd} is the power supply voltage, and f_c is the chip clock frequency. For succeeding technology generations, C_a and f_c increase sharply, so V_{dd} is reduced to keep P_d within tolerable limits. This is particularly critical for high-performance logic, where P_d is typically considerably larger than P_{st} [12]. Reducing V_{dd} with scaling is also beneficial for controlling short channel effects, improving reliability, and for general device scaling. In contrast to P_d, P_{st} is dependent on the transistor leakage current:

$$P_{st} = N_{\text{off}} W I_{\text{off}} V_{dd} \qquad (2)$$

where N_{off} is the average number of transistors that do not switch during a clock cycle (N_{off} is generally a substantial fraction of the total number of transistors per chip listed in Table 1), W is the average transistor width, and I_{off} is the total transistor leakage current (in units of amperes per micron of transistor width). Since N_{off} generally increases with scaling, while W and V_{dd} generally decrease, it is difficult to generalize about the behavior of the product, ($N_{\text{off}} W V_{dd}$), with scaling. Nevertheless, any increase of I_{off} with scaling is constrained by the need to keep P_{st} within tolerable limits.

The 2001 ITRS utilizes the transistor intrinsic switching frequency, f_i, as the figure of merit for transistor speed. f_i is defined as the reciprocal of the transistor intrinsic delay, τ_i:

$$\tau_i = 1/f_i = (C_g V_{dd})/I_{\text{on}} \qquad (3)$$

where C_g is the gate capacitance per micron of transistor width for a MOSFET of gate length L_g, and I_{on} is the transistor saturation drive current (in units of amperes per micron of transistor width). τ_i can be understood as approximately the time required for the MOSFET to charge or discharge the gate of an identical MOSFET through a potential difference of V_{dd} (in this approximation, the S/D junction capacitance of the driving MOSFET is ignored). Alternatively, utilizing the explanation in Taur and Ning [13], τ_i is approximately the switching delay of an inverter for a capacitive load (C_L) equal to the gate capacitance of one transistor, (i.e., $C_L = W C_g$, units are farads). (Note that, for real gates such as ring oscillators, the measured delay [τ_{RO}] is several times larger than τ_i, because the load capacitance of a ring oscillator [$C_{L,RO}$] is several times larger than C_L, and $\tau_{RO}/\tau_I \sim C_{L,RO}/C_L$.) As seen from Eq. (3), f_i is dependent only on the transistor parameters and characteristics, and to maximize f_i, I_{on} must be maximized. f_i is a good figure of merit for the transistor intrinsic speed in dense logic circuits, where the interconnect wire lengths are short and load capacitance is dominated by the gate capacitance of the transistors in the fan out path. In contrast, for logic circuits with long interconnect wires, load capacitance is dominated by the wiring capacitance.

As a result, the switching frequency is a function of both the driving transistor(s) layout and characteristics and the interconnect layout and characteristics. The interconnect dominated case is beyond the scope of this paper, but transistors optimized to maximize f_i for dense logic should also be reasonably optimal for interconnect dominated logic.

In the 2001 ITRS, the following approach was used to scale the MOSFETs. Simplified models were developed to capture the essentials of the impact of such key MOSFET parameters as V_{dd}, gate dielectric equivalent oxide thickness *(EOT)*, physical gate length (L_g), etc., on the important transistor electrical output characteristics such as I_{off}, I_{on}, etc. These models were then embedded into a commercial spreadsheet. Key output characteristics calculated by the spreadsheet included the MOSFET intrinsic delay, $\tau_i = (C_g V_{dd})/I_{\text{on}}$, and $f_i = 1/\tau_i$. In the expression, C_g includes the ideal gate capacitance per micron of width (defined as $[\varepsilon_{ox} L_g/EOT]$, where ε_{ox} is the dielectric constant of silicon dioxide) plus both parasitic gate overlap and fringing capacitance per micron transistor width. To determine the projected MOSFET parameter values for the tables, a target was set for either the MOSFET leakage current or f_i. Then an initial set of MOSFET parameters was chosen based on scaling rules, engineering judgment, and physical device principles. Using the spreadsheet, the input parameters were iteratively varied until the target was met, and the final values of the parameters were then used for the projected MOSFET parameter values in the tables.

For high-performance logic, an average 17% per year rate of increase in f_i was targeted, which matches the historic rate of device performance improvement. The iterative approach described above was used. From the previous discussion, since I_{on} must stay relatively high to maximize f_i, the projected NMOSFET I_{on} stays constant at 900 μA/μm until the 65 nm technology node in 2007, and then it increases somewhat in the later years. (The PMOSFET I_{on} is 40 – 50% of the NMOSFET I_{on} because of the lower mobility of holes.) V_{dd} is decreasing with device scaling, both for control of power dissipation and for reliability, as discussed above. Since I_{on} is strongly dependent on the overdrive, $(V_{dd}-V_t)$, where V_t is the effective threshold voltage, V_t must be reduced along with V_{dd} to keep I_{on} at the specified values. However, $I_{sd,leak}$ increases exponentially as V_t decreases [14], where $I_{sd,leak}$ is the source-drain subthreshold leakage current of the MOSFET, and is a component of I_{off}. The scaling of V_{dd} and V_t is shown in Figure 2. Both V_{dd} and V_t scale with succeeding years, and V_t becomes very close to zero for the long-term years, beyond 2007. The overall scaling results for f_i and $I_{sd,leak}$ are shown in Figure 3. f_i increases at the targeted rate of ~17% per year, while $I_{sd,leak}$ increases rapidly and becomes particularly large (at greater than 1μA/μm) for the long-term years beyond 2007. As a result of the increasing $I_{sd,leak}$, the static power dissipation per device increases with device scaling despite the reduced V_{dd} and device dimensions with scaling. This increased static power dissipation has a major impact, which will be discussed later, on the approaches to controlling chip power dissipation. (An important point is that, for the parameters that are discussed in this section, such as I_{on} and $I_{sd,leak}$, the ITRS projects nominal values to characterize the technology. Worst-case values, which are of great practical importance for any IC technology, are generally specific to the details of the MOSFET design, the process flow, and the process control techniques and capabilities, and even reflect the product goals, and hence are left to the individual companies to determine.)

Finally, $I_{sd,leak}$ is only one of three components of the total transistor leakage current, I_{off}. The other components are the gate leakage current and the S/D junction leakage current, including band-to-band tunneling. A reasonable restriction imposed in the 2001 ITRS is that each of these other components of I_{off} must be less than $I_{sd,leak}$.

For high-performance chips, the rapid increase in $I_{sd,leak}$ (and hence in I_{off}) with scaling must be dealt with to keep the chip static power dissipation within tolerable limits. This is illustrated graphically in Figure 4, where the relative total allowable chip power dissipation from Table 1 is plotted, as well as the relative total chip dynamic and static power dissipation with scaling. The dynamic and static dissipation were calculated using Equations (1) and (2), respectively, with the parameters

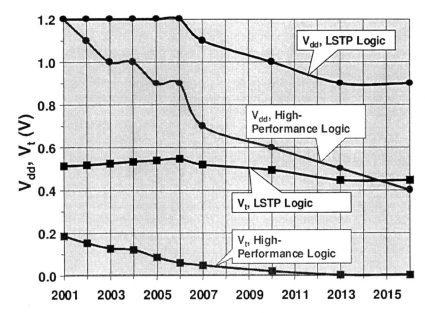

Fig. 2. Scaling of V_{dd} and V_t for high-performance and low standby power (LSTP) logic, data from 2001 ITRS.

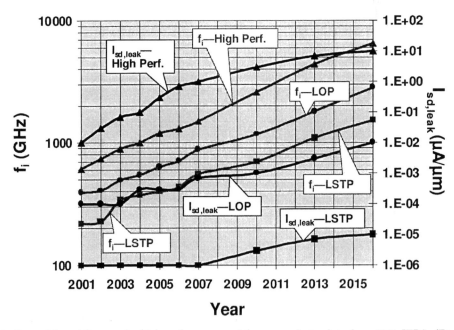

Fig. 3. Projections of f_i and $I_{sd,leak}$ for high-performance and low-power logic, data from 2001 ITRS. (Figure us from Zeitzoff et al.[7], ©2002 IEEE)

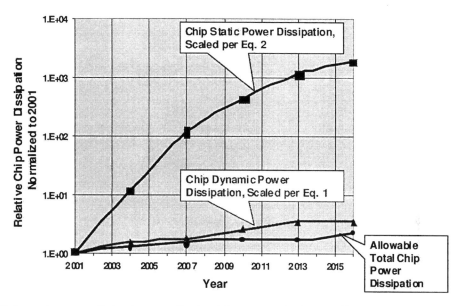

Fig. 4. Relative chip power dissipation normalized to 2001 versus calendar year. Data from 2001 ITRS is utilized.

scaled per the ITRS. In this figure, an unrealistic assumption is made to illustrate the impact of the high $I_{sd,leak}$ values noted above: it is assumed that all transistors on the chip are the high-speed, low V_t, high leakage MOSFETs discussed in the previous paragraph. The dynamic power dissipation scales up at about the same rate as the total allowable chip power dissipation, with an increase of about three times by 2016, which is acceptable. However, the static power dissipation scales up much more rapidly, with an increase of over three orders of magnitude by 2016. Clearly, this rate of increase of the static power dissipation is so large that, unless measures are taken to reduce it, the total allowable chip power dissipation limit will be exceeded within several technology nodes. An approach that is becoming common is to fabricate multiple transistor types on the chip in addition to the high-speed, low V_t, high leakage device described above. The other MOSFET(s) have higher V_t and larger EOT to reduce I_{off} [15], [16], [17], but they also have lower I_{on} and lower f_i than the high-speed devices. The high-speed MOSFETs are used only in critical paths or in circuits which are constantly switching, while the lower leakage devices are used everywhere else. This approach can significantly reduce the chip static power dissipa-

tion without seriously degrading chip performance, while also increasing the chip flexibility for diverse, system-on-a-chip (SoC) functions. There are other techniques utilized to curtail static power dissipation, and additional ones are being investigated for future utilization. These include well-biasing, use of electrically or dynamically adjustable V_t devices, and circuit/architectural techniques such as pass gates to cut off access to power/ground rails to power down circuit blocks [18]. Hence, a realistic picture of scaled high-performance ICs is that the static power dissipation will be controlled by utilizing the high-speed, high leakage MOSFETs sparingly, only when needed for performance, and by utilizing sophisticated device and IC design and architectural techniques for power conditioning.

For low-power logic, which is focused on mobile applications, the ITRS targets the transistor leakage current to ensure reasonable battery life for mobile applications. Two types of low-power logic are considered: low operating power *(LOP)* and low standby power *(LSTP)*. *LOP* chips are used for higher performance mobile applications such as notebook computers, where the battery capacity is relatively large. In contrast, *LSTP* chips are used for consumer type applications such as cellular telephones, where the performance requirements and

8

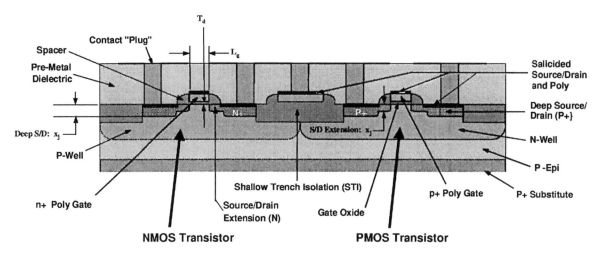

(Not to scale)

Fig. 5. Simplified cross section of a typical PMOSFET and NMPSFET. (Figure is from Zeitzoff et al[7], ©2002 IEEE.)

the battery capacity are both lower than for LOP chips. The leakage current targets are thus most stringent for $LSTP$. In the 2001 ITRS, the LOP logic target for $I_{sd,leak}$ (and using the same reasoning as with the high-performance logic, the maximum allowed gate leakage or S/D junction leakage current) starts at $100\,\mathrm{pA}/\mu m$ in 2001 and rises steadily in succeeding years. For $LSTP$ logic, the $I_{sd,leak}$ target starts at $1\,\mathrm{pA}/\mu m$ in 2001 and stays at that level until 2007, after which it rises slowly. L_g and EOT scale down with succeeding years, but the actual values of L_g lag behind those used in high- performance logic by two years (i.e., L_g is 65 nm for high performance logic in 2001, but it is 65 nm for LOP and $LSTP$ logic in 2003). EOT scaling also lags behind that for high-performance logic. Figure 3 shows the results of using the iterative approach described previously to meet the $I_{sd,leak}$ targets. The average rate of increase in f_i of about 14% per year for both LOP and $LSTP$ logic is smaller than the 17% per year rate obtained with high-performance logic. As expected, $I_{sd,leak}$ and f_i are lowest for $LSTP$, intermediate for LOP, and highest for high-performance logic, with $I_{sd,leak}$ for high-performance logic ranging from four to six orders of magnitude larger than for $LSTP$. Figure 2 shows the scaling of V_{dd} and V_t for $LSTP$ logic,

as well as for high–performance logic. One key issue is the slow scaling of V_{dd} for the $LSTP$ transistors. This is a result of the slow scaling of V_t to meet the low and relatively constant targets for $I_{sd,leak}$. V_{dd} follows V_t in scaling slowly for two reasons: first, for reasonable device performance I_{on} must be sustained, and hence the overdrive, $(V_{dd}-V_t)$, must remain relatively large. Secondly, for adequate circuit switching noise margins, V_{dd} must be larger than at least $2 \times V_t$. (The curves of V_{dd} and V_t versus year for LOP logic are not shown in the figure, but they are similar to those for $LSTP$, with slightly smaller values.

For LOP and especially for LSTP logic, another important issue is caused by the sharp increase in the lateral electric field, $E_{lat} \sim V_{dd}/L_g$, with device scaling. This increase will result in difficulty in controlling short channel effects and possibly in significant reliability problems in the long-term years, particularly for $LSTP$ logic. A final critical issue is with gate leakage current. As will be discussed in more detail later, for $LSTP$ logic, given the EOT and V_{dd} in 2005, the gate leakage current target cannot be met using oxynitride because of direct tunneling. Hence, it is projected that high-k gate dielectric will be initially required for $LSTP$ ICs in 2005.

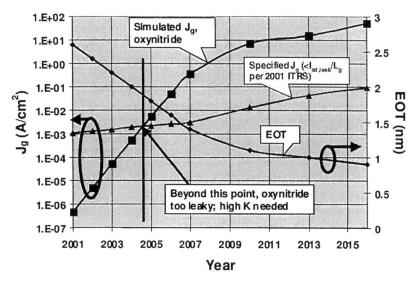

Fig. 6. 2001 ITRS projections versus si,ulations of gate leakage current density for low standby power (LSTP) logic. (Figure is from Zeitzoff et al[7], ©2002, IEEE.)

3. Planar Bulk CMOS: Front-End Process Challenges and Potential Solutions

As mentioned above, the 2001 ITRS projects that all the MOSFET dimensions are scaling down rapidly with succeeding years. In the last section, key issues related to overall circuit and MOSFET scaling, including performance and power dissipation, were discussed under the assumption that scaled transistors with acceptable characteristics (including reliability, statistical control of key characteristics, electrical performance, etc.) can be successfully fabricated. However, serious material and process limits are being approached that will make it difficult to fabricate highly scaled planar bulk transistors with all the desired properties, and these limits and potential material and process approaches to dealing with them form the subject of this section.

A simplified cross section of a PMOS and an NMOS transistor typical of current leading-edge technology is shown in Figure 5, where L_g, gate dielectric physical thickness, T_d, and the junction depths, x_j, of the deep source/drain (S/D) and S/D extension are defined. (Traditionally, the gate dielectric is silicon dioxide, also called oxide.) Important features illustrated here include:

(a) shallow trench isolation *(STI)*, (b) deposited, anisotropically etched spacers, and (c) dual (N+ for NMOS, P+ for PMOS) polysilicon (*"poly"*) gate electrodes. In addition, *"salicide"* (self-aligned silicide) for the poly gate electrodes and the deep S/D's (which are also known as the S/D *contacts*) are shown, as well as the shallow S/D extensions.

For all the logic technologies, several problems are associated with the rapid scaling of *EOT*. The first problem, boron penetration, is the uncontrolled diffusion of boron from the heavily doped P+ poly gate of the PMOSFETs through the gate oxide and into the MOSFET channel. The result is an uncontrollable shift in V_t in a positive direction. Adding a relatively small amount of nitrogen to the gate oxide to suppress the boron diffusion through the oxide is generally effective in dealing with this problem,[19], [20] and is the standard technique used in current leading edge technologies. The other major problem is the rapid increase in gate leakage current with scaling. This is due to direct tunneling, which increases exponentially with decreasing thickness for thin oxides. As noted before, the gate leakage current, I_g (units are amperes/micron of width), must be less than the S/D subthreshold leakage current, $I_{sd,leak}$, which is projected for the various types of

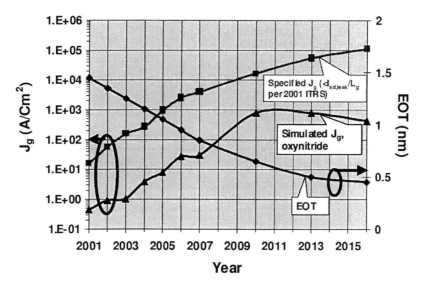

Fig. 7. 2001 projections of maximum allowed gate leakage current density (J_g) versus simulations for high-performance logic[24].

logic chip in Figure 3. One approach to reducing I_g is to use a "high-k" gate dielectric, i.e., one with a value for its relative dielectric constant, K, that is significantly higher than the 3.9 value for thermal silicon dioxide (SiO_2). For a high-k film with physical thickness, T_d, the equivalent oxide thickness, EOT, is:

$$EOT = T_d(3.9/K) \qquad (4)$$

To first order, a transistor with such a high-k gate dielectric has the same gate capacitance per unit area (C_{ox}) and hence the same electrical performance as an otherwise identical transistor with a thermal silicon dioxide film of thickness EOT. ($C_{ox} = \varepsilon_{ox}/EOT$, where ε_{ox} is the dielectric constant of thermal SiO_2.) With K larger than 3.9, T_d is larger than EOT. Since direct tunneling decreases exponentially with T_d, it is expected that I_g will be smaller for the high-k gate dielectric compared to the thermal oxide gate dielectric with the same EOT. However, another important factor impacting tunneling is the barrier height between the conduction band of silicon and that of the high-k material. Tunneling increases sharply as the barrier height decreases, and as the K of different materials increases, the energy gap and the barrier height tend to decrease [21]. Hence, in evaluating

candidate high-k materials, barrier height should be examined as well as K, and I_g should be experimentally determined to see whether the material is advantageous for use as a gate dielectric. Extending current approaches using lightly nitrogen doped silicon dioxide (called oxynitride), heavily nitrogen doped oxynitride or oxynitride/silicon nitride stacks have been reported with K up to ~6.5 and I_g reduced by about two orders of magnitude relative to thermal oxide with the same EOT [20], [22], [23]. In Figure 6, the 2001 ITRS projections for EOT and maximum allowed gate leakage current density, $J_g = I_g/(L_g)$, are plotted for $LSTP$ logic, which has the most stringent targets for gate leakage.

Also plotted are simulations of the expected J_g using heavily nitrogen-doped oxynitride, as calculated in the 2001 ITRS [24]. By 2005, J_g due to direct tunneling through the oxynitride, as deduced by simulation, is well beyond the projection in the 2001 ITRS. Hence, the introduction of high-k gate dielectric (with K larger than the ~6.5 value of heavily nitrogen doped oxynitride) is expected to be driven by $LSTP$ logic in 2005, with a projected EOT~ 1.8 nm and $I_g = 1$ pA/μm [25]. Some of the high-k materials currently being considered are hafnium oxide, aluminum oxide, their silicates, and quarternary combinations[26]. Active research and development programs are being carried out globally on high-k gate dielectrics and reported on at the VLSI Symposia on Technology (for example, in the 2001 Symposium see the papers by Choi et al.[27], Lee et al.[28] and others), the IEDM (for example, in the 2001 IEDM, see the papers by Kim et al.[29], Onishi et al.[30] and others), and other conferences and workshops.

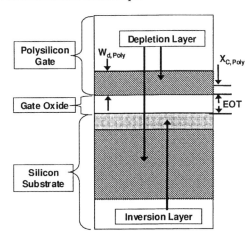

Fig. 8. Illustration of polysilicon depletion for MOSFET. A schematic cross-section through the middle of a turned-on MOSFET is shown.

In contrast to *LSTP* logic, for *LOP* and especially for high-performance logic, the maximum allowed gate leakage current is relatively large and increases significantly with scaling. For high-performance logic, Figure 7 has plots of the 2001 ITRS projections for maximum allowed gate leakage current density, J_g, as well as simulations from the ITRS of the expected J_g for heavily nitrogen doped oxynitride [24]. From the figure, heavily nitrogen doped oxynitride or oxynitride/silicon nitride stacks appear to be adequate to meet the ITRS gate leakage projections through the end of the Roadmap in 2016; the same appears to be true for *LOP* logic. However, there are serious questions about oxynitride's reliability for the thin films and large gate direct tunneling currents projected in the ITRS, particularly for high-performance logic [21]. (Note that the maximum allowed J_g is 10^3 A/cm^2 in 2005 and 10^5 A/cm^2 in 2016 for high-performance logic, which are exceedingly high val-

ues.) Also, achieving acceptable levels of thickness control for these thin films will become extremely difficult, particularly since I_g increases by an order of magnitude with a decrease of $\sim$0.2 nm in EOT [5]. Finally, it may be useful to reduce the gate leakage current below the targets in the ITRS to help reduce the chip static power dissipation. Hence, a likely scenario is that high-k gate dielectric will be implemented for high-performance and *LOP* logic after several years of learning on *LSTP* chips [31], in 2007 or beyond.

Another major front-end issue is polysilicon depletion in the gate electrode. When a MOSFET is turned on, a depletion region of thickness $W_{d,Poly}$ forms in the polysilicon adjacent to the polysilicon-dielectric interface. Consequently, the centroid of the charge is not at the interface, but at a distance $x_{c,poly}$ from the interface, where $x_{c,poly}$ is proportional to $W_{d,Poly}$ (in particular, $x_{c,poly} = W_{d,Poly}/2$ for uniform doping.). (See Figure 8 for a schematic illustration.) The effective electrical thickness of the gate dielectric, $EOT_{elec} = EOT + \Delta_{poly}$[32], where $\Delta_{poly} = (x_{c,poly}K_{ox}/K_{si}) = (x_{c,poly}/3)$. In this expression, $K_{ox} = 3.9$ is the relative dielectric constant of oxide and $K_{si} = 11.9$ is the relative dielectric constant of silicon[33]. According to Wilk et al.[21], Δ_{poly} can be as much as 0.4 nm. For a MOSFET with a given gate voltage applied, the increase in EOT_{elec} due to the non-zero Δ_{poly} results in a reduction in the inversion layer charge and hence a corresponding reduction in I_{on} compared to the case where there is no polysilicon depletion (i.e., where $EOT_{elec} = EOT$). The impact of polysilicon depletion becomes more severe in succeeding technology generations because of EOT scaling. Increased doping of the polysilicon can ameliorate this effect since $W_{d,Poly}$ and hence $x_{c,poly}$ decrease with doping. However, this solution will eventually become inadequate because of the limited solubility of the dopants, particularly boron, even if poly-SiGe, which has higher solid solubility for boron, is used[34]. An alternate solution, metal gate electrodes, is being pursued because metal gates have virtually no depletion, no boron penetration, and the gate electrodes have very low sheet resistance. However, CMOS optimization requires different work functions for PMOS and

NMOS devices (∼5.0 eV [near the silicon valence band edge] and ∼4.1 eV [near the silicon conduction band edge], respectively[35]) to set the desired symmetric threshold voltages of 0.2 to 0.5 V for NMOS and –0.2 to –0.5 V for PMOS.

For poly gates, this is done with P+ doping of the gate electrode for PMOSFETs and N+ doping of the gate electrodes for NMOSFETs. For metal gates, different metals with the required different work functions could be used, at the cost of difficult process integration problems (particularly including the thermal stability of the metal electrode) and increased chip processing complexity and cost. For NMOSFETs, TaSiN[36], W/TiN[37], and other combinations are being investigated, while TaN[38], RuO$_2$[39], TiAlN[40], and others are being investigated for PMOSFETs. An alternate approach that promises to reduce the added process complexity and cost is to use one material system for the metal electrodes, and to vary the work function for PMOS and NMOS by varying the alloy composition or by doping. Zhong et. al.[41] describe a RuTa binary alloy in which they varied the alloy composition to obtain work functions appropriate for PMOSFETs (Ta < 20%) and for NMOSFETs (40% < Ta < 54%). Q. Lu et al.[42] used (110)-Mo for the PMOSFET metal electrode, and they doped the Mo with nitrogen to shift the work function toward the NMOSFET required value.

Another important effect is due to the displacement away from the silicon-dielectric interface of the inversion charge centroid in the silicon substrate. Following the same reasoning as in the case of polysilicon depletion, this displacement of the inversion charge gives rise to an increase, Δ_{inv}, in EOT_{elec}. Taking account of quantum mechanical effects, $\Delta_{inv} \sim 0.3-0.6$ nm[21], [43] when the transistor is fully turned on. It should be noted that polysilicon depletion and quantum effects in the substrate limit the effectiveness of gate dielectric scaling. $\Delta_{poly} + \Delta_{inv} \sim 0.8$ nm in inversion, and since $EOT_{elec} = EOT + \Delta_{poly} + \Delta_{inv}$, it becomes difficult to scale EOT_{elec} much below 1.0 nm, no matter how much the EOT of the dielectric is reduced. Furthermore, Δ_{inv} cannot be significantly reduced, so even if metal gate electrodes are used to make $\Delta_{poly} \sim$ zero, the ability to scale EOT_{elec} is still limited by Δ_{inv}[21].

The series S/D parasitic resistance, $R_{sd,series}$, has increasingly stringent limits with scaling. These requirements are critical, because high series parasitic resistance degrades the saturation current drive, I_{on}. $R_{sd,series}$ is made up of components from both the S/D extension regions and the S/D contact regions[44] (see Figure 5).

For the S/D extension regions, very shallow junctions are required with scaling, mainly to minimize short channel effects[12], [45], [46]. However, the shallow profiles tend to lead to high sheet resistance, R_s, which increases $R_{sd,series}$[44]. This tradeoff between x_j and R_s will become increasingly difficult with scaling. Another requirement is that the S/D extension lateral doping profile in the vicinity of the extension-channel junction must become increasingly abrupt with scaling to minimize $R_{sd,series}$ and short channel effects[12], [44], [47]. The highest dose implant consistent with meeting the junction depth target will be used to minimize the series resistance. For the PMOSFET, where boron is the dopant, it is expected to be especially difficult to meet the junction depth and lateral abruptness requirements with scaling. Since boron is a light atom, it has a large implantation projected range and diffuses rapidly, and it strongly exhibits transient enhanced diffusion (TED)[48]. Ultra-low energy implants (0.5 keV or below at the 90 nm technology generation) are expected to be utilized to reduce the projected range of the boron as-implanted profile. In addition, rapid thermal processing (RTP) will be used extensively to minimize the x_j and to improve the lateral abruptness, since RTP is effective at reducing both the thermal budget and TED. Particularly with the implementation of "spike annealing", which has high ramp rates, these enhanced RTP solutions will probably suffice through the 90 nm node in 2004. However, beyond the 90 nm node, novel doping schemes such as atomic layer epitaxy, plasma doping, or co-implantation[49] and novel annealing schemes such as laser thermal annealing, microwave annealing, or low temperature annealing[46], [50] will eventually be required[24]. The ultimate solution may be deposited layers[5], [46].

For the deep S/D regions (also known as the S/D contact regions), low energy implantation

Elevated S/D with Selective (Epitaxial) Silicon and Post Implant

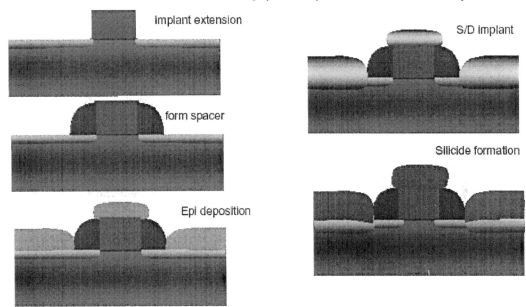

Fig. 9. Schematic illustration of elevated S/D process (courtesy of E. Graetz, Infineon).

followed by RTP for anneal is the current mainstream process[24]. After the RTP anneal, self-aligned silicide (salicide) is formed selectively on all deep S/D regions and the polysilicon electrodes (i.e., on all exposed silicon) through deposition of the metallic component of the silicide (i.e., titanium for titanium silicide or cobalt for cobalt silicide or nickel for nickel silicide) followed by low-temperature RTP. During the low temperature RTP, the metal combines with the exposed silicon to form the silicide, and during that process a portion of the silicon is consumed by converting it to silicide. After the RTP, the unreacted metal on the regions covered with silicon oxide or silicon nitride is removed by etching. Finally, a somewhat higher temperature RTP is done to reduce the sheet resistance, R_s, of the silicide. To further minimize the R_s, a thicker silicide layer must be formed, which is accomplished by depositing a thicker metal layer. However, this results in more silicon being consumed. If the consumed silicon depth gets too large, excessive junction leakage results because the silicide-silicon interface gets too close to the p-n junction. This excessive leakage

is caused by junction depletion region encroachment upon the silicide-silicon interface, which is quite rough and hence has many defects that can serve as leakage current generation centers[51]. To maintain junction leakage within tolerable limits, the silicon consumption depth is limited to half of the S/D contact junction depth. For previous mainstream process flows, titanium salicide was used, but cobalt salicide, which has a more favorable tradeoff between silicon consumption and low R_s[51], is becoming predominant at the 130 nm node. Eventually, cobalt salicide may be replaced by nickel salicide, which has an even more favorable tradeoff between silicon consumption and low R_s[52], [53]. These approaches appear to be adequate through the 65 nm technology generation in 2007. Beyond that, novel schemes such as elevated S/D may be needed, possibly including SiGe or Ge materials[24]. In elevated S/D, illustrated in Figure 9, selective epitaxial growth builds up just the deep S/D regions. As a result, more silicon consumption and hence lower R_s are obtainable without the excessive junction leakage that would ensue in the absence of elevated S/D.

A final issue is the specific contact resistance (also known as the contact resistivity) between the silicide and the deep S/D, which is an important component of $R_{sd,series}$[54]. The contact resistance must decrease sharply with scaling to meet the overall requirements on $R_{sd,series}$. Starting with the 65 nm technology generation in 2007, where the specific contact resistance is projected to be as low as 1.1×10^{-7} ohm-cm^2, there is no known manufacturable solution for meeting these requirements. Potential solutions being examined include selectively deposited silicide and selectively deposited metals [55], as well as selectively deposited, heavily doped SiGe contacted by tungsten, which results in as low as $\sim 2 \times 10^{-8}$ ohm-cm^2 specific contact resistance for P+ material[54]. Finally, dual low barrier silicide films (platinum silicide for PMOSFETs and erbium silicide for NMOSFETs, where platinum silicide has a Schottky barrier height, ϕ_{b0p}, of 0.24 V to P+ silicon and erbium silicide has a Schottky barrier height, ϕ_{b0n}, of 0.32 V to N+ silicon) show promise of reducing the specific contact resistance to acceptable levels for 20 nm and below MOSFETs[54], [56].

4. Non-Classical CMOS

As CMOS technology is scaled to the 65 nm node (in 2007) and beyond, planar bulk CMOS technology will face increasing difficulties in meeting all of the transistor requirements (high I_{on}, low I_{off}, reasonable short channel effects, good noise margins, adequate statistical control, etc.), even with the utilization of high-k gate dielectric, metal electrodes, novel annealing schemes, and the other material and processing type potential solutions discussed above. These problems are particularly acute with high-performance logic, which scales more rapidly than low-power logic. Key difficulties include controlling short channel effects adequately, the impact of the high channel doping required for such small devices, the increasing relative impact of quantum effects and statistical variation, and others. As a solution, non-classical CMOS devices such as enhanced mobility strained silicon/silicon-germanium channels, ultra thin

body SOI, and various types of multiple-gate MOSFET may be utilized [57], [58]. Furthermore, use of these non-classical CMOS devices may facilitate a slowdown in the scaling of some of the parameters such as EOT, thus allowing a delay in the implementation into manufacturing of some of the solutions such as high-k gate dielectric for high-performance logic.

For high-performance logic, an improvement in the effective mobility beyond that of standard silicon is needed to meet the targets for I_{on} and f_i, particularly in the long-term years beyond 2007. To a lesser extent, the same need for enhanced mobility holds true for LOP and $LSTP$ logic. An approach to realizing this enhanced mobility is to utilize an epitaxially grown layer of relaxed silicon-germanium (SiGe) on a silicon substrate with an epitaxially grown surface layer of strained silicon on top. The silicon surface layer is the channel of the MOSFETs, and this channel is under strain due to its having a smaller lattice constant than the underlying relaxed SiGe layer. For NMOSFETs, tensile strain splits the degeneracy in the silicon conduction band so that the minima with lower effective mass are lower in energy. These lower mass minima are therefore preferentially populated with carriers when the device is turned on and an inversion layer forms. As a result, since mobility, $\mu = q\tau/m^*$, where q = electronic charge, τ = mean free time against scattering, and m^* is the effective mass [59], the effective mobility of those carriers is increased [60], [61], [62]. For PMOSFETs, either tensile or compressive strain lifts the degeneracy of the valence band and reduces the effective mass of the bands, thereby increasing the hole effective mobility[62], [63]. In addition, for both NMOSFETs and PMOSFETs, interband scattering is reduced, which also contributes to increasing the effective mobility[64], [65]. The details of the SiGe layers, including the Ge concentration, the layer thickness, the layer doping, and grading, are different for the NMOSFET and PMOSFET to separately optimize the mobility enhancement for each. Cross sections of a PMOSFET and an NMOSFET with strained silicon channels are shown in Figure 10[65], [66]. Rim et al.[67] utilized strained silicon on relaxed SiGe for NMOSFETS and found that the effective mobility was increased by a factor of about

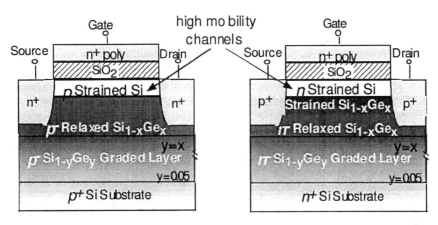

Fig. 10. Cross sections of PMOS and NMOS devices with strained silicon enhanced mobility channel (courtesy J. Hoyt, MIT).

1.8 times compared to unstrained silicon control samples for values of the vertical effective electric field, E_{eff}, up to 1 MV/cm. An improvement by a factor of 1.8 in bulk PMOSFETs[65] and a factor of 1.45 in SOI PMOSFETs[60] has been found. An interesting approach was taken by J. Alieu et. al.[68]. They used one type of multiple SiGe well for both NMOSFETs and PMOSFETs to form strained PMOS and NMOS channels. They realized a PMOSFET transconductance gain of 50% and an NMOSFET transconductance gain of 10% compared to unstrained Si. Use of enhanced mobility channels is a promising technique, but numerous issues need to be addressed, including such process integration issues as how to deposit the different SiGe layers needed for the NMOS and PMOS devices and the thermal stability of the SiGe and strained silicon layers during subsequent IC processing operations. Another important issue is how to establish and maintain an optimally strained layer in an SOI device with ultra-thin body.

Aside from the enhanced mobility issues discussed above, planar bulk CMOS transistors will face additional difficult issues as they are scaled to the 65 nm technology generation (in 2007) and beyond. Perhaps the most difficult issue in scaling planar bulk CMOS devices to the 65 nm technology generation ($L_g = 25$ nm) and especially beyond is achieving adequate control of short channel effects (SCEs). The difficulty is caused by electrostatic coupling between the charge in the

source and drain regions and depletion charge in the MOSFET channel. This is illustrated schematically in the bulk MOSFET of Figure 11, where field lines from some of the charges in the source and drain regions terminate on charges in the channel depletion region, as shown schematically by the lines connecting the S/D and the depletion charges. The field lines from the rest of the depletion charge (not drawn) originate from the charge in the gate electrode. The SCEs are caused by the increasing proportion of the channel depletion charge that is electrostatically coupled to charge in the source and drain, rather than to charge in the gate, as the gate length (L_g) becomes shorter. As a result gate control of the channel is compromised and SCEs such as threshold voltage roll off (reduction in $|V_t|$ as gate length decreases), drain induced barrier lowering *(DIBL)*, punchthrough, and increased subthreshold slope, S, beyond the long-channel value all become more predominant as the gate length is scaled down[69]. *DIBL* is defined as the reduction in $|V_t|$ due to increased drain voltage. For long channel MOSFETs, *DIBL* is approximately zero, but it increases sharply as L_g becomes small. Punchthrough occurs for very short channel devices when the drain depletion region extends all the way to the source end of the channel region for relatively high drain voltage (typically equal to or approaching V_{dd}). As a consequence, I_{ds} is large, even for zero gate voltage. Finally, S is measured in the subthreshold region of operation ($|V_{gs}| < |V_t|$), and is defined as the slope of the log $(I_{ds}) - V_{gs}$ curve there, in units of mV/decade. It increases sharply as L_g becomes small.

The techniques to control SCEs include:

(i) scaling the vertical device dimensions such as EOT, S/D extension x_j, channel depletion region width (W_{ch}) (specifically, x_j and W_{ch} must be significantly less than L_g[24])

(ii) super steep retrograde *(SSR)* profiles[70], [71] (vertical profiles relatively lightly doped near the silicon/dielectric interface with a peak in the interior)

(iii) halo implant profiles[72], [73], [74], where heavy, laterally asymmetric doping is implanted in the channel near the source and drain.

The extremely sharp, asymmetric profiles needed for SSR and halo implant profiles at the 65 nm technology node and beyond will be difficult to successfully fabricate.

A major issue is the impact of high doping in the channel. Scaled technologies will require high channel doping (well over 10^{18} cm^{-3}), both to set the threshold voltage to the desired levels ~ 0.2 to 0.5 V (NMOS) and $\sim$ -0.2 to -0.5 V (PMOS) and to keep W_{ch} significantly less than L_g to adequately control short channel effects. At the 65 nm technology node, $\sim 3 \times 10^{18}$ cm^{-3} doping is required to meet the V_t requirement, and $\sim 2 \times 10^{19}$ cm^{-3} doping is required to meet the W_{ch} requirement, and these increase with further scaling[24]. Doping levels of 3×10^{18} cm^{-3} and higher lead to reductions in the effective mobility in the channel[75] and hence to reduced I_{on}, and to possible problems with band-to-band leakage tunneling between the transistor drain and body, thereby increasing I_{off} [76], [77]. Another important issue is the impact of statistical variation on the control of the transistor characteristics. With scaling, the total number of dopant atoms, N_T, in the depletion region of a MOSFET's channel is significantly reduced. For example, $N_T \sim 650$ atoms at the 130 nm technology node, where $L_g = 65$ nm, W is taken to be 3 L_g, the channel doping $\sim 2 \times 10^{18}$ cm^{-3}, and the depletion region width ~ 25 nm. On the other hand, for the 22 nm technology node, where $L_g = 9$ nm, the same type of analysis yields $N_T \sim 60$ atoms.

N_T fluctuates stochastically for different transistors[78] with a standard deviation $\Delta N_T = N_T^{0.5}$. Hence, $(\Delta N_T / N_T) = 1/N_T^{0.5}$, which is about 0.04 (4%) for the 130 nm node, but is a quite large 0.13 (13%) for the 22 nm node. This increase in the percent fluctuation in the number of dopant atoms in the depletion region of the MOSFETs can give rise to increases in the statistical fluctuations of V_t, which is undesirable, particularly because these fluctuations in V_t cannot be reduced by improved process control of L_g, EOT, etc. Following Taur and Ning[79] and using the numbers quoted above, the standard deviation of V_t (denoted ΔV_t) due to the fluctuations in N_T is ~ 15mV for the 130 nm node and ~ 60 mV for the 22 nm node, where $L_g \sim 10$nm. Since the maximum expected statistical variation is $\sim 3\Delta V_t$, at the 22 nm node, where $3\Delta V_t \sim 180$mV, the expected statistical variation is significantly larger than the 50 mV total statistical variation allowed for 10 nm gate length MOSFETs per the analysis in Ren et al.[80]. Alternatively, the maximum variation in $I_{sd,leak}$ can be estimated using a typical 90 mV/decade subthreshold swing and setting the maximum V_t variation to $+3\Delta V_t$. The result is an unacceptably large two orders of magnitude difference between the mean and the maximum value of $I_{sd,leak}$.

Active research and development are being carried out on ultra-thin body SOI *(UTB SOI)* and various types of double-gate *(DG) MOSFETs* as a means to deal with many of the above problems for devices with L_g of 25 nm (characteristic of the 65 nm technology node in 2007) and below. UTB SOI and DG SOI MOSFETs are shown schematically in Figure 11. The UTB SOI has a very thin silicon film on top of a relatively thick buried oxide *("BOX")*, which itself is on top of the silicon substrate. The thin film contains the active area of the MOSFET, and its thickness, T_{si}, is optimally $< L_g/3$[81] to adequately control SCEs, and in all cases, it must be sufficiently small that the depletion region extends throughout the silicon thin film when the device is turned on. Such a device is called a fully depleted SOI *(FDSOI) MOSFET*, and it has a number of advantages. V_t can be controlled by the work function of the gate electrode, rather than by the substrate doping. The substrate doping is then relatively light compared to bulk pla-

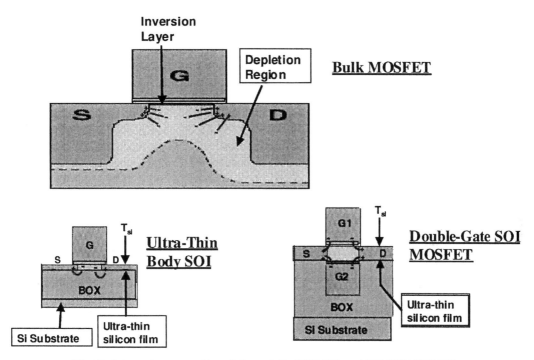

Fig. 11. Schematic cross section of planar bulk, UTB SOI, and DG SOI MOSFETs.

nar MOSFETs, which will ameliorate the problems due to heavy channel doping discussed above. Furthermore, since the channel doping is a minor factor in controlling V_t, the stochastic variation of N_T will not be reflected in the statistical variation of V_t. Another advantage of the UTB SOI MOSFET is that the parasitic junction capacitance is significantly reduced from that of the bulk MOSFET because the bottom of the deep S/D is directly on top of the thick BOX, which is inherently low capacitance. However, there are several shortcomings with the UTB SOI device. One is the higher cost of SOI substrates compared to standard bulk substrates. Another is the difficulty of establishing acceptable control of SCEs in highly scaled MOSFETs. This difficulty is due to the electrostatic coupling between charge in the source and drain and charge in the silicon body's depletion region. This is similar to the electrostatic coupling for planar bulk MOSFETs discussed above, except that the thick BOX strengthens the coupling and thus worsens the problem of controlling SCEs for highly scaled UTB SOI devices (see Figure 11 for an illus-

tration of this coupling). As mentioned above, in order to ameliorate the SCEs, optimally T_{si} should be less than $L_g/3$, which results in very small values for T_{si}. In particular, for the 65 nm technology node or beyond, this amounts to silicon film thickness of 8 nm or less. Fabricating such thin layers of silicon with adequate process control is a difficult challenge. However, encouraging data on UTB SOI devices with $L_g = 50$ nm and $T_{si} < 30$ nm has been reported[82]. The transistor electrical characteristics are generally good, with PMOSFET I_{on} of 0.65 mA/μm and $I_{off} = 9$ nA/μm, subthreshold slope, S, of 75 mV/decade, and reduced $DIBL < 50$ mV/V. The NMOSFET results, for devices with $L_g = 65$ nm, are I_{on} of 1.18 mA/μm and $I_{off} = 60$ nA/μm, subthreshold slope, S, of 75 mV/decade, and $DIBL$ of 45 mV/V. For both devices, $V_{dd} = 1.3$ V. Y. K. Choi et al.[83] reported on $L_g = 30$nm, $T_{si} = 4$ nm PMOSFET devices with a selective Ge elevated S/D. They found generally good transistor electrical characteristics: $S \sim 75$ mv/decade, $DIBL < 50$ mV/V, and $I_{on} \sim 250$ μA/μm for $V_{dd} = 1$ V. They also found that the SCEs got significantly worse if T_{si} was increased to ~ 8 nm; for example, S increased from about 75 to almost 100 mV/decade.

A simplified cross section of a DG SOI MOSFET is shown in Figure 11. This is an UTB SOI MOSFET with a bottom gate in the BOX. The bottom gate is fully self-aligned with the top gate, and the equivalent oxide thickness of the gate dielectric for the bottom gate, as well as the bottom gate electrode material, are usually the same as those of the top gate. Because of the complex structure, particularly the self-aligned top and bottom gates, the process for fabricating such devices is complicated and challenging. Typically, the DG SOI MOSFET is operated in the fully depleted mode, and the top and bottom gates are electrically tied together. The DG SOI MOSFET has the same advantages as the UTB SOI MOSFET, but in addition, it has an enhanced immunity to SCEs for highly scaled MOSFETs, and hence for optimal control of SCEs, $T_{si} \leq L_g/2$ [80]. This enhanced SCE immunity is because the majority of the charge in the source and drain regions is electrostatically coupled to charge on the two gate electrodes, rather than to depletion charge in the transistor channel region (i.e., the two gates tend to shield the channel from charge in the source and drain regions, as illustrated in Figure 11). Furthermore, the subthreshold swing, S, can be close to the ideal value of 60 mV/decade. Another advantage is that, since both the top and bottom gates can induce inversion charge in the silicon channel region, I_{on} can be significantly enhanced in a DG MOSFET compared to a similar single-gate SOI MOSFET. Clearly, from the point of view of device design and electrical characteristics, the DG MOSFET is an optimal structure for highly scaled MOSFETs down to $L_g \sim 10$ nm[76], [80], [84], [85]. However, series S/D parasitic resistance is of particular concern, because of the difficulty in making a low resistance S/D extension and a low resistance contact with the very thin silicon film. Heavy doping and a very abrupt lateral doping profile for the S/D extension as well as elevated S/D are required to deal with this issue[80], [85]. Finally, as with single-gate SOI MOSFETs, fabricating the thin silicon film with adequate thickness control is a difficult challenge.

Reports of such devices are in Neudeck et al.[86] and Guarini et al.[87] Guarini et al. used $\sim$20 nm thick silicon film to fabricate MOSFETs with Lg as small as 50 nm, and found that the subthreshold slope, S, for both NMOSFETs and PMOSFETs is very close to the ideal 60 mV/decade if the back gate and front gate bias is equal. They also report that both the electron and hole effective mobilities match the universal curves for planar bulk silicon CMOSFETs from Takagi[88], and they demonstrate working logic circuits.

There are several other types of DG MOSFET that have been discussed in the literature: these include the vertical MOSFET and the FinFET. These have the same basic structure and the same optimal electrical characteristics as the DG SOI device, but the details of the layout and structure are different. A simplified cross-section of the vertical MOSFET is shown in Figure 12. This device is built above the silicon substrate using epitaxial growth techniques, and the polysilicon gate is formed to surround the device. The fabrication process is complicated and challenging. The current flow direction is vertical, as shown, and this functions like a double gate device with the two gates on either side, as shown in Figure 12. The width, X, is the analog of the film thickness, T_{si}, for the DG SOI. A number of reports on experimental vertical MOSFETs have appeared in the literature[77], [89], [90], [91]. The smallest L_g achieved so far is about 50 nm, but the devices are not fully depleted because X is not small enough[89]. The electrical characteristics are generally acceptable, with NMOSFET $I_{on} \sim 1000$ μA/μm, and high-k gate dielectrics using HfO_2 or Al_2O_3 have been successfully integrated. Schultz et al. report on fully depleted MOSFETs with 70 nm gate length and $X \sim 50$ nm[77]. The SCEs are well under control, with negligibly small DIBL.

The other type of DG MOSFET is the FinFET[92], [93], [94], [95], schematically illustrated in Figure 13. In this device, very thin vertical silicon fins are etched in SOI films. Poly gates are formed surrounding these fins, so the double gates are on either side of the fin, and the MOSFET inversion layers are formed along the vertical edges of the fin. The current flow is from source to drain

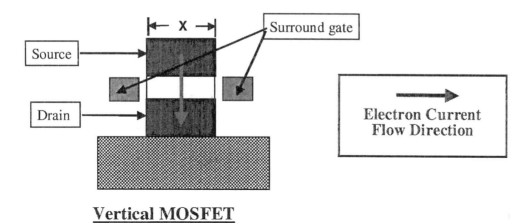

Vertical MOSFET

Fig. 12. Simplified cross section of vertical MOSFET.

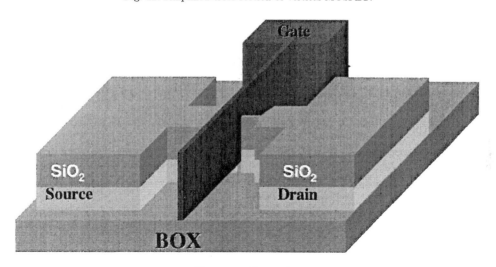

a. Perspective view

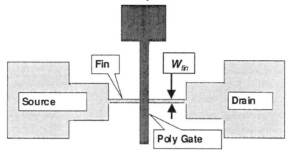

b. Top view

Fig. 13. a. Simplified perspective view of FinFET (courtesy T-J King, UC-Berkeley). b. Top view of FinFET.

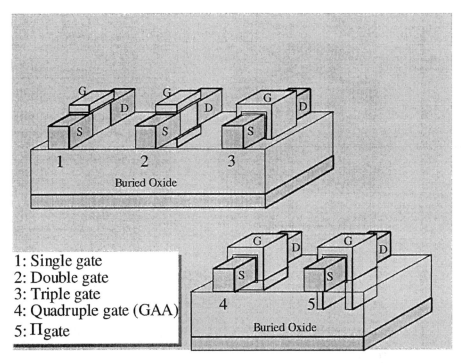

Fig. 14. Schematic illustrations of various multiple-gate MOSFET structures (courtesy: J. P. Colinge, UC-Davis).

in the figure. The horizontal width of the fin in the direction perpendicular to the current flow direction (W_{fin}, see Figure 13b) is the equivalent of T_{si} in the DG SOI MOSFET. An advantage of the FinFET is that the process flow and process tools are relatively compatible with current techniques. Partly for this reason, the FinFET appears to be the current leader in DG MOSFET structures. However, optimally $W_{fin} \leq L_g/2$ to control SCEs, as discussed previously, and defining and controlling such small fin widths is a major challenge. Recent results for fully depleted, 20 nm gate length, 10 nm W_{fin} PMOS and NMOS devices have been reported[92]. The electrical characteristics were generally good; for example, the subthreshold swing was around 70 mV/decade (NMOS) and 80 mV/decade (PMOS) for L_g down to 20 nm, and the Vt roll off was acceptable. The impact of W_{fin} was explored, and it was found that an increase in W_{fin} (to 26 nm) significantly degraded the SCEs such as V_t roll off, DIBL, and increased subthreshold swing for short L_g devices. A main issue for all types of DG MOSFETs is how to fabricate these

devices: a cost-effective, manufacturable process flow is still to be determined, and any process flow that emerges seems likely to be complicated and difficult.

Even more complicated multiple-gate MOSFETs, with gates on three or four sides of the silicon channel, are being explored[96] (see Figure 14 for schematic illustrations of several of these structures). They appear to have even more optimal electrical characteristics for highly scaled devices, particularly the SCEs, compared to DG MOSFETs. This is because, with the gates on more sides of the MOSFET, the charge in the source and drain regions is more effectively shielded by the gates. However, because of their more complex structure, they tend to be more complicated to fabricate than DG MOSFETs.

In summary, the IC industry is pursuing highly scaled, 25 nm or less gate-length devices (at the 65 nm technology node and beyond) along two parallel paths: planar bulk MOSFETs and non-classical MOSFETs such as UTB SOI or multiple-gate MOSFETs. Bulk planar MOSFETs are being aggressively scaled, and a recent paper reports on data from 15 nm gate-length devices with 0.8 nm EOT gate dielectric using a nitride/oxynitride stack[53]. The electrical results are reasonable but not optimal, with NMOS $I_{on} = 615$ $\mu A/\mu m$ and PMOS $I_{on} = 265$ $\mu A/\mu m$, $I_{off} < 500$ nA/μm, and $V_{dd} = 0.8$ V. Very fast gate delays (CV/I) of 0.29 ps for NMOSFET and 0.68 ps for PMOSFET were obtained. However, the SCEs are a major issue, with DIBL, for example, well over 100 mV/V. Clearly, both planar bulk CMOS and non-classical CMOS are candidates for the technology nodes toward the end of the Roadmap. Given the difficulties of scaling planar bulk CMOS into that regime, however, non-classical CMOS (particularly multiple-gate) is preferred from a device point of view, but difficult challenges, particularly in processing, manufacturability, and fabrication cost, remain to be solved for non-classical CMOS. The ultimate CMOS device with L_g of 10 nm or less is likely to be a DG or other type of multiple-gate MOSFET with strained silicon high-mobility channel and with advanced materials and processes such as high-k gate dielectric, metal gate electrodes, elevated source/drains, and advanced annealing.

5. Conclusions

IC technology is continuing to scale according to Moore's Law, with the overall chip circuit requirements driving the MOSFET device and process integration requirements and optimal choices. In the 2001 ITRS the driver for the high performance logic is maximizing MOSFET intrinsic speed, while the driver for low power logic is minimizing MOSFET leakage current. In the front end, high-k gate dielectric is an important issue, and will apparently be needed initially for low standby power logic in 2005, with subsequent implementation for high-performance logic. Other important front-end issues include polysilicon depletion in the gate electrode, which will probably eventually drive the industry to metal gates, and series source/drain parasitic resistance, which will probably drive the industry to various new solutions for source/drain

processing. Finally, the adequacy of planar bulk CMOS devices to meet the MOSFET requirements beyond 2007 appears questionable, and the industry is exploring alternatives in non-classical CMOS structures to deal with this issue.

References

[1] G. E. Moore, "Lithography and the Future of Moore's Law," *Proceedings of VIII th Optical/Microlithography Conference*, SPIE Vol. **2440**, pp. 2 – 17, February 1995.

[2] Semiconductor Industry Association (SIA), *International Roadmap for Semiconductors 2001 edition*, Austin, TX: International SEMATECH, 2001, Executive Summary Chapter. (This is available for printing and viewing from the internet, with the following URL: http://public.itrs.net.)

[3] G.E. Moore, "Progress in Digital Integrated Electronics," *IEDM Tech. Digest*, pp. 11-13, Dec. 1975.

[4] G.E. Moore, "Cramming More Components onto Integrated Circuits," *Electronics*, Vol. **38,** pp. 114 - 117, 1965.

[5] H. R. Huff et al., "Sub-100 nm Gate Stack/Ultrashallow Junction Integration Challenges," *ECS PV* **2001-09**, 263-297 (2001). [Originally presented at the ECS *RTA and Other Short Time Processing Technologies II* Symposium, March 2001.]

[6] P. K. Vasudev and P. M. Zeitzoff, "ULSI Semiconductor Technology Trends and Challenges in the Coming Decade," *IEEE Circuits and Devices Magazine,* **14**, p. 19 (March 1998).

[7] P. M. Zeitzoff and J. E. Chung, "Weighing in on Logic Scaling Trends," *IEEE Circuits and Devices Magazine*, Vol. **18**, pp. 18 – 27, (March 2002).

[8] Semiconductor Industry Association (SIA), *International Roadmap for Semiconductors 2001 edition*, Austin, TX: International SEMATECH, 2001. (This is available for printing and viewing from the internet, with the following URL: http://public.itrs.net.)

[9] Semiconductor Industry Association (SIA), *International Roadmap for Semiconductors 1999 edition*, Austin, TX: International SEMATECH, 1999. (This is available for printing and viewing from the internet, with the following URL: http://public.itrs.net.)

[10] J. D. Meindl et al., "Limits on Silicon Nanoelectronics for Terascale Integration," *Science*, Vol. **293**, pp. 2044 – 2049 (14 Sept. 2001).

22

[11] Semiconductor Industry Association (SIA), *International Roadmap for Semiconductors 2001 edition*, Austin, TX: International SEMATECH, 2001, Front End Process and Lithography Chapters. (This is available for printing and viewing from the internet, with the following URL: http://public.itrs.net.)

[12] S. Thompson et al., "MOS Scaling: Transistor Challenges for the 21st Century," 3rd Quarter 1998 on Intel's® website. URL is http://www.intel.com/technology/itj/q31998/articles/art_3.htm.

[13] Y. Taur and T. H. Ning, *Fundamentals of Modern VLSI Devices,* Cambridge University Press, New York, 1998, pp. 228-230.

[14] Y. Taur and T. H. Ning, *Fundamentals of Modern VLSI Devices,* Cambridge University Press, New York, 1998, pp. 173 – 175.

[15] S. Thompson et al., "An Enhanced 130 nm Generation Logic Technology Featuring 60 nm Transistors Optimized for High Performance and Low Power at 0.7 – 1.4 V," *IEDM Tech. Digest,* pp. 257 – 260, Dec. 2001.

[16] S. F. Huang et al., "High Performance 50 nm CMOS Devices for Microprocessor and Embedded Processor Core Applications," *IEDM Tech. Digest,* pp. 237 - 240, Dec. 2001.

[17] S. Parihar, "A High Density $0.10\mu m$ CMOS Technology Using Low K Dielectric and Copper Interconnect," *IEDM Tech. Digest,* pp.249 - 252, Dec. 2001.

[18] Semiconductor Industry Association (SIA), *International Roadmap for Semiconductors 2001 edition*, Austin, TX: International SEMATECH, 2001. Process Integration, Devices, and Structures chapter and Design chapter. (This is available for printing and viewing from the internet, with the following URL: http://public.itrs.net.)

[19] B.Maitl et al., "High Performance 20 Å NO Oxynitride for Gate Dielectric in Deep Sub-Quarter Micron CMOS Technology," *IEDM Tech. Digest,* pp. 651- 654, Dec. 1997.

[20] B. Yu et al., "Scaling Toward 35nm Gate Length CMOS," *Symposium on VLSI Technology, Digest of Tech. Papers,* pp. 9 – 10, June 2001.

[21] G. D. Wilk et al., "High-k Gate Dielectrics: Current Status and Material Properties Considerations," *Journal of Applied Physics,* **89**, pp. 5243 – 5275 (May 2001).

[22] X. Guo and T. P. Ma, "Tunneling Leakage Current in Oxynitride: Dependence on Oxygen/Nitrogen Content," IEEE Elec. Dev. Letters, EDL-**19**, pp. 207-209 (June 1998).

[23] A. Karamcheti et al., "Silicon Oxynitride Films as a Segue to the High-K Era," *Semiconductor FABTECH Magazine,* **12th ed.**, pp. 207 – 214 (June 2000).

[24] Semiconductor Industry Association (SIA), *International Roadmap for Semiconductors 2001 edition*, Austin, TX: International SEMATECH, 2001. Front End Process chapter. (This is available for printing and viewing from the internet, with the following URL: http://public.itrs.net.)

[25] Semiconductor Industry Association (SIA), *International Roadmap for Semiconductors 2001 edition*, Austin, TX: International SEMATECH, 2001. Process Integration, Devices, and Structures chapter, Memory and Logic Potential Solutions section. (This is available for printing and viewing from the internet, with the following URL: http://public.itrs.net.)

[26] H. R. Huff et al., "Integration of High-k Gate Stack Systems into Planar CMOS Process Flows," *Proceedings of International Workshop on Gate Insulators (IWGI)*, Tokyo, Japan, November 2001.

[27] R. Choi et al., "High-Quality Ultra-thin HfO2 Gate Dielectric MOSFETs with TaN Electrode and Nitridation Surface Preparation," *Symposium on VLSI Technology, Digest of Tech. Papers,* June 2001.

[28] S. J. Lee et al., "Performance and Reliability of Ultra Thin CVD HfO_2 Gate Dielectrics with Dual Poly-Si Gate Electrodes," *Symposium on VLSI Technology, Digest of Tech. Papers,* June 2001.

[29] Y. Kim et al., "Conventional n-channel MOSFET devices using single layer HfO_2 and ZrO_2 as high-k gate dielectrics with polysilicon gate electrode," *IEDM Tech. Digest,* pp. 455-458, Dec. 2001.

[30] K. Onishi et al., "Reliability Characteristics, Including NBTI, of Polysilicon Gate HfO_2 MOSFET's," *Tech. Digest of 2001 IEDM,* pp. 659-662, Dec. 2001.

[31] Semiconductor Industry Association (SIA), *International Roadmap for Semiconductors 2001 edition*, Austin, TX: International SEMATECH, 2001. Process Integration, Devices, and Structures chapter. (This is available for printing and viewing from the internet, with the following URL: http://public.itrs.net.)

[32] Y. Taur and T. H. Ning, *Fundamentals of Modern VLSI Devices,* Cambridge University Press, New York, 1998, pp. 75 – 78.

[33] S. M. Sze, *Physics and Technology of Semiconductor Devices,* 2nd ed., Wiley and Sons, New York, 1981, Appendixes H and I.

[34] S. Gannavaram et al., "Ultra-Shallow P^+-N Junctions for 35-70 nm CMOS Using Selectively Deposited Very Heavily Boron-Doped Silicon-Germanium Films, *Rapid Thermal Processing and Other Short-Time*

Processing Technologies I, (F. Roozeboom et al., eds.) ECS PV 00-9, pp. 73 – 82 (2000).

[35] H. Zhong et al., "Properties of Ru-Ta Alloys as Gate Electrodes for NMOS and PMOS Silicon Devices," *IEDM Tech. Digest,* pp. 467-470, Dec. 2001.

[36] Y. Suh et al., "Electrical Characteristics of TaSixNy Gate Electrodes for Dual Gate Si CMOS Devices," *VLSI Symposium Dig. Of Tech. Papers,* pp.47-48, June 2001.

[37] D. Park et al., "Novel Damage-Free Direct Metal Gate Process Using Atomic Layer Deposition," *VLSI Symposium Dig. Of Tech. Papers,* pp. 65-66, June 2001.

[38] Y. H. Kim et al., "High Quality CVD TaN Gate Electrode for Sub-100nm MOS Devices," *IEDM Tech. Digest,* pp. 667-670, Dec. 2001.

[39] H. Zhong et al., "Characterization of RuO_2 electrode on Zr silicate and ZrO_2 dielectrics," *Appl. Phys. Lett.,* **78,** pp. 1134-1136 (2001).

[40] D. Park et al., "Robust Ternary Metal Gate Electrodes for Dual Gate CMOS Devices," *IEDM Tech. Digest,* pp. 671-674, Dec. 2001.

[41] H. Zhong et al., "Characterization of RuO_2 electrode on Zr silicate and ZrO_2 dielectrics," *Appl. Phys. Lett.,* **78,** pp. 1134-1136 (2001).

[42] Q. Lu et al., "Metal Gate Work Function Adjustment for Future CMOS Technology," *VLSI Symposium Dig. Of Tech. Papers,* pp. 45-46, June 2001.

[43] Y. Taur and T. H. Ning, *Fundamentals of Modern VLSI Devices,* Cambridge University Press, New York, 1998, pp. 194 – 200.

[44] K. W. Ng and W. T. Lynch, ""The impact of intrinsic series resistance on MOSFET scaling", *IEEE Transactions on Electron Devices,* ED-**34**, p.502 (March 1987).

[45] Y. Taur and T. H. Ning, *Fundamentals of Modern VLSI Devices,* Cambridge University Press, New York, 1998, pp. 164 - 173.

[46] L. Larson and H. Huff, "ITRS front-end process roadmap," *Solid State Technology,* pp. 32-34, January 2002.

[47] Y. Taur and T. H. Ning, *Fundamentals of Modern VLSI Devices,* Cambridge University Press, New York, 1998, pp. 214 – 219.

[48] S. W. Crowder et al., "The Effect of Source/Drain Processing on the Reverse Short Channel Effect of Deep Sub-Micron Bulk and SOI NMOSFETs," *IEDM Tech. Digest,* pp. 427 - 430, Dec. 1995.

[49] H-J. Li et al., "Boron Diffusion and Activation in the Presence of Other Species," *IEDM Tech. Digest,* pp. 515 - 518, Dec. 2000.

[50] K. Tsuji et al., "High Performance 50-nm Physical Gate Length pMOSFETs by using Low Temperature Activation by Re-Crystallization Scheme", *VLSI Symposium Dig Of Tech. Papers,* pp. 9 - 10, June 1999.

[51] W. P. Maszara, "Very thin $CoSi_2$ films by Co sputtering", *Appl. Phys. Lett.,* **62**, 961 (March 1993).

[52] S. Inaba et al., "High Performance 35 nm Gate Length CMOS with NO Oxynitride Gate Dielectric and Ni SALICIDE," *IEDM Tech. Digest,* pp. 641 - 644, Dec. 2001.

[53] B. Yu et al., "15 nm Gate Length Planar CMOS Transistor," *IEDM Tech. Digest,* pp. 937- 939, Dec. 2001.

[54] S. D. Kim et al., "Advanced Model and Analysis for Series Resistance in Sub-100nm CMOS Including Poly Depletion and Overlap Doping Gradient Effect," *IEDM Tech. Digest,* pp. 723 – 726, Dec. 2000.

[55] Semiconductor Industry Association (SIA), *International Roadmap for Semiconductors 2001 edition,* Austin, TX: International SEMATECH, 2001. Front End Process chapter, Figure 30: Doping Potential Solutions. (This is available for printing and viewing from the internet, with the following URL: http://public.itrs.net.)

[56] J. Kedzierski et al., "Complementary silicide source/drain thin-body MOSFETs for the 20nm gate length regime," *IEDM Tech. Digest,* pp. 57- 60, Dec. 2000.

[57] Semiconductor Industry Association (SIA), *International Roadmap for Semiconductors 2001 edition,* Austin, TX: International SEMATECH, 2001. Process Integration, Devices, and Structures chapter, Emerging Research Devices section. (This is available for printing and viewing from the internet, with the following URL: http://public.itrs.net.)

[58] J. Hutchby et al., "Extending the Road Beyond CMOS," *IEEE Circuits and Devices Magazine,* March 2002.

[59] Y. Taur and T. H. Ning, *Fundamentals of Modern VLSI Devices,* Cambridge University Press, New York, 1998, Appendix 4.

[60] T. Mizuno et al., "Advanced SOI-MOSFETs with Strained-Si Channel for High Speed CMOS Electron/Hole Mobilty Enhancement," *VLSI Symposium Dig Of Tech. Papers,* pp.210 - 211, June 2000.

[61] Y. Taur and T. H. Ning, *Fundamentals of Modern VLSI Devices,* Cambridge University Press, New York, 1998, pp. 285 – 286.

[62] M. V. Fischetti and S. D. Laux, "Band Structure, Deformation Potentials, and Carrier Mobility in Strained Si, Ge, and SiGe Alloys, *Electron. Lett.,* **17**, pp. 503 – 504 (1996).

24

[63] Y. Taur and T. H. Ning, *Fundamentals of Modern VLSI Devices,* Cambridge University Press, New York, 1998, pp. 285 – 286.

[64] K. Rim et al., "Strained Si NMOSFETs for High Performance CMOS Technology," *IEDM Tech. Digest,* pp. 59 - 60, *VLSI Symposium Dig Of Tech. Papers,* pp. 59 - 60, June 2001.

[65] K. Rim et al., "Enhanced Hole Mobilities in Surface-channel Strained-Si *p*-MOSFETs," *IEDM Tech. Digest,* pp. 517 – 520, Dec. 1995.

[66] J. Welser et al., "Strain Dependence of the Performance Enhancement in Strained-Si *n*-MOSFETs," *IEDM Tech. Digest,* pp. 373 – 376, Dec. 1994.

[67] K. Rim et al., "Transconductance Enhancement in Deep Submicron Strained-Si n-MOSFETs," *IEDM Tech. Digest,* pp. 707 – 710, Dec. 1998.

[68] J. Alieu et al., "Multiple SiGe Well: a New Channel Architecture for Improving Both NMOS and PMOS Performances, *VLSI Symposium Dig Of Tech. Papers,* pp.130 - 131, June 2000.

[69] Y. Taur and T. H. Ning, *Fundamentals of Modern VLSI Devices,* Cambridge University Press, New York, 1998,, pp. 139 – 149.

[70] S. Song et al., "Design of Sub-100nm CMOSFETs: Gate Dielectrics and Channel Engineering," *VLSI Symposium Dig Of Tech. Papers,* pp.190 - 191, June 2000.

[71] Y. Taur and T. H. Ning, *Fundamentals of Modern VLSI Devices,* Cambridge University Press, New York, 1998, pp. 189 – 194.

[72] H. Wakabayashi et al., "45-nm Gate Length CMOS Technology and Beyond Using Steep Halo," *IEDM Tech. Digest,* pp. 49 - 52, Dec. 2000.

[73] Y. Taur, "25 nm CMOS Considerations," *IEDM Tech. Digest,* pp. 789 - 792, Dec. 1998.

[74] Y. Taur and T. H. Ning, "CMOS Devices Below 0.1 μm: How High Will Performance Go?" *IEDM Tech. Digest,* pp. 215 - 218, Dec. 1997.

[75] S. Takagi et al, "On Universality of Inversion-Layer Mobility in Si MOSFET's: Part I—Effects of Substrate Impurity Concentration," *IEEE Trans. on Elec. Devices,* **41,** 2357 – 2362 (December 1994).

[76] L. Chang et al., "Gate Length Scaling and Threshold Voltage Control of Double-Gate MOSFETs," *IEDM Tech. Digest,* pp. 719 - 722, Dec. 2000.

[77] T. Schultz et al., 50-nm Vertical Sidewall transistors With High Channel Doping Concentrations," *IEDM Tech. Digest,* pp.61 - 64, Dec. 2000.

[78] Y. Taur and T. H. Ning, *Fundamentals of Modern VLSI Devices,* Cambridge University Press, New York, 1998,, pp. 200 – 202.

[79] Y. Taur and T. H. Ning, *Fundamentals of Modern VLSI Devices,* Cambridge University Press, New York, 1998,, pp. 200 – 202, Eq. 4.64.

[80] Z. Ren et al., "Examination of Design and Manufacturing Issues in a 10 nm Double Gate MOSFET Using Nonequilibrium Green's Function Simulation," *IEDM Tech. Digest,* pp. 107 - 110, Dec. 2001.

[81] Y. K. Choi et al., "30 nm Ultra-Thin-Body SOI MOSFET with Selectively Deposited Ge Raised S/D," 58th Device Research Conf. (DRC), pp. 23 – 24.

[82] R. Chau et al., "A 50nm Depleted-Substrate CMOS Transistor (DST)," *IEDM Tech. Digest,* pp. 621 - 624, Dec. 2001.

[83] Y-K. Choi et al., "Ultra-Thin Body PMOSFETs with Selectively Deposited Ge Source/Drain," *VLSI Symposium Dig Of Tech. Papers,* pp.19 - 20, June 2001.

[84] D. T. Frank et al., "Monte Carlo Simulation of a 30 nm Dual-Gate MOSFET: How Short Can Si go?", *IEDM Tech. Digest,* pp. 553 - 556, Dec. 1992.

[85] D. Hisamoto, "FD/DG-SOI MOSFET—a Viable Approach to Overcoming the Device Scaling Limit," *IEDM Tech. Digest,* pp.429 - 432, Dec. 2001.

[86] G. W. Neudeck et al, "Novel Silicon Epitaxy for Advanced MOSFET Devices," *IEDM Tech. Digest,* pp. 169 - 172, Dec. 2000.

[87] K. W. Guarini et al., "Triple-Self-Aligned, Planar Double-Gate MOSFETs: Devices and Circuits, *IEDM Tech. Digest,* pp. 425 - 428, Dec. 2001.

[88] S. Takagi et al., "On Universality of Inversion-Layer Mobility in n- and p-Channel MOSFETs, *IEDM Tech. Digest,* pp. 398 – 401, Dec. 1988.

[89] J. M. Hergenrother et al., "50nm Vertical Replacement Gate (VRG) nMOSFETs with ALD HfO$_2$ and Al$_2$O$_3$ Gate Dielectrics, *IEDM Tech. Digest,* pp.51 - 54, Dec. 2001.

[90] S-H. Oh et al., "50 nm Vertical Replacement-Gate pMOSFETs," *IEDM Tech. Digest,* pp.65 - 68, Dec. 2000.

[91] J. M. Hergenrother et al., "The Vertical Replacement Gate (VRG) MOSFET: a 50-nm Vertical MOSFET with Lithography-Independent Gate Length," *IEDM Tech. Digest,* pp.75 - 78, Dec.1999.

[92] Y-K Choi et al., "Sub-20nm CMOS FinFET Technologies," *IEDM Tech. Digest,* pp. 421 - 424, Dec.2001.

[93] J. Kedzierski et al., "High-Performance Symmetric-Gate and CMOS-Compatible Vt Asymmetric-Gate FinFET Devices," *IEDM Tech. Digest,* pp. 437 - 440, Dec.2001.

[94] D. Hisamoto et al., "FinFET—A Self-Aligned Double-Gate MOSFET Scalable to 20 nm," *IEEE Trans. on Electron Devices*, **47**, pp. 2320 – 2325 (2000).

[95] X. Huang et al., "Sub 50-nm FinFET: PMOS," *IEDM Tech. Digest*, pp. 67 - 70, Dec.1999.

[96] J. T. Park and J. P. Colinge, "Multiple-Gate SOI MOSFETs: Device Design Guidelines'", *IEEE Trans. on Electron Devices,* Vol. 49, pp. 2222 – 2229, Dec. 2002.

Nano and Giga Challenges in Microelectronics
Greer at al (Editors)
© *2003 Elsevier Science B.V. All rights reserved*

Electronics Below 10 nm

Konstantin Likharev

State University of New York, Stony Brook, NY 11794-3800

Abstract

This chapter reviews prospects for the development and practical introduction of ultrasmall electron devices, including nanoscale field-effect transistors (FETs) and single-electron transistors (SETs), as well as new concepts for nanometer-scalable memory cells. Physics allows silicon FETs to be scaled down to ~3 nm gate length, but below ~10 nm the devices are extremely sensitive to minute (sub-nanometer) fabrication spreads. This sensitivity may send the fabrication facilities costs (high even now) skyrocketing, and lead to the end of the Moore Law some time during the next decade. Lithographically defined SETs can hardly be a panacea, since the critical dimension of such transistor (its single-electron island size) for the room temperature operation should be below ~1 nm. Apparently, the only breakthrough that would allow to make 1-nm-scale electron devices practical, would be the introduction of "CMOL" hybrid integrated circuits that would feature, in addition to an advanced CMOS subsystem, a layer of ultradense molecular electron devices. These devices would be fabricated by chemically-assisted self-assembly from solution on few-nm-pitch nanowire arrays connecting them to the CMOS stack. Due to the finite yield of molecular devices and their sensitivity to random charged impurities, this approach will require a substantial revision of integrated circuit architectures, ranging from defect-tolerant versions of memory matrices and number crunching processors to more radical solutions like hardware-implemented neuromorphic networks capable of advanced image recognition and more intelligent information processing tasks.

Key words: nanoelectronics, electron devices, memory cells, logic

1. Introduction

The phenomenal success of semiconductor electronics during the past three decades was based on scaling down of silicon field-effect transistors (MOSFETs) and the resulting increase of density of logic and memory chips. The most authoritative industrial forecast, the International Technology Roadmap for Semiconductors [1] predicts that this exponential ("Moore-Law") progress of silicon MOSFETs and integrated circuits will continue at least for the next 15 years. By the end of this period, devices with 10-nm minimum features (transistor gate length L) should become commercially available.

The prospects to continue the Moore Law beyond the 10 nm frontier are more uncertain, and it is very important to understand them. This chapter is an attempt at a review of the recent research results related to the scaling prospects of silicon MOSFETs and possible alternatives to this technology. I will begin the review with a discussion (Sec. 2) of advanced field effect transistors, with a focus on their most prospective variety: double-gate SOI MOSFETs. In Sec. 3, single electron transistors and other Coulomb-blockade-based devices will be discussed. (A brief review of single- and few-electron memory cells and some other new memory ideas is also included in that section.) This will naturally lead us to a discussion of molecular devices and prospects of their self-assembly and hybridization with CMOS technology (Sec. 4). Finally, in Conclusion (Sec. 5) I will try to summarize the basic problems facing future integrated circuits beyond the 10 nm frontier, and possible ways of their solution.

2. Advanced Field Effect Transistors

2.1. *Bulk and SOI MOSFETs*

Bulk silicon MOSFETs (for their detailed description, see, e.g., Refs. 2-5) are extremely versatile electron devices combining a (relatively) easy fabrication with very high performance in a broad variety of logic and memory circuits. Moreover, the devices are scalable to deep-submicron range. This powerful combination has allowed the bulk MOSFET devices to serve as the work horse of the leading electronic technology, CMOS, for more than 30 years. However, as the bulk MOSFETs enter the sub-100-nm range, their further scaling runs into several problems, including short-channel effects and gate oxide leakage – see Ref. 6 for an extensive review of these issues. Despite the recent experimental demonstrations of several bulk transistors with gate length below 20 nm [7-14], performance of these prototypes is far from perfect.

There is a growing consensus (see, e.g., Chapter 1 of this collection) that reaching high performance (good saturation at high ON current and high ON/OFF ratio) below 20 nm will require the use of advanced FETs, primarily double-gate MOSFETs with thin, undoped silicon-on-insulator (SOI) channel connecting highly doped source and drain. The main reasons in favor of this choice are as follows:

1 Such device is a close approximation to what may be called the "ultimate MOSFET", because two gates allow a very effective control of the electrostatic potential of the channel, and hence the carrier transport. (Similar devices with single-gate [15, 16] loose to double-gate devices in scalability, though are certainly preferable to bulk MOSFETs.)

2 Although the fabrication of double-gate transistors is certainly more complex than that of the usual bulk MOSFETs, they have already been implemented in various geometrical versions, including planar [17-24], fin-type [25-33] and vertical [34-37] geometries.

Because of these reasons, the double-gate MOSFETs have become a focus of recent theoretical efforts to understand MOSFET scaling laws and limits [38-61]. These analyses are based on a variety of models and calculation techniques, but give results in the same ballpark. I will present the most recent results [59, 61] based on a model (Fig. 1) that I believe provides the best trade-off of analysis simplicity and result accuracy.

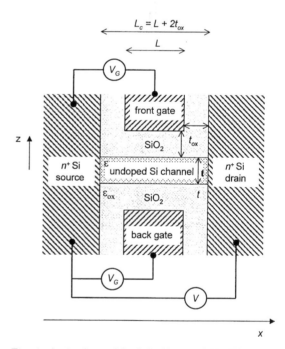

Fig. 1. A simple model of double-gate MOSFETs with ultrathin intrinsic channel. Notice the difference between the gate length L and channel length L_c.

2.2. Model

In this model of a nanoscale n-MOSFET, the channel thickness t is assumed to be so small that the lateral electron confinement (in direction z perpendicular to the channel plane) considerably increases the effective potential energy

$$U_{ef}(x,y) = E_z - e\langle\Phi\rangle(x,y) \qquad (1)$$

of 2D motion of electrons in the channel ($e > 0$). Here

$$\langle\Phi\rangle = \int\limits_{-t/2}^{+t/2} \Phi(x,y,z)\,|\psi(x,y,z)|^2\,dz \qquad (2)$$

is the effective 2D value of the 3D electrostatic potential Φ. Calculations [86] show that if $t \lesssim 3$ nm, while the channel length L_c and width W are much larger than t, the 3D electron wavefunction ψ may be factored as

$$\psi \approx \frac{1}{\sqrt{2t}} \cos\frac{\pi z}{2t} \sum_{E_x,k_y} \psi(x)e^{ik_y y}, \qquad (3)$$

so that the confinement energy

$$E_z = \frac{\pi^2\hbar^2}{2m_h t^2}, \qquad (4)$$

where m_h (close to m_0 for silicon) is the heavy electron mass. For the value $t = 2$ nm, accepted in most illustrations below, E_z is quite considerable (~ 0.1 eV). (The channel is assumed to be in the [102] direction of the silicon crystal, so that the only two valleys with heavy mass in z direction participate in transport; for four other valleys with light mass $m_l \approx 0.19\ m_0$ in z direction, the confinement energy is $m_h/m_l \approx 5$ times higher and for reasonable values of applied voltages these valleys are not populated.)

The functions $\psi(x)$ and $\Phi(x,y)$ (if $W \gg L_c$, the y-dependence of Φ is negligible) may be found by a solution of, respectively, the 1D Schrödinger equation

$$-\frac{\hbar^2}{2m_l}\frac{\partial^2\psi(x)}{\partial x^2} + U(x)\psi(x) = E_x\psi(x), \qquad (5a)$$

$$U(x) \equiv U_{eq}(x,y) - \frac{\hbar^2 k_y^2}{2m_l}, \qquad (5b)$$

and the 2D Poisson equation

$$\frac{\partial^2\Phi}{\partial x^2} + \frac{\partial^2\Phi}{\partial y^2} = \begin{cases} en(x,z)/\varepsilon\varepsilon_0, & |z| \le t/2, \\ 0, & t/2 \le |z| \le t/2 + t_{ox}. \end{cases} \qquad (6)$$

Since $n(x,z)$ is the full 3D density of electrons in the channel:

$$n(x,z) = |\psi(x,y,z)|^2 = \frac{1}{2t}\cos^2\frac{\pi z}{2t}\sum_{E_x,k_y}|\psi(x)|^2, \qquad (7)$$

equations (5ab) and (6) are generally coupled and should be solved self-consistently. After that, drain current density (per unit channel width) may be calculated as

$$J = 4\frac{e\hbar}{m_l}\,\mathrm{Im}\sum_{E_x,k_y}\psi^*(x)\frac{d}{dx}\psi(x) \qquad (8)$$

in any cross-section. (The factor 4 is the product of the number of silicon electron valleys participating in the transport by the spin degeneracy.)

Equation (5) neglects electron scattering inside the channel, i.e. describes ballistic electron transport. This assumption is sometimes questioned on the basis of experiments with doped silicon channels, that indicate considerable scattering. However, results of recent experiments [62] indicate that electron mobility in undoped SOI layers as thin as 5 nm may be very high (> 400 cm^2/V-s), i.e. essentially the same as in good bulk MOSFETs [63] in the equivalent perpendicular field (pressing the electron wavefunction "centroid" equally close to the Si/SiO$_2$ interface). If the similar surface roughness can be sustained down to $t \sim 2$ nm, the mobility should be about 200 cm^2/V-s (decreasing by a factor of two because of electron scattering at both interfaces). This mobility (μ) corresponds to an elastic scattering time $\tau = m_l\mu/e$ close to 25 fs. (The inelastic relaxation time for relevant energies is much longer, of the order of 100 fs [64].) For the most important transport region, where the effective potential $U(x)$ is close to its maximum, the electron kinetic energy is of the order of thermal energy $k_B T \sim 25$ meV, i.e. their average speed $v \approx (3k_B T/m_l)^{1/2}$ is close to 2.5×10^7 cm/s; for this speed the elastic mean free path $l = \tau v$ should be about 6 nm. (This estimate is consistent with results of the recent direct measurements of l in bulk

30

MOSFETs [65].) While this value of l may be lower than the total channel length L, it is still larger than that of the transport bottleneck, where electrons overcome the potential barrier maximum. This estimate means that the ballistic transport may be a reasonable approximation for MOSFETs with high-quality ultrathin undoped channels.

Equations (5ab)-(7) should be solved with appropriate boundary conditions. For the model shown in Fig. 1, an electron leaving the channel has much more chances to be scattered into the bulk drain than back into the channel [66]. As a result, one can use "completely absorbing" boundary conditions that neglect the backscattering completely. (These conditions are frequently used for the analysis of transistor with thin source and drain, but here this assumption is much more questionable, because backscattering may be quite substantial [67].) In order to make these conditions self-consistent, one should assume that occupation of each particular mode of electron propagation in the channel, described by parameters k_y and E_x, is equal to the equilibrium Fermi function of energy $E = E_x + \hbar^2 k_y^2/2m_l + E_z$ in the source (for the few electrons traveling is the back direction, in the drain). This function depends on the level of doping of source and drain. If device-to-device fluctuation of transistor parameters has to be relatively small (this is necessary for acceptable yield of integrated circuit fabrication), the average number N of dopants in electrode regions immediately adjacent to the channel has to be is much larger than one. For sub-10-nm devices the volume V of these regions is of the order of 0.3×10^{-18} cm^3; hence to keep device-to-device fluctuations below 10% the doping rate N/V should be at least as high as 3×10^{20} cm^{-3}, corresponding to deeply degenerate silicon. The numerical results shown below correspond to this value, since higher doping degrades the MOSFET performance.

2.3. Results

Figures 2-4 show typical results of the numerical solutions of the equations of the model discussed above. One can see that for relatively long devices ($L = 10$ nm) the characteristics are close to ideal:

at positive gate voltage the current rapidly saturates at a level considerably larger than the industrial standard (for n-MOSFETs, 600 μA/μm, i.e. 6 A/cm [1]), while the subthreshold curve slope is close to the perfect, thermally-determined value 60 mV/decade. However, as soon as the gate length L is reduced below approximately 5 nm (channel length L_c, below $\sim$ 8 nm), transistor performance starts to degrade. In particular, the saturation becomes less pronounced and is achieved at higher source-drain voltage V. The subthreshold curve slope (Fig. 3) becomes considerably lower than the perfect and shows increasing dependence of V (the so-called drain-induced barrier lowering, or "DIBL") as L decreases. Moreover, the subthreshold curves (plotted on semi-log scale) start bending upward for large negative values of V_G, due to the contribution from quantum-mechanical source-to-drain tunneling along the channel.

The overall degradation of the transistor can be characterized by voltage gain, i.e. the derivative $G_V = \partial V/\partial V_G$ taken at a fixed drain current J. In good MOSFETs, $G_V \to \infty$ at saturation, so this is not a very popular engineering figure-of-merit. (In the sub-threshold region, the notion of DIBL that is essentially G_V^{-1}, is used instead.) However, as the transistor degrades, the voltage gain becomes an important characteristic, since digital logic circuits fundamentally require $G_V > 1$ for their operation. Figure 4 shows G_V as a function of the gate voltage V_G, at fixed drain-source voltage $V = 0.5$ V. One can see that as the gate length is reduced, G_V rapidly decreases, its maximum approaching unity at $L \approx 2$ nm, i.e. at the channel length $L_c \equiv L + 2t_{ox}$ about 5 nm.

2.4. Discussion

Detailed analysis of the results presented above shows that two effects contribute comparably to the device degradation at $L \to 0$: a loss of electrostatic control of the bottleneck potential U_{max} by the gate voltage V_G, and source-to-drain tunneling along the channel.

In order to estimate the electrostatic degradation analytically, one can use the 2D Laplace equation to find electrostatic field distribution for a

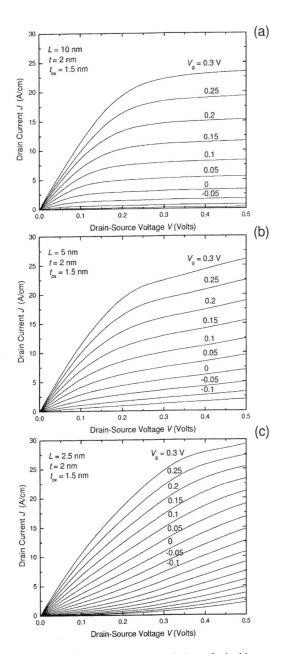

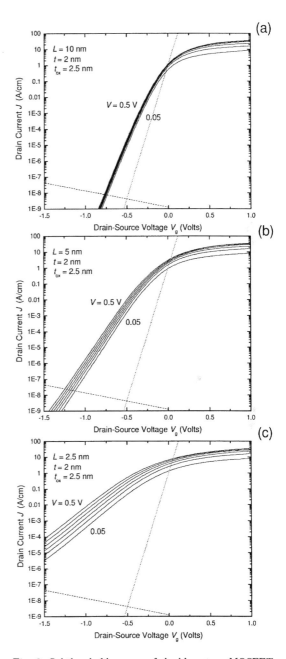

Fig. 2. Drain $I - V$ characteristics of double-gate n-MOSFETs for 3 values of gate length L, numerically calculated [59, 61] using the model shown in Fig. 1. Here and in Figs. 3, 4, 6 and 7 below, the electrode doping level is 3×10^{20} cm^{-3}.

Fig. 3. Subthreshold curves of double-gate n-MOSFETs (Fig. 1) for 3 values of gate length L, each for 10 values of drain-source voltage V (with 50-mV steps) [61]. The dashed lines show the estimated gate oxide leakage current. Notice that the gate oxide is thicker than in Fig. 2. (Subthreshold curves are mostly important for memory applications where oxide leakage should be small). Dotted lines show the ideal slope of 60 mV/decade.

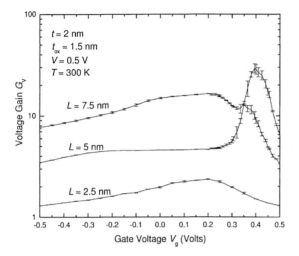

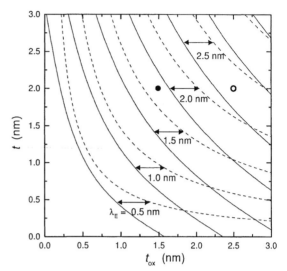

Fig. 4. Voltage gain $G_V \equiv \partial V / \partial V_{G|J=const}$ of nanoscale MOSFETs as a function of gate voltage for $V = 0.5$ V and several values of gate length L [59, 61].

Fig. 5. The contours of fixed characteristic electrostatic length λ_E on the plane of channel thickness t and oxide thickness t_{ox} for Si double-gate MOSFETs ($\varepsilon = 12$, $\varepsilon_{ox} = 3.9$). Solid lines: Eq. (9); dashed lines: the parabolic approximation given by Eq. (11). Solid and open points show the parameter sets used for Fig. 2 ($t_{ox} = 1.5$ nm) and Fig. 3 and 4 ($t_{ox} = 2.5$ nm), respectively.

simplified model of the device [43] in the depletion regime ($n = 0$). The assumption of exponential dependence of the electrostatic potential on x, $\Phi(x, z) = \Phi(z) \exp(\pm x / \lambda_E)$, readily yields the following equation for λ_E:

$$(\varepsilon / \varepsilon_{ox}) \tan(t / 2\lambda_E) \tan(t_{ox} / \lambda_E) = 1. \quad (9)$$

Figure 5 shows levels of constant λ_E on the $[t, t_{ox}]$ plane, calculated using this equation. It is evident that λ_E decreases with both t and t_{ox}. For small values of the $\varepsilon_{ox} / \varepsilon$ ratio, and close values of t and t_{ox}, the following relations are approximately satisfied:

$$(\varepsilon_{ox} / \varepsilon)^2 t_{ox} / \varepsilon_{ox} \ll t / 2\varepsilon \ll t_{ox} / \varepsilon_{ox}. \quad (10)$$

In this case, Eq. (9) is reduced to a simple expression,

$$\lambda_E = (\varepsilon t t_{ox} / 2\varepsilon_{ox})^{1/2}, \quad (11)$$

following also from the so-called "parabolic approximation" [68, 69]. Equation (11) allows a very simple interpretation: this is the standard length $(C_s / C_p)^{1/2}$ of field penetration into a 1D line of series capacitances $C_s = t\varepsilon\varepsilon_0$ (representing the longitudinal capacitance of silicon channel per unit length and width) shunted by parallel capacitances $C_p = 2\varepsilon_0\varepsilon_{ox} / t_{ox}$ (the specific transversal capacitance of two oxide layers in parallel).

One can expect the transistor electrostatics to degrade seriously if the channel length is reduced below $\sim \pi\lambda_E$. For the parameters accepted for Fig. 3 ($t = 2$ nm, $t_{ox} = 1.5$ nm, see the solid point in Fig. 5) this gives the channel length limitation $L_c \approx 5.5$ nm, i.e. $L \approx 2$ nm (Fig. 1), while the increase of t_{ox} to 2.5 nm (the open point in Fig. 5) increases the minimum L_c to ~ 8 nm, i.e. L to about 3 nm. These estimates are in a reasonable agreement with numerical results shown in Fig. 2 and 3, so that one may use Fig. 5 to estimate the transistor electrostatics degradation for other parameters. It is commonly assumed that SiO_2 gate oxide with acceptable leakage and reliability [70-73], and SOI layers [74, 75] with acceptable roughness may be both ultimately thinned to 1 nm. In this case Fig. 5 shows that the electrostatic limit $\pi\lambda_E$ on the channel length is around 3 nm. Some further (probably, modest) improvement of electrostatics can apparently be achieved using the thinned source and drain and more complex structures with nonuniform ("graded") channel and/or gates – see, e.g., Refs. 76-80. Finally, the use of new, high-ε dielectrics such as ultradense silicon oxynitride [81], hafnium oxide [37, 82], zirconium silicate [83], or aluminum oxide [84] (see also Chapter 4 of this collection) may allow the channel length limit to be pushed down a little bit further, possibly to about 2 nm.

However, all these efforts may be inadequate, because of the second important limitation on L_c imposed by source-to-drain tunneling. Let us give a simple estimate of this effect. Since for very short MOSFETs (with channel length of the order of $\pi\lambda_E$) the potential distribution along the channel is rather smooth, a reasonable estimate of tunneling importance may be obtained from the famous Kemble formula (see, e.g., Ref. 85):

$$D(E_x) = \frac{1}{1 + \exp\left[2\pi\frac{U_{\max}-E_x}{\hbar\omega}\right]}, \qquad (12)$$

for the WKB transparency of electron tunneling under an inverted quadratic potential $U(x) = U_{max} - m_l\omega^2(x-x_0)^2/2$. By coincidence, this dependence of barrier transparency on energy E_x has exactly the same functional form as the Fermi distribution of the incident electrons. Using this fact, it is straightforward to show that quantum tunneling under such a barrier dominates over "thermionic" charge transfer over the barrier if the physical temperature T is lower than the so-called "inversion temperature"

$$T_{inv} \equiv \hbar\omega/2\pi k_B. \qquad (13)$$

For the effective potential $U(x) = E_z - e\langle\Phi\rangle$ equal to the Fermi level ε_F of source and drain at $(x-x_0) = \pm L_c/2$, and peaking at U_{max} over that level, we obtain that tunneling dominates at $L_c < L_T$, where

$$L_T \equiv (\hbar/\pi k_B T)(2U_{max}/m_l)^{1/2}. \qquad (14)$$

For $U_{max} = 0.05$ eV (which is the typical barrier height in the transistors discussed above) and $T = 300$ K, Eq. (14) yields $L_T \approx 7$ nm. This estimate agrees well with the numerical results shown in Fig 2-4. Shorter devices, like those shown in Figs. 2c and 3c, operate essentially as "tunnel transistors" were gate voltage controls electron tunnel-

ing through the barrier. [1] For devices with channel length $L_c \lesssim L_T$, this control still may be effective, tunneling corresponding crudely to an increase of effective temperature from T to $T_{inv} \propto 1/L_c$. This conclusion is (at least qualitatively) confirmed by first experiments [7, 8, 15, 16] with sub-10-nm transistors.

The control is, however, virtually lost as soon as $k_B T_{inv}$ becomes comparable to U_{max}, because the tunnel barrier becomes almost completely transparent. Formula (13) shows that this happens at

$$L_c \approx L_{min} = \hbar/(2m_l U_{max})^{1/2}. \qquad (15)$$

U_{max} can hardly be larger than half the bandgap (otherwise Zener tunneling [2] begins), giving for silicon $L_{min} \approx 2$ nm. Notice, however, that in ultrathin channels the conduction band edge rises by E_z given by Eq. (4), while the valence band edge lowers according to a similar formula, but with the hole rather than electron mass. As a result, the bandgap grows (by as much as ~0.5 eV at $t = 1$ nm), so as a matter of principle the length limitation due to source-to-drain tunneling may be decreased even a little bit further.

2.5. Parameter sensitivity

The theoretical predictions made above seem very optimistic: to summarize, they indicate that physics allows FET channel length to be scaled down as deeply as to at least 2 nm, still enabling the performance necessary for operation of logic and memory circuits. However, these results also indicate two major challenges on the way to approaching these limits in commercial practice.

The first problem the rapidly increasing power dissipation [49, 59]. For relatively long transistors $(L_c \gg l)$ that are well described by the drift-diffusion model, the specific power P_0 per unit

---

[1] Earlier, tunnel transistors were discussed mostly in the context of structures with metallic or silicide source and drain which create additional Schottky barriers at channel interfaces – see, e.g., Sec. 9.7 of Ref. 2 as well as recent publications [87-89] and references therein. These barriers cause additional reduction of device transparency; as a result, in most cases the tunnel transistor transconductance and ON current were rather low.

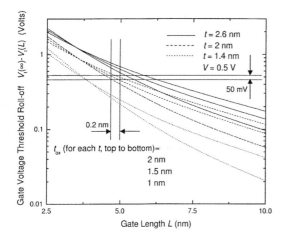

Fig. 6. Threshold voltage shift (relative to that for $L \to \infty$) as a function of gate length L, for all combinations of 3 values for oxide thickness t_{ox} and 3 values for channel thickness t [61].

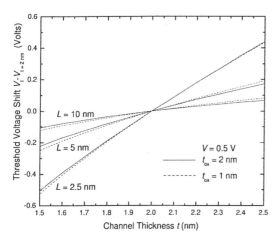

Fig. 7. Threshold voltage dependence on the channel thickness t[61].

channel width decreases with L_c, because drawing a certain ON current density requires approximately the same longitudinal electric field E, but lower drain voltage $V \sim EL_c$. (The real relation of P_0 and V is more complex and depends on many details [3-6], but this does not affect the result qualitatively.) However, as soon as L_c becomes comparable with the mean free path, the power decrease stalls, because the ON current is now limited by the source electron supply exhaustion [39, 40] rather than channel mobility. Moreover, at $L_c \lesssim 10$

nm the transistor degradation due to electrostatics and source-to-drain tunneling leads to increase of P_0 with the channel length reduction, even at the optimum choice of the power supply voltage V_{DD} [49, 59]. This increase would further exacerbate the power management problem that is very severe even now [1, 90, 91].

However, even more significant is the second problem, the rapidly increasing sensitivity of transistor characteristics to unavoidable random variation of geometrical parameters, due to fabrication uncertainties. For example, Fig. 6 shows the decrease ("roll-off") of the threshold gate voltage V_t (defined as the value of V_G providing a certain small drain current) with a reduction of the gate length. The plot shows that small changes in the length definition may lead to large variations of V_t. These variations should be compared with the minimum power supply voltage V_{DD} necessary for driving the device.

For example, consider a relatively long device with physical gate length $L = 9$ nm, planned by the ITRS [1] for commercial introduction in approximately the year 2016. The same document (see, e.g., Table 57) predicts that by then the critical dimension control accuracy (at the 3σ level) will reach $\sim$0.7 nm. Figure 6 shows that this control would enable variations of V_t to be kept within approximately 50 mV. On the other hand, Fig. 2a shows that in order to keep ON current density at the standard 600 μA/μm level [1], the voltage swing V_{DD} should be above 300 mV; hence the 50 mV variation seems acceptable.

Now consider a shorter device with $L = 5$ nm, $t = 2$ nm, and $t_{ox} = 1.5$ nm ($L_c = 8$ nm). According to Fig. 6, in order to keep fluctuations of V_t below 50 mV, the critical dimension should be controlled better than $\sim$ 0.2 nm, much tighter than the farthest ITRS projections for even the most advanced lithographic techniques such as EUV [97, 98], currently in the stage of laboratory development.

Moreover, Fig. 7 shows that V_t is even more sensitive to the channel thickness t, mostly because of the strong dependence of the quantum confinement energy E_z on t – see Eq. (4). For example, in order to keep fluctuations of V_t at the same level (below 50 mV), t should be controlled better than $\sim$0.1 nm, a very hard task indeed – see, e.g., Chapter 3 of this collection.

The necessity to ensure such tight control of device dimensions will lead to a rapid increase of fabrication cost facilities, that may reach the point of diminishing economic returns. As a result, the Si-MOSFET-based exponential Moore-Law progress may stop at $L \sim 10$ nm, i.e., long before fundamental physical limits have been reached.

2.6. *Alternative FET materials*

In view of this situation, it is important to take a second look at the numerous suggestions to replace silicon with another channel material. Among those, single-wall carbon nanotube FETs [92-96] seem to be most promising. (Results were obtained for multi-wall carbon nanotubes [96] are less impressive.) A possible advantage of these structures could be smaller interface roughness that may allow to implement small effective values of channel thickness t ($\sim$1.5 nm) while still sustaining reasonable mobility. However, for sub-10-nm transistors the requirement of very high mobility fades away, because current is more limited by source exhaustion. (Since electrode material is different from that of the nanotube, current may be also substantially limited by contact resistances.)

In addition, present day production methods give nanotubes with random helicity, which strongly affects the parameters of longitudinal electron transport, effectively changing them from metals all the way to semiconductors with considerable bandgap. Finally, carbon nanotubes (or any other channel material) cannot help solve the key problem of transistor parameter sensitivity to channel length – see Sec. E above. As a result, I do not believe that an alternative channel material may be a universal remedy against the sub-10-nm transistor woes.

3. Single Electron Devices

3.1. *Motivation*

The problem outlined in the end of the last section motivates a search for new nanoscale electronic devices based on different physics. General guidelines for such a search may be provided by the following arguments [99] based on the famous particle-wave duality. Quantum mechanics says that electrons may behave either as discrete particles or continuous de Broglie waves, depending on experimental conditions. (Surprisingly enough, a sufficiently clear understanding of these conditions for conduction electrons in solids was achieved not so long ago, in the 1980s – for reviews see, e.g., Refs. 100-102.)

Consider for example a generic situation where two parts of a conductor are separated by some interface, and ask whether the electric charge of each conductor is a multiple of the fundamental charge e at any instant (this is natural for the particle picture) or may be continuous (the wave picture allows this, because the wavefunction of each electron may be split between the two parts). The answer to this question turns out to be dependent on whether the effective tunnel resistance R of the interface is larger or smaller than the natural quantum unit of resistance

$$R_Q \equiv \hbar/e^2 \approx 4.1 k\Omega. \qquad (16)$$

If the resistance is low, $R \ll R_Q$, the charge of each conductor may be continuous, but in the opposite limit it may be only a multiple of e.

This relation may be derived and explained in numerous ways; perhaps the simplest interpretation is as follows. In a closed ("Hamiltonian") quantum system, the characteristic energy of quantum fluctuations per degree of freedom is $E_Q \sim \hbar\omega/2$, where ω is a characteristic frequency. In contrast, each part of the conductor we are discussing, concerning its electric charge degree of freedom, is an "open" system, strongly interacting with its environment (in classics, corresponding to an RC relaxator, rather than an LC oscillator). For such a system, $\hbar\omega$ should be replaced by $\hbar/\tau$, with $\tau = RC$, where C is the capacitance between the two conducting parts. Transfer of a single electron between the parts causes an electrostatic energy change of the order of $E_C \sim e^2/2C$. Comparing E_C and E_Q we notice that C cancels, and see that if $R \ll R_Q$, quantum fluctuations smear out the electrostatic energy difference which tries to keep the electric charge of each part constant.

To comprehend the importance of this result, let us combine it with the so-called Landauer formula for the interface conductance $G \equiv 1/R$ [100-102]:

$$G = (e^2/\pi\hbar)\Sigma_i D_i \qquad (17)$$

where D_i is the interface transparency (i.e. the probability of electron transmission) for a particular transversal mode of electron propagation; the sum is over all the modes. Comparison of Eqs. (16) and (17) shows that if the conductor cross-section is so narrow that quantum confinement makes only one propagating mode possible, the condition of electric charge discreteness takes a simple and natural form: $D \ll 1$. However, for devices with a larger cross-section A, the restriction on the average transparency is much more severe: $D \ll 1/N$, where N is the transversal mode number. (For a degenerate conductor, N is of the order of λ_F^2/A where λ_F is the Fermi wavelength, typically of the order of 1 nm.)

An FET transistor is a good example of a device where the number n of electrons in the channel is never quantized, because the boundaries between the electrodes and channel are typically highly transparent ($D \sim 1$). As a result, these transistors do not exhibit single-electron charging effects even if the *average* number of electrons in the channel is small. An adequate understanding of such devices may be achieved using the wave language, used in particular in Sec. 2 above.

Theoretically, for nanoscale "wave" devices ($R \ll R_Q$), the FET-type control of transport is not the only possible mode of operation: the effects of quantum interference of electron de Broglie waves can be, as a matter of principle, used for this purpose as well. In the 1980s and early 1990s, much attention was focused on such "quantum electronic devices" – see, e.g., Refs. 103-105. However, later the prospects for their practical applications have been recognized as rather poor, mostly for the following reason. In contrast to optical phonons (which obey Bose statistics), charge carriers in solids are fermions and in particular obey the Pauli principle: each of them must have a different energy, and hence a different de Broglie wavelength. Hence, high-contrast interference patterns require operation with either a single transversal mode or a small number of modes. The accuracy δL of size

definition of the nanostructures that single out such a mode from a continuum, and then handle its interference, should be much better than the de Broglie wavelength $\lambda = h/(2mE)^{1/2}$ of the used electrons. Simultaneously, the electron energy E should be well above the thermal fluctuation scale (typically, $\sim k_B T$) in order to avoid interference pattern smearing by thermal fluctuations. Combining these two requirements, and plugging in fundamental constants, we may see that for room temperature δL should be, as in nanoscale MOSFETs, well below 1 nm. (This is only natural, because λ has the same order of magnitude as L_T defined by Eq. (14)). This simple estimate show that quantum interference devices do not have any substantial advantage over FETs for sub-10-nm scaling. [2]

The above discussion pertains to *spatial* quantum interference; one can also consider using *temporal* quantum coherence of electrons for information processing. The most prominent ideas put forward in this field are those of quantum encryption and quantum computing – see, e.g., [113]. Presently, the former goal seems to be much closer than the latter; however, both of them represent rather narrow application niches. For most tasks faced by digital electronics, quantum computing does not seem to offer significant advantages over

---

[2] I am aware of two important exceptions of this conclusion. First, if the only critical dimension of a device may be defined by film thickness, it may be readily controlled with sub-nm accuracy. This is the case of vertical resonance diodes [106, 107]. If such devices with acceptably large peak-to-valley ratio are implemented in CMOS-compatible technology (see, e.g., Ref. 108) there will be hope for their practical introduction. Unfortunately, the range of possible applications of these two-terminal devices is probably limited to fast semiconductor memories (challenging the current SRAM technology). The second important exception are Cooper pairs in superconductors, that are bosons rather than fermions. As a result, many Cooper pairs may have exactly the same wavefunction, enabling the so-called macroscopic quantum interference effects in structures much larger than λ - see, e.g., Refs. 109, 110. These effects are used, in particular, in fast, ultra-low-power Rapid Single-Flux-Quantum (RSFQ) logic circuits – see, e.g., recent reviews [111, 112]. Unfortunately, these circuits need deep refrigeration, the fact that has so far has hindered their wide practical introduction.

the usual ("classical") computing, while being much harder to implement. Because of this reason, I will abstain from discussing this issue in this chapter which deals with potentially practicable digital technologies. (A brief review of quantum computing may be found in the last chapter of this collection.)

To summarize, in the category of "wave" electron devices ($R \ll R_Q$) we are left with not much more than the field effect transistors. This motivates us to move to nanoscale structures with high impedance ($R \gg R_Q$), dominated by single-electron charging effects.

3.2. Single-electron box

By now, the basic physics of single-electron devices has been developed quite deeply, but is known within a much narrower circle than that of MOSFETs. Therefore I will give its brief review. (More detailed reviews may be found, e.g., in Refs. 114-117.) Let us start with a generic single-electron device, frequently called the "single-electron box" - Fig. 8a. [3]

The device consists of just one small conductor ("island") separated from an external electrode by a tunnel barrier with high resistance,

$$R \gg R_Q. \qquad (18)$$

An external electric field may be applied to the island using a capacitively coupled gate electrode. The field changes the local Fermi level of the island and thus determines the conditions of electron tunneling. Elementary electrostatics shows that the energy of the system may be presented as

$$W = Q^2/2C_\Sigma - (C_g/C_\Sigma)QV_g + const, \qquad (19)$$

where $Q = -ne$ is the island charge (n is the number of uncompensated electrons), C_g is the island-gate capacitance, while C_Σ is the total capacitance of

---

[3] The basic physics of this device was understood by Lambe and Jaklevic [118] as early as in 1969, on the basis of their experiments with disordered granular structures, while the first quantitative theory of the box was developed by Kulik and Shekhter [119]. The first experiments with an individual box were, however, carried out much later [120], after key experiments with other, more complex, single-electron devices.

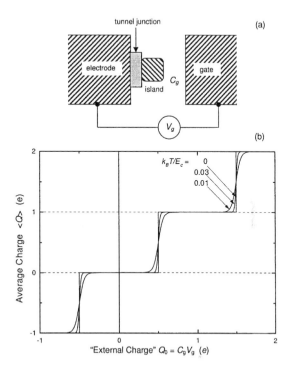

Fig. 8. Single-electron box: (a) schematics, and (b) the "Coulomb staircase", i.e. the step-like dependence of the average charge Q of the island on the gate voltage V_g, for several values of relative temperature.

the island (including C_g). It is convenient to rewrite Eq. (19) expression in another form,

$$W = (Q_0 - ne)^2/2C_\Sigma + const, \quad Q_0 \equiv C_g V_g, \quad (20)$$

where parameter Q_0 is usually called the "external charge". From its definition, it is evident that in contrast with the discrete total charge Q of the island, the variable Q_0 is continuous, and may be a fraction of the elementary charge e.

At sufficiently low temperatures,

$$k_B T \ll E_C, \quad E_C \equiv e^2/C_\Sigma, \qquad (21)$$

the stationary number n of electrons in the island corresponds to the minimum of W; an elementary calculation using Eq. (20) shows that Q is a step-like function of Q_0, i.e. of the gate voltage (Fig. 8b), jumping by e when

$$Q_0 = e(n + \frac{1}{2}), \quad n = 0, \pm 1, \pm 2, \ldots \qquad (22)$$

38

If the temperature is increased to $k_B T \sim E_c$, the system has non-vanishing probability p_n to be in other states as well, with

$$p_n = \frac{\exp\{-W(n)/k_B T\}}{\sum\limits_n \exp\{-W(n)/k_B T\}}. \qquad (23)$$

A straightforward calculation of the average charge $\langle Q \rangle = -\sum\limits_n enp_n$ yields the pattern shown in Fig. 8b: the step-like dependence of charge on gate voltage is gradually smeared out by thermal fluctuations. This is typical for all single-electron devices, so that the operation temperature of most of them should satisfy Eq. (21). (A notable exception are single-electron temperature standards that operate at $k_B T \sim E_C$ – see, e.g., Refs. 121, 122.)

The physics of this "Coulomb staircase" is very simple: increasing gate voltage V_g tries to attract more and more electrons to the island. The discreteness of electron charge, provided by low-transparency barriers (with $R \gg R_Q$) ensures that the changes may be only discrete. Notice that in this sense the box (and all other single-electron devices) are not really a "quantum electron device", or at least much less so than the usual field-effect transistor.

3.3. Single-Electron transistor

A simple modification of the single-electron box, splitting its electrode into two parts (source and drain), so that voltage V may be applied between them, turns it into a very important device, the single-electron transistor (Fig. 9a). This device, that was first suggested in 1985 [123, 124] and first implemented two years later [125], is clearly reminiscent of an FET, but with a small conducting island limited by two tunnel barriers, instead of the usual channel, connecting the source and drain.

Figure 9b shows a typical set of dc $I - V$ curves of such transistor, for several values of the "external charge" Q_0, defined in the same way as in the single-electron box – see the second of Eqs. (20). One can see that at small drain-to-source voltage V, there is virtually no current, besides the special values of Q_0 given by Eq. (22). The physics of this phenomenon (the "Coulomb blockade") is easy to

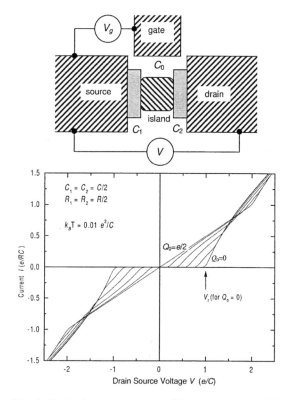

Fig. 9. Single-electron transistor: (a) schematics, and (b) a typical set of source-drain $I - V$ curves of a symmetric transistor for several values of the "external charge" Q_0, i.e. of the gate voltage V_g, calculated using the "orthodox" theory of single-electron tunneling.

understand: even if $V > 0$, and thus it is energy-advantageous for a electron to go from source to drain, on its way the electron has to tunnel into the island first, and change its charge Q it by $\Delta Q = - e$. Such charging would increase the electrostatic energy W of the system

$$W = (Q_0 - ne)^2/2C_\Sigma$$
$$- eV(n_1 C_2 + n_2 C_1)/C_\Sigma + const, \qquad (24)$$
$$C_\Sigma \equiv C_g + C_1 + C_2,$$

(where n_1 and n_2 are the numbers of electrons passed through the tunnel barriers 1 and 2, respectively, so that $n = n_1 - n_2$), and hence at low enough temperatures ($k_B T \ll E_C$) the tunneling rate is exponentially low.

At a certain threshold voltage V_t the Coulomb blockade is overcome, and currents starts to grow

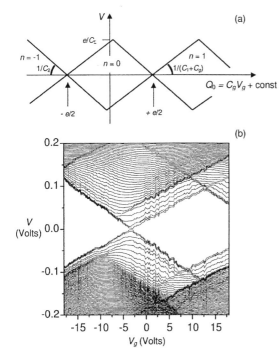

Fig. 10. Coulomb blockade threshold voltage V_t as a function of Q_0: (a) theoretical dependence at $T \to 0$, and (b) experimental contour plots of current on the $[V, V_g]$ plane for an aluminum SET with $E_c \approx 100$ meV at $T = 4.2$ K [126].

with V. The most important property of the single-electron transistor is that V_t is a periodic function of V_g, vanishing in special values of gate voltage, given by Eq. (22) – see Fig. 10. The reason for these so-called "Coulomb blockade oscillations" is evident from the above discussion of the single-electron box: in the special points (22), one electron may be transferred to the island from either drain or source without changing the electrostatic energy of the system even at $V = 0$. Hence, an electron can tunnel from the source to the island and then to the drain even at negligible V, so that $V_t = 0$. As can be readily shown from Eq. (24), at low temperatures the dependence of V_t on V_g is piecewise-linear, with its lower and upper branches forming the so-called "diamond diagram" (Fig. 10) [123].

Since the $I - V$ curves of the transistor are continuous (Fig. 9b), if a small current is fixed by an external circuit, V is close to V_t and also follows

the diamond diagram – see, e.g., Fig. 10b. Thus the voltage gain and transconductance of single-electron transistors may change sign depending on the gate voltage – an important difference in comparison with usual field-effect transistors. On the other hand, the same diamond diagram shows that the voltage gain is limited by a capacitance ratio: $(G_V)_{max} = C_g/C_2$ [124]. It may be higher than unity (see, e.g. experiments [127-129]), but hardly much higher than that, especially in room-temperature transistors.

3.4. Single-electron trap

Another key device, the "single-electron trap", may also be understood as a generalization of the single-electron box. Let us replace the single tunnel junction in Fig. 8a with a one-dimensional array of $N > 1$ islands separated by tunnel barriers - Fig. 11a. [4] The main new feature of this system is its internal memory, i.e., bi- or multi-stability: within certain ranges of applied gate voltage V_g the system may be in one of two (or more) charged states of its edge island (Fig. 11b).

The reason for this multi-stability stems from the peculiar properties of an electron inside a 1D array: an electron located in one of the islands of the array extends its field to a certain distance [133, 134] and hence may interact with the array edges (is attracted to them). As a result, the electrostatic self-energy of the electron has a maximum in the middle of the array. By applying sufficiently high gate voltage $V_g = V_+$ the energy profile may be tilted enough to drive an electron into the edge island; if the array is not too long, other electrons feel its repulsion and do not follow. If the gate voltage is subsequently decreased to the initial level $(V = 0)$, the electron is still trapped in the edge island, behind the energy barrier. In order to remove the electron from the trap, the voltage has to be reduced further, to $V_g < V_- < 0$. As a result, the $n(V_g)$ dependence exhibits regions of bi-

[4] This device was first discussed explicitly in 1991 [130, 131], but in fact it may be considered just a particular operation mode of a more complex device, the single-electron turnstile, invented earlier [132].

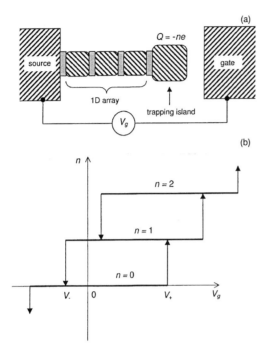

Fig. 11. (a) Single-electron trap and (b) its static characteristic (schematically).

or multi-stability, in which the charge state of the trap depends on its prehistory (Fig. 11b).

The retention time of a certain charge state within the multi-stability region is fundamentally limited by the thermal activation over the energy barrier, and a higher-order quantum process, cotunneling [135]. The first effect is exponentially low in E_c/k_BT, while the second effect falls exponentially with the array length N. As a result, electron retention time may be very long; experimentally, single electron trapping for more than 12 hours has been demonstrated [136-138].

Note that a similar multi-stability may be also achieved in the simpler device, single-electron box (Fig. 8a) if the tunnel barrier is so thick that the reciprocal tunneling rate Γ^{-1} is longer than the measurement time scale. However, in order to suppress this barrier and ensure fast box recharging, the energy eV_g available from the applied gate voltage should be comparable with the barrier height, typically of the order of a few electron-volts. On the other hand, in the trap the necessary energy is of the order of E_C, and in low temperature ex-

periments may be much lower. However, for room-temperature devices, E_C should be also of the order of 1 eV or higher (see Fig. 13 and its discussion below), and the advantage of the trap over the box fades away.

3.5. Single-electron parametron

The last key single-electron device, the so-called parametron [140] is essentially a short segment of a 1D array of islands, galvanically detached from both electrodes. The simplest version of the device uses three small islands separated by two tunnel barriers (Fig. 12a). For simplicity, I will describe its operation for the case when the system is charged by one additional electron, though such pre-charging is in fact unnecessary [141].

Let the parametron be biased by a periodic "clock" electric field $E_c(t)$, oriented vertically, for example, by a slight vertical shift of the central island. (Of course, the same effect may be achieved in a strictly linear array, using a special gate located closer to the central island [141].) This field keeps an extra electron in the central island during a part of the clock period. At some instant, the field E_c reaches a certain value E_t at which electron transfer to one of the edge islands becomes energy advantageous. If the system were completely symmetric, the choice between the two edge islands would be random, i.e. the system would undergo what is called the "spontaneous symmetry breaking". However, even a small additional field E_s applied by a similar neighboring device(s) may determine the direction of electron tunneling at the decision-making moment. Once the energy barrier created by the further change of the clock field has become large enough, the electron is essentially trapped in one of the edge islands, and the field E_s may be turned off. Now the device itself may serve as a source of the dipole signal field E_s for the neighboring cells. The sign of this field (i.e. of the electric dipole moment of the device) presents one bit of information.

Figure 12b shows the phase diagram of the parametron [141], that gives a quantitative description of the properties described above. Each of "ON" states (with the extra electron in one

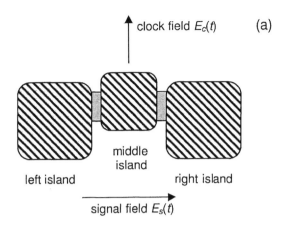

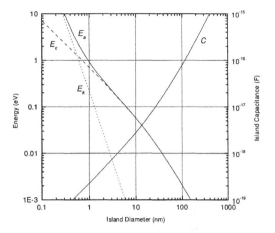

Fig. 13. Single-electron addition energy E_a (solid line), and its components: charging energy E_C (dashed line) and electron kinetic energy E_k (dotted line), calculated for a simple model of a single-electron island [117].

device.

3.6. Scaling and implementation

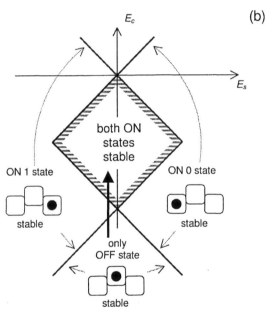

Fig. 12. (a) Single-electron parametron and (b) its phase diagram for the case of precharging by one extra electron. The bold arrow shows the evolution described in the text.

of the edge islands) is stable within an angle-limited region. These regions overlap providing the (hatched) diamond-shaped region of cell bistability. The route from "OFF" state (with the electron in the central island) to one of the ON states, that was described above, in shown by the bold arrow.

Recently, low-temperature prototypes of the single-electron parametron were experimentally demonstrated by two groups [142, 244, 309], albeit one of them prefers to use a different name for this

Before discussing the current and prospective applications of single-electron devices, we should consider their scaling. The most important conditions for operation of such devices are given by Eqs. (18) and (21). The former condition, $R \gg \hbar/e^2 \approx$ 10 kΩ, is relatively easy to satisfy experimentally, since resistance R of a tunnel barrier grows exponentially with its thickness. On the other hand, this relation shows that the output resistance of single-electron devices is unavoidably high, practically above $\sim$ 100 kΩ. This feature is frequently cited as a drawback. However, what is really important for applications (e.g., the interconnect recharging speed) it the maximum current density per unit width. For a room-temperature SET with $V \sim V_t \sim E_C/e \sim 1$ Volt, the total current would be below 10 μA. However, since width of such transistor has to be very small, $\sim$ 1 nm (see below), the available current density may be well above 1,000 μA/μm, i.e., even higher than that of standard silicon MOSFETs [1].

The latter condition (21) makes the practical implementation of single-electron transistors operating at room temperature rather problematic. Figure 13 shows the energy E_a necessary to put an additional electron on a transistor island of a certain radius, calculated within the framework of a simple model [117]. This energy is crudely a sum of the electrostatic contribution $E_C = e^2/C$, where C is the island capacitance, and quantum confinement energy E_k. While E_C dominates for relatively large islands, for 1-nm-scale islands with size comparable with the electron de Broglie wavelength, E_k becomes substantial. [5]

Both theory and experiment show that single-electron tunneling effects (i.e. some current modulation by gate voltage in single-electron transistors) become visible at $E_a \sim 3k_BT$. It means that in order to notice these effects at $T \sim 100$ mK (the temperature routinely reached in standard helium-dilution refrigerators), E_a should be above ~ 25 μeV, corresponding to the island capacitance $C \sim 5 \times 10^{-15}$ F and island size of the order of 1 micron, with tunnel junction area $\sim 0.1 \times 0.1$ μm^2. Such dimensions can be routinely reached by several methods, including notably the now-classical technique of metal evaporation from two angles through hanging resist mask (most typically formed by direct e-beam writing on a double-layer resist). This method was used, in particular, for the first experimental demonstration of a stand-alone single-electron device [124]; later this technique, complemented with subsequent island oxidation, has been advanced to increase E_a to ~ 0.1 eV in certain samples – see, e.g., Fig. 10b [126]. This is sufficient to see a slight current modulation even at room temperature.

However, for reliable operation of most digital single-electron devices, the single-electron addition energy should be approximately 100 times larger than k_BT. This means that for room temperature operation, E_a should be as large as ~ 3 eV. According to Fig. 13, this value corresponds to island size about 1 nm. Reproducible fabrication of integrated circuits with features so small presents quite a challenge. Most claims of success in this direction have been based on results from evidently irrepro-

ducible structures, for example, arrays of nanoparticles deposited between rather distant source and drain. (By chance, one of these particles may be tunnel-coupled to its neighbors much more weakly than are others, forming a single-electron transistor island, while strongly coupled particles effectively merge, forming its source and drain.) Of course, the parameters of such transistors are unpredictable, and there is no hope of using them in integrated circuits.

A more interesting option is fabrication of discrete transistors with scanning probes, for example by nano-oxidation of metallic films [145, 146] or manipulation with carbon nanotubes [147-149]. For the former devices, single-electron addition energies as high as 1 Volt have been reached [146], though unfortunately the current was very low (below 10^{-11} A). Moreover, the scanning probe techniques are so slow that there is no hope for their use in the fabrication of circuits of any noticeable integration scale, though for discrete devices this approach may be promising.

Several methods which are closer to standard CMOS technology have also been used to fabricate single-electron transistors, mostly by the oxidation of a thin silicon channel until it breaks into one or several tunnel-coupled islands – see the pioneering work [150] and recent results [151-154]. The highest values of E_a reached in those efforts (~ 250 mV [154]) are still not high enough for logic applications, but for some memory cells (see below) this value is almost sufficient. A problem with this approach is that the parameters of the resulting transistors are still highly irreproducible, and there is little chance of changing this situation without the development of patterning technologies with a sub-nm resolution – a very distant goal indeed (see, e.g., the next chapter of this collection).

A radical way to overcome the reproducibility is to use natural 1-nm-scale objects of exact size and shape: molecules. Starting from the mid-1990s, several groups have managed to measure electron transport through a single molecule (or a few molecules in parallel) placed between two

---

[5] Notice that at $E_k \stackrel{>}{\sim} E_C$, the basic ("orthodox") theory of single-electron tunneling [115] should be modified [142, 143] (see also review [144]), although the device implications do not change significantly.

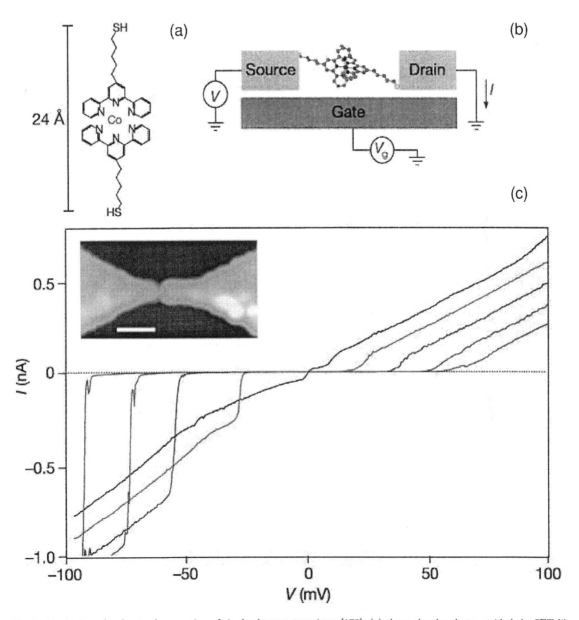

Fig. 14. A single-molecular implementation of single-electron transistor [172]: (a) the molecules that provided the SET-like characteristics shown in panel (c) for several gate voltages. Panel (b) shows the scheme of the experiment, while a SEM picture of the used nanoelectrodes is shown in the inset.

metallic electrodes [6] [155-178, 314, 315]. During this period, the accuracy and reproducibility of such experiments has been improved dramatically, and some of the observed phenomena have already been understood, at least on semi-quantitative basis. The first attempts at qualitative explanation of the transport data [179-184] (see also Chapter 6 of this collection) have been only partly successful so far, apparently because the chemistry of molecule-to-metal interfaces still has to be understood in more detail for each particular method of structure formation.

Besides some very short molecules (see, e.g., Refs. 172, 173, 176) that feature high low-voltage conductance $G \sim e^2/h$ (that is interpreted, in terms of Eq. (17), as high transparency D of the molecule for at least one transversal mode), most samples show $I - V$ curves reminiscent of those shown in Fig. 9b – see, e.g., Fig. 14b. This suppression of current at low voltages is typically interpreted as follows: if the molecule is long enough, it would typically feature a part where electron orbitals are not too much affected by metallic electrodes. These orbitals typically feature considerable discreteness, and hence a substantial energy gap Δ between its higher occupied molecular orbital (HOMO) and lowest unoccupied orbital (LUMO). If the applied voltage is small, Fermi levels of both electrodes have a good chance to be inside this gap, so the conductance is small. (If $\Delta \gg k_B T$, the only mechanism is quantum-mechanical tunneling through the molecule as a whole; this tunneling is very small is the molecule length is above a few nanometers – cf. the discussion of MOSFETs in Sec. 2.3) As the applied voltage increases, either HOMO or LUMO is eventually aligned with one of the Fermi levels, and electrons may pass through the molecule in two hops (from one electrode to the inner orbital and then to another electrode), resulting in considerable increase of current.

This physics is very close to that of single-electron transistor (essentially identical to it, if the discreteness of inner energy levels is taken into account [142, 143]). Consequently, one might expect that if the electrostatic potential of the molecule interior could be changed by voltage V_g applied a galvanically-insulated gate electrode, the Coulomb blockade threshold voltage would be changed in the piece-linear fashion shown in Fig. 10a. This conclusion has been confirmed in first experiments [156, 165, 167, 169, 171-173, 314] where such gate has been arranged – see Fig. 14b. Hence, one may state that single-molecule, single-electron transistors have already been implemented.

So far, these devices have been formed by techniques excluding practical fabrication of integrated circuits. However, there are very good prospects for chemical synthesis of special molecules that would combine the structure suitable for single-electron tunneling with the ability to self-assembly from solution on prefabricated nanostructures, with acceptable yield – see, e.g., the discussion in Ref. 185. Then a way to generically inexpensive fabrication of VLSI circuits would be open. We will come back to this opportunity in Sec. 4 below.

3.7. Random background charge problem

Besides fabrication, single-electronics faces one more serious problem. Let a single charged impurity be trapped in the insulating environment, say on the substrate surface, at a distance r from a single-electron island, comparable to its size a. The impurity will polarize the island, creating on its surface a polarization ("image") charge of the order of $e(a/r)$, that is effectively subtracted from the external charge Q_0 – see, e.g., Eq. (20). This charge affects all characteristics of single-electron devices, for example, in the single-electron transistor it shifts the Coulomb blockade threshold V_t – see Fig. 10. For $r \sim a$, this shift may be large, of the order of $(V_t)_{max}$, even from a simple impurity. Using even the most optimistic estimate compatible with experimental data, 10^9 cm^{-2} [186], for the minimum concentration of charged impu-

---

[6] The list of employed configurations includes scanning tunnel microscopes (STM) [155, 156, 158, 164-166, 168, 169], "break junctions" (mechanically controlled cracks in narrow metallic wires) [156, 160, 162, 170, 174], crossings of two narrow wires [159, 163,], nano-orifices in a thin film [161, 175], gaps in nanowires, either narrowed [167] or created [172, 173, 314] by electromigration, and even junctions fabricated on an STM tip [171].

rities, and assuming 1-nm island size, we can estimate that at least 10^{-3} part of the chip area would be "poisoned", so that the same fraction of single-electron devices will have an unacceptably large background charge fluctuation, $\delta Q_0 \gtrsim 0.1 \; e$.[7]

A possible way to circumvent this problem is the use single-electron transistors with resistive (rather than capacitive) coupling [124, 187-188], which are insensitive to background charge. These devices require "Ohmic" resistors with very high resistance (above ~ 1 MΩ) and quasi-continuous ("sub-electron") transfer of charge that would provide the compensation for the fractional part of the random background charge. Such resistors have been indeed demonstrated (see, e.g., Ref. 190), however, their implementation for room temperature operation presents a problem. In fact, theoretical analyses [191-193] show that in order to provide the continuous transfer of charge a diffusive conductor has to be much longer than the electron-phonon interaction length. For most materials at room temperature, this length is well above 10 nm (see, e.g., Ref. 64), i.e. much larger than the desirable size of the whole device. Moreover, the stray capacitance of such a resistor would be much larger than that of the island itself, reducing its single-electron charging energy and making room temperature operation impossible. As a matter of principle, electron hopping in quasi-insulators may ensure higher resistance at smaller resistor length; however, results of a recent analysis of quasi-continuous transport in this regime [194-196] also leads to pessimistic conclusions concerning the resistor size. For these reasons, prospects for room-temperature operation of any resistively-coupled single-electron devices do not look encouraging.

---

[7] It would be unfair to say that such poisoning is the specific problem of single-electronics. The electrostatic potential created by a single charged impurity in a typical dielectric is of the order of 1 Volt, the typical voltage scale for most electron devices including field-effect transistors. This is why this effect is important for *all* nanoscale devices. The only reason why it was noted in single-electron devices first is that these devices retain high charge sensitivity even if their islands are much larger (this is acceptable for low-temperature operation).

3.8. *Electrometry*

Now we are ready to discuss applications of single-electron devices. Unfortunately, due to space and time restrictions, I will not be able to discuss their use in fundamental physical experiments, and also in such interesting but narrow areas of electronics as single-electron spectroscopy (for a review, see, e.g., Ref. 144), dc current standards, and temperature standards [121, 122]. (For general reviews of analog and metrological applications of single-electron devices see, e.g., Refs. 131, 197 and 198.) I have to mention, however, one analog application which may become important for future integrated circuits, namely using single-electron transistors as ultrasensitive electrometers.

If the source-drain voltage V applied to a single-electron transistor is slightly above its Coulomb blockade threshold V_t, source-drain current I of the device is extremely sensitive to the gate voltage V_g. In fact, Figs. 9, 10 show that even the changes δV_g corresponding to sub-single-electron variations of the external charge lead to measurable variations of I. Calculations based on the orthodox theory have shown [199, 200] that the optimized charge sensitivity of such an electrometer is limited by its white (combined Johnson-Nyquist and shot) noise at the level

$$\delta Q \approx (k_B T R \Delta f)^{1/2} \times \begin{cases} 5.4C, & C_i << C, \\ 2.7C_i, & C_i >> C, \end{cases}$$

$$(25)$$

where $R = R_1 = R_2$ and $C = C_1 = C_2$ ($\ll e^2/k_B T$) are resistance and capacitance of each tunnel junction of the transistor, C_i is the effective output capacitance of the signal source (including the capacitance of the wires connecting it to the transistor), and Δf is the measurement bandwidth. This sensitivity is not at all impressive if C_i is large on the scale of C, which is typically very low (Fig. 13). On the other hand, if the source is so small and so close to the single-electron transistor that C_i of the order of C, the white noise limits the charge sensitivity only at an extremely low level of the order of $10^{-6}e/\sqrt{\text{Hz}}$. This is some 7 orders of magnitude better than sensitivity of the best commercially available instruments, and about 4 orders of magnitude more sensitive than specially designed low-temperature MOSFETs.

Tunnel barriers and electrostatic environment of single-electron devices always contain electron trapping centers and other two-level systems, each capable of producing "telegraph noise" – discrete, random low-frequency variations of the barrier conductance. An ensemble of these variations, with exponentially broad distribution of parameters, produces excess $1/f$-type noise; in single-electron transistors such noise may be very high, typically limiting the charge resolution at the level of the order of $10^{-4}e/\sqrt{\text{Hz}}$ for a-few-Hertz signal frequencies. The sensitivity may be reduced radically (to a level below $10^{-5}e/\sqrt{\text{Hz}}$ [201]) using special stacked geometry in which the single-electron island is lifted over the substrate that apparently hosts most of $1/f$ noise sources. Another way to reach a similar sensitivity is to modulate the transistor parameters, and pick out its output, at a GHz-range frequency [202, 203]. Such modulation cannot beat the white-noise-imposed limitation of sensitivity [204], but helps to avoid most of $1/f$ noise. In digital circuits it is easy to avoid the $1/f$ noise by digital modulation, thus approaching the fundamental noise limit given by Eq. (25).

3.9. Single- and few-electron memories

The trade-off of advantages and drawbacks of single-electron devices is most favorable for memory applications (see, e.g., Ref. 205), because of the following reasons:

(i) for memories, the bit density is the most important single figure-of-merit,

(ii) low voltage gain of single-electron transistors may be tolerated,

(iii) simple rectangular-matrix architecture of the memory banks makes the exclusion of bad bits (say, with thresholds shifted by a single charged impurity) possible – see below.

The published suggestions for single-electron memory cells are based mainly on two approaches:

- using various modifications of the single-electron trap (Fig. 11) with either MOSFET or single-electron-transistor readout, and

- direct scaling down the cells of usual nonvolatile

memories [206].[8]

Physics of these two approaches is not much different: in both of them, insertion and extraction of a single electron to the trapping island ("floating gate"), i.e. write and erase operations, are achieved by the field suppression of the potential barrier separating the island from the electron source (word line). The only difference is that in the former case the barrier is created by the Coulomb repulsion of electrons in a short 1D tunnel junction array, while in the latter case the barrier physics is the usual conduction band offset a single thick tunnel barrier (e.g., $\sim$8 nm of SiO_2).

Numerous experiments with single-electron cells of both types were useful for the development of the field; in particular, room temperature operation of single cells has been demonstrated by several groups [26, 216-218]. However, because of the background charge randomness (see Sec. F above), these cells can hardly be sufficiently reproducible. In fact, a single charge impurity near the floating gate has an effect equivalent to an addition of (positive or negative) external charge $\Delta Q_0 \sim e$, and thus shifts the threshold for both write/erase and readout operations from their nominal values rather considerably. In the trap-type cells the same effect, in addition, may change the array proper-

---

[8] The latter approach includes an interesting proposal [207, 208] (see also Refs. 210-217) for relatively large (multi-electron) memory cells, using many (N) nanometer-scale silicon crystals rather than a single floating gate, in usual nonvolatile memories. The main advantage of this idea is that a single leaking defect in the tunneling barrier would not ground the whole stored charge, but only its minor ($\sim 1/N$) part. This may allow to use very thin tunnel barriers (which would be unreliable in the ordinary case) and thus to decrease the characteristic time of Fowler-Nordheim tunneling, which presents the lower bound for the write/erase cycle. A potential drawback of nanocrystalline floating gates is that the electric field of the surface of each crystal, which determines the Fowler-Nordheim tunneling rate, depends on the size and exact shape of the crystal and is basically unpredictable. This may provide an undesirable broadening of the statistics of the electric field at the surface of nanocrystal surfaces, and hence of write/erase thresholds, especially at any attempt to scale the cells below 10 nm, where the number N would be relatively small.

ties randomly [196].[9]

Due to the regular structure of memory arrays, several ways of avoiding the random background charge effects are available. The idea suggested first [221] was to use the periodic character of the threshold characteristic of single-electron transistors for the cell contents readout. In this approach the memory structure may be very simple – see Fig. 15a. Binary 1 is stored in a relatively large floating gate in the form of a positive charge $Q = Ne$, with $N \sim 10$, while binary 0 is presented by a similar negative charge. (Since $N \neq 1$, the effect of random background charge on the floating gate is negligible.) Write/erase process is achieved by Fowler-Nordheim tunneling through a barrier separating the floating gate from the word line. Readout is destructive, and combined with WRITE 1 operation: if the cell contents was 0, during the WRITE 1 process the injected electrons ramp up the electric potential U of the floating gate, so that the external charge Q_0 of the readout single-electron transistor is ramped up by $N'e$, with $1 < N' < N$. Due to the fundamental periodicity of the transconductance (Fig. 10), this ramp-up causes N' oscillations of the transistor current. (If the initial charge of the floating gate corresponded to binary 1, the transistor output is virtually constant.) The current creates oscillations of voltage between two bit lines connected to SET source and drain. These oscillations are picked up, amplified, and rectified by an FET sense amplifier; the resulting signal serves as the output. The main idea behind this device is that the random background charge will cause only an unpredictable shift of the initial phase of the current oscillations, which does not affect the rectified signal.

This concept has been verified experimentally [222] using a low-temperature prototype of the memory cell. A very attractive feature of such SET/FET hybrid approach is a relatively mild minimum feature requirement: room temperature operation is possible with an electron addition en-

---

[9] Memory cells based on single- (or few-) electron trapping in grains of nanocrystalline MOSFET channels [219, 220] have even larger threshold spreads due to random locations of the grains and random transparency of tunneling barriers.

ergy of about 250 meV. Figure 13 shows that this level requires a minimum feature (SET island) size of about 3 nm, i.e. much larger than that required for purely single-electron digital circuits. The reason for this considerable relief is that in this hybrid memory the single-electron transistor works in the essentially analog mode, as a sense preamplifier/modulator, and can tolerate a substantial rate of thermally activated tunneling events. One drawback of this memory is the need for an FET sense amplifier/rectifier. However, estimates show that since its input signals have already been preamplified with the SET, one FET amplifier may serve several hundred memory cells within the ordinary NOR architecture [205, 206], and hence the associated chip real estate loss per bit is minor. The next drawback is more essential: the signal charge swing $\Delta Q_0 = N'e$ at the single-electron island should exceed e. Since the gate oxide should not bee too leaky to sustain acceptable retention time of the floating gate, its thickness should be at least a few nanometers. For silicon dioxide this gives a specific capacitance of approximately 1 $\mu F/cm^2$, corresponding to ~ 10 nm^2 area per one electron (at 1 Volt). This prevents scaling of the island well below its maximum size (~ 3 nm).

Both these drawbacks may be avoided using a more complex memory cell (Fig. 15b, adapted from Ref. 223) in which the random background charge Q_0 of the island is compensated by weak capacitive coupling with the additional ("compensating") floating gate. The necessary few-electron charge of this gate may be inserted from an additional ("compensation") word line before the beginning of memory operation, when special peripheral CMOS circuit measures Q_0 of each single-electron transistor (this may be done in parallel for all bits on a given word line) and develops an adequate combination of bit line signals. During the actual memory operation, the compensating charge is constant, but may be adjusted periodically if necessary. Due to the compensation, the SET may be biased reliably at the steep part of its control characteristic (Fig. 10) and hence used for nondestructive readout of the cell state. Moreover, the signal charge swing may be relatively small, $\Delta Q_0 \sim (k_B T/E_a)e$, conveniently decreasing with the island size. This may allow the single-electron

48

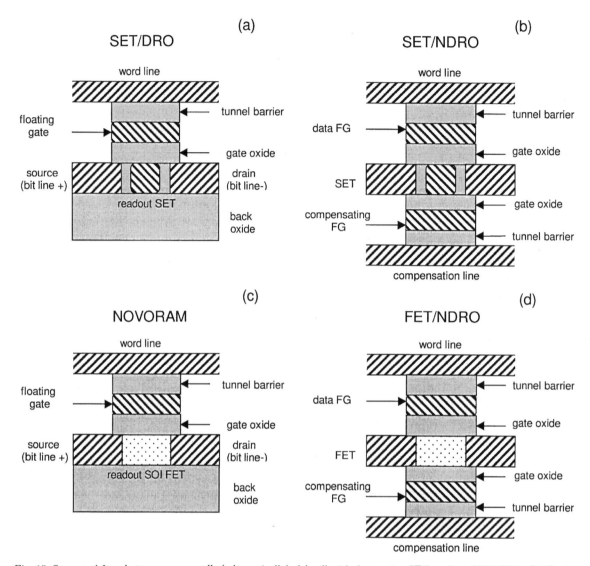

Fig. 15. Suggested few-electron memory cells (schematically): (a) cell with destructive SET readout (SET/DRO, [221]); (b) SET-readout cell with background charge compensation, making non-destructive readout possible (SET/NDRO, [223]), (c) cell with nondestructive FET readout which may also be used in NOVORAM [226], and (d) a similar cell with compensation of FET parameter variations [99].

transistor island to be scaled down to ~1 nm, with the ~3 nm floating gate storing just 2 to 3 electrons.

An alternative approach to few-electron memories is to scale down the generic structure of nonvolatile memory cells [205, 206] using a nanoscale MOSFET for readout – see Fig. 15c. As was already mentioned in Sec. 2, advanced MOSFETs with ultrathin silicon-on-insulator channel may be scaled down to ~10 nm sustaining high performance and reproducibility. Further scaling down would lead to large random fluctuations of the threshold voltage, that may be again corrected using a back floating gate (Fig. 15d).

Estimates of the maximum density of the SET/FET hybrid memories may be carried out under the assumption that the SET islands are fabricated by some self-aligned method, so their size is independent of the wiring line half-pitch F. In this case the area of the cells shown in Fig. 15 is essentially independent of the island size and ranges between $6F^2$ and $8F^2$, where F is the minimum feature size of a given technology. The cell type determines the possible limits of this scaling: for the SET/DRO (Fig. 15a) it is confined in a narrow range around 3 nm, while for the SET/NDRO (Fig. 15b) F may be scaled down to ~1.5 nm (line patterning permitting). For the FET/NDRO cells shown in Fig. 15c,d, the range of F does not have an upper bound; for the simple cell shown in Fig. 15c the lower bound for F is about 5 nm, while the more complex cell with background charge compensation (Fig. 15d) can be scaled down to $F \sim 2$ nm, i.e. to the cell area about 30 nm^2. This should allow memories with density well beyond 10^{12} bits/cm^2, enabling chips with multi-terabit integration scale. These exciting prospects are, however, contingent on the development of nm-scale fabrication technologies.

Another problem with all the memory cells described above is the slowness of the write/erase operations. In fact, they rely on the process of Fowler-Nordheim tunneling, similar to that used in the ordinary nonvolatile, floating-gate memories [206]. The standard 8-to-10-nm SiO$_2$ barriers used in such memories do not change transparency fast enough to allow floating gate recharging than in ~1 μs even if the applied electric field is as high as

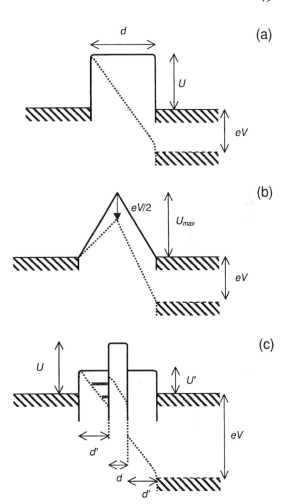

Fig. 16. Tunnel barriers options: (a) usual, uniform barrier; (b) ideal, triangular crested barrier; and (c) trilayer crested barrier [224]. Thick horizontal lines show (schematically) the subbands formed at large values of applied voltage, and enabling resonant tunneling through the barrier.

~10 MV/cm, close to the breakdown threshold. (A reduction of the barrier thickness or height makes retention time too short.) Such a long write/erase cycle is acceptable for typical applications of flash memories [206], but is too long to replace DRAM in bit-addressable memories.

This problem can be solved using special layered ("crested") barriers [224] (Fig. 16). Calculations show [224-226] that trilayer crested barriers may combine a 1-ns-scale write time with a ~10-year retention time, at apparently acceptable electric

50

fields (about 10 MV/cm) – see Fig. 17. This gives hope that such barriers may have extremely high endurance necessary for RAM applications. Moreover, ratio of the necessary write voltage (V_W) to the highest retention voltage (V_R) for such barriers may be well below 3. This condition ensures that the disturb effects on self-selected cells are small enough even in the simplest memory architectures [226].

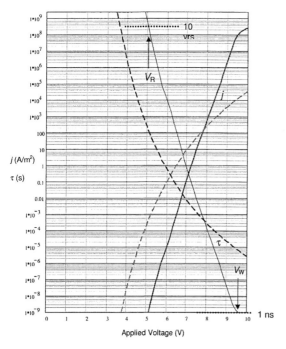

Fig. 17. Calculated current density j and floating gate recharging time constant τ as functions of the applied voltage V for a uniform tunnel barrier (10 nm of SiO_2, dashed lines) and a trilayer crested barrier (5 nm of thermally-grown and post-annealed Al_2O_3 sandwiched between two plasma-grown 2.5-nm Al layers, solid lines) [99].

Thus, if CMOS-compatible crested barriers are implemented, the few-electron cells shown in Fig. 15 may be used as random-access memories, replacing the generically unscalable DRAM. Moreover, crested barriers may enable the so-called Nonvolatile Random Access Memory (NOVORAM [224-226]) with a very simple cell structure (Fig. 15c) and architecture (Fig. 18), which may be able to compete with DRAM and possibly SRAM even at the current technology level (mini-

mum feature size F of the order of 100 nm), and be scaleable all the way down to $F \sim 3$ nm. Some further decrease of F is possible by dropping the nonvolatility requirement and organizing cell refresh with period ~ 1 s, because in this case the crested barrier and gate oxide may be thinned by $\sim 50\%$.

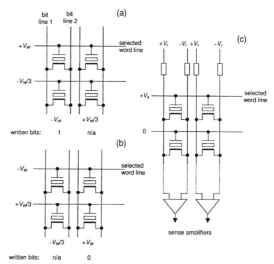

Fig. 18. Possible architecture of NOVORAM and applied voltages necessary for: (a) WRITE 1, (b) WRITE 0, and (c) READ operations [99].

An alternative way to speed up the write/erase operation in floating gate memories is to replace the tunnel barrier with a transistor. Such two-transistor (2-T) memory cells were suggested long ago [227]; for a while they could not compete with DRAM because of the much larger cell size. The apparent impossibility to scale DRAM much below 100 nm has led to a recent revival of R&D work in this direction, and several new versions of the 2-T memories have been suggested (and some explored experimentally):

1. Single-electron trap with an additional gate controlling the potential profile high of the array, and readout using either a single-electron-transistor [228], or a nanoscale MOSFET [229-232]. Unfortunately, neither approach addresses the problem of random background charge. (Single-electron arrays, in addition, are themselves sensitive to this effect [196, 233].) Notice that in experiments with semiconductor single-

electron arrays the Coulomb blockade mechanism of electron transport control coexists (and was repeatedly confused) with the usual field-effect mechanism. In this case we are speaking about the effective replacement of the single-electron array by an MOSFET with very thin (and hence not very uniform) channel.

2. The authors of a recent work [234] have made this replacement conscientiously, arguing that their 2-nm thick silicon channel should have lower parasitic source-to-drain leakage, because quantum confinement increased the effective bandgap. Indeed, their MOSFETs exhibited leakage current as low as 10^{-19} A, sufficient to keep the cell retention time above 100 ms, i.e., the typical DRAM refresh time. A memory similar in structure, but with a special "stacked" vertical FET [235] for fast write/erase is called Phase-State Low Electron Drive Memory, or PLEDM [236] – see also [237]. The channel of such a transistor incorporates three horizontal silicon nitride tunnel barriers partitioning the channel into several parts connected in series. Such separation improves transport control of the channel potential (and hence of the electron transport) by the surrounding gate; the considerable loss of ON current caused by these barriers is tolerable because the MOSFET should only recharge a very small capacitance of the charging node. Unfortunately, in such vertical MOSFETs, with their relatively large channel cross-section, getting acceptably small leakage current (and hence retention time) may be a difficult problem.

Though the approach [234] looks very interesting, I believe that the successful implementation of silicon-compatible crested barriers [224] will make NOVORAM, with its simple structure and small cell area, a more promising option.

The single- and few-electron memories will have to compete not only with each other, but with several other prospective memory concepts.

In *ferroelectric memories*, information is stored as a sign of electric polarization of a layer of a ferroelectric material. This polarization may be read out either destructively (as in DRAM) or nondestructively (e.g., if it controls a readout FET) [238]. Strong features of these memories include a simple cell structure (and, as a result, small cell area), potential nonvolatility, and fast write/erase

time (some materials have internal re-polarization time well below 1 ns). In terms of immediate practical introduction, the complexity of ferroelectric materials and their compatibility with the generic CMOS process is the main challenge. However, in the long term, scalability of these memories may be a larger problem. With the decrease in area, the height of the energy barrier $\Delta U \approx P_s E_c V$ separating two polarization states (where P_s is the saturation polarization, E_c the coercive electric field, and V the ferroelectric layer volume) decreases and should finally become comparable with the thermal fluctuation scale $k_B T$, resulting in random cell switching. For typical ferroelectric film parameters $P_s \sim 50 \ \mu C/cm^2$ and $E_c \sim 5 \ V/\mu m$ [206], such spontaneous switching should become rather noticeable ($\Delta U \sim 300 \ k_B T$) at $V \sim 20 \times 20 \times 2 \ nm^3$, even if P_s and E_c do not degrade with size. Thus, if no new breakthroughs are made, ferroelectric memory cells can hardly compete for terabit applications, which is the main promise of single- and few-electron memories.

In *magnetic memories* [238, 240], bits are stored in the form of thin film magnetization. These memories share almost all the advantages and drawbacks of ferroelectric cells listed above (low 0/1 output signal value is an additional issue). However, because of the essentially similar physics, scaling of such cells to nanoscale may again be a problem. In this case, $\Delta U \sim B_s H_c V$ (where $B_s \sim 2 \ T$ is the saturation magnetization and $H_c \sim 10$ Oe the coercive magnetic field) drops below 300 $k_B T$ at even larger volume ($\sim 30 \times 30 \times 10 \ nm^3$).

Memory cells based on *structural phase transition* include notably the "Ovonics Unified Memory" (*OUM*) [241, 242]. In an OUM cell, a chalcogenide alloy (GeSbTe) is switched from a conductive crystalline phase to a highly-resistive amorphous phase under the effect of heating by current passed through a special heater. Though the chalcogenide materials are relatively complex, considerable progress in their deposition has been made in the course of development of CD-RW and DVD-RW technologies. As a result, OUM cells with surprisingly high endurance (up to 10^{13} cycles) have been demonstrated. OUM problems include relatively long write/erase time (reportedly, close to 100 ns, i.e., considerably longer than that

52

of DRAM). Unfortunately, I am not aware of any published experimental data or reliable theoretical results sufficient to evaluate the dependence of the retention time on the storage region volume, and thus evaluate prospects of OUM scaling into the terabit range.

Finally, *single-molecular memories* may be based on various background physics. For example, they may be just molecular implementation of single-electron memories discussed above (Fig. 15). However, there is an alternative possibility: to employ molecular conformation changes [159, 163, 315]. In this case the molecule has internal bistability that manifests itself as a hysteretic region on the molecule as a two-terminal device. Very recently, this approach allowed the first demonstration of an 8×8 memory matrix, where each bit was stored in a state of many similar molecules, sandwiched in parallel between two crossing 40-nm metallic wires [245]. Since the technical details of this work have not yet been published, it is too early to evaluate the possible speed, retention time, and reliability of such memory cells. However, I will discuss this general approach more in Sec. 4 below.

3.10. *Electrostatic data storage*

A combination of single-electron transistors with crested barriers may be used not only in terabit-scale memories, but also in ultra-dense electrostatic data storage systems [224] – see Fig. 19. In this "ESTOR" design, a single-electron transistor, followed by a MOSFET amplifier, a few microns apart, would be fabricated on a tip-shaped chip playing the role of a READ/WRITE head. The data bits are stored as few-electron charge trapped in a group of nanoscale conducting grains deposited on top of a crested tunnel barrier. It is important that since each bit is stored in a few (~10) grains, their exact shape and location are not important, so the storage medium production does not require any nanofabrication.

WRITE is performed by the application of the same voltage V_W to both input terminals, relative to the conducting ground layer of the moving substrate. The resulting electric field of the tip in-

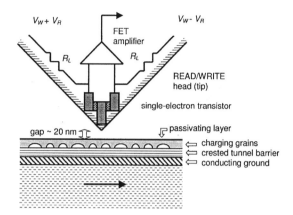

Fig. 19. Proposed electrostatic data storage system ("ESTOR") with hybrid SET/FET readout [224].

duces rapid tunneling of electrons from the ground through the crested barrier into a ~30-nm-wide group of grains. For READ, the single-electron transistor is activated by source-drain voltage $2V_R \geq V_t$. In this state it is very sensitive to the electric field created by the group of charged grains it is being flown above.

Simple estimates show that with a 20-nm tip-to-substrate distance (close to those already implemented in the best present-day magnetic storage systems), the electrostatic system is capable of a density ~3 Terabits per square inch, i.e. at least an order of magnitude than the best prospects for the magnetic competition of which I am aware. The use of crested barriers may provide a very broad bandwidth of both WRITE and READ operations, up to 1 Gbps per channel, possibly quite adequate even for this enormous bit density.

Notice that for this particular application the difficulties of fabrication of room-temperature single-electron transistors, outlined above, are not a major concern, because one would need just one (or a few) transistors per system, and slow fabrication techniques (like the scanning probe oxidation) may be acceptable. Moreover, like in single-electron memories, the transistor would work as an analog amplifier, so that the single-electron island size of ~3 nm would be sufficient.

Recent experiments [246] may be considered as the first step toward the implementation of this idea.

3.11. *Logic circuits*

Most suggestions of logic circuits based on single-electron devices may be referred to one of two groups.

1). In circuits of the first, *"voltage state"* group, single-electron transistors are used in CMOS-like circuits. This means that the single-electron charging effects are confined to the interior of the transistor, while externally it looks like the usual electronic device switching persisting currents, with binary unity/zero presented with high/low dc voltage levels (physically not quantized). This concept simplifies the circuit design which may ignore all the single-electron physics particulars, except the specific dependence of the drain current I on the drain-to-source voltage V and gate-to-source voltage V_g - see Figs. 9 and 10.

Analyses of this opportunity has shown that due to the specific shape of this dependence (oscillating transconductance), both resistively-coupled [124] and capacitively-coupled [247] single-electron transistors allow a very simple implementation of CMOS-type inverters, without a need for two types of transistors (like n-MOSFETs and p-MOSFETs in the standard CMOS technology). On the other hand, peculiarities of functions $I(V, V_g)$ makes the exact copying of CMOS circuits impossible, and in order to get substantial parameter margins, even simple logic gates have to be re-designed. Such circuits (Fig. 20) may operate well within a relatively wide window of parameters R, C, and V_{DD} [248, 249]. Even after such optimization, the range of their operation with acceptable bit error rate shrinking under the effect of thermal fluctuations as soon as their scale $k_B T$ reaches approximately $0.01 E_a$. (For other suggested versions of the voltage state logic [247, 250], the temperature range apparently is even narrower). The maximum temperature may be somewhat increased by replacing the usual single-island single-electron transistors with more complex devices with 1D-array "channels" and distributed gate capacitances [251]. However, this would increase the total transistor area, at the given minimum feature size.

A disadvantage of voltage state circuits is that neither of the transistors in each complementary

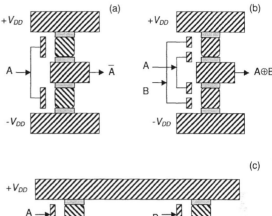

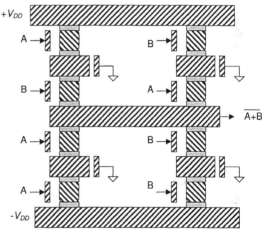

Fig. 20. Basic gates of the complementary SET logic family using capacitively-coupled single-electron transistors [249]: (a) inverter; (b) XOR, and (c) NOR/NAND.

pair is closed too well, so that the static leakage current in these circuits is fairly substantial, of the order of $10^{-4} e/RC$ [248]. The corresponding static power consumption is negligible for relatively large devices operating at helium temperatures. However, at the prospective room-temperature operation this power becomes on the order of 10^{-7} Watt per transistor. Though apparently low, for the hypothetical circuits which would be dense enough ($>10^{11}$ transistors per cm^2) to challenge the prospective CMOS technology, this number gives an unacceptable static power dissipation density (>10 kW/cm^2).

2) The power dissipation problem may be avoided, to a large extent, by using *"charge state"* logic circuits in which single bits of information are presented by the presence/absence of single

electrons at certain conducting islands throughout the whole circuit. In this case the static currents and power virtually vanish. This approach has been explored theoretically since 1987 [252], and several few families of charge state logic circuits have been suggested and analyzed [139, 140, 253-283]. In most suggestions, an electron is confined in a cell consisting of one or a few islands, while the logic switching is achieved via electrostatic (or spin [259, 264, 271]) coupling of the cells. Another classification of single electron logics may be based on where they take the energy necessary for logic operations: from dc power supply [252, 261, 269], ac power supply (also playing the role of global clock) [138, 140, 260, 262, 263, 274, 278-283], or just from the energy of an external signal [254-259, 265, 267, 271, 272]. [10]

Only a few of these concepts have been analyzed in detail, especially at finite temperatures. To my knowledge, the most robust charge-state logic circuits suggested till now are those based on the single-electron parametrons (see Sec. 3.5 above). Figure 21 shows a possible shift register based on the parametrons [139]. The direction of the shift of the central island of each next device is shifted by $\pi/3$ within the yz plane. The circuit is driven by electric field $\boldsymbol{E}_c(t)$ rotating in the same plane and providing the periodic switching on the SET parametrons, with an appropriate phase shift. As a result, each digital bit (one per three cells) is being shifted by 3 cells along the structure each clock period. Majority logic gates, sufficient for arbitrary logic circuits, may be implemented in the same way

[140]. Geometric modeling and numerical simulation of these circuits within the framework of the orthodox theory have shown that they may operate correctly within approximately $\pm20\%$ deviations from the optimal clock amplitude. Estimates show that the maximum operation temperature of these logic circuits is of the order of that of voltage mode circuits, i.e. of the order of $0.01E_c/k_B$, if the bit error rate is in the practically acceptable range (below $\sim10^{-20}$).

A new, potentially useful feature of the charge state logics is the natural internal memory of their "logic gates" (more proper terms are "finite-state cells" or "timed gates"), thus combining the functions of the combinational logic gates and latches. This feature makes natural the implementation of deeply pipelined ("systolic") and cellular automata architectures. The back side of this advantage is the lack of an effective means of transferring a signal over large distances: crudely speaking, this technology does not allow passive wires, just shift registers.

3) Within the framework of our classification, the so-called *phase-mode logic* [284-287] should be placed in a separate category, because in these circuits the information-keeping cell are coupled by rf signals, carrying the binary information coded by either of two possible values of the rf signal phase. The elementary cell of this logic is essentially a relaxation oscillator generating SET oscillations with frequency $f_{SET} = I/e$ [123, 286], phase locked by an external reference with frequency $2f_{SET}$. Such "subharmonic" phase locking allows two possible phases of oscillations, which differ by π; the oscillation state with each phase has exactly the same amplitude and is locally stable, i.e. can be used to code a single bit of information. Moreover, the oscillator can impose its phase upon a similar, adjacent oscillator that had been turned off temporarily (e.g., by turning off its power supply current I), and is now being turned on. Thus information may be transferred along a row of oscillators, working as a shift register, very much similar to that consisting of single-electron parametrons – see Fig. 21. Majority gates can be organized similarly, enabling the implementation of arbitrary logic functions.

A theoretical advantage of this logic family is

---

[10] The last category includes the so-called Quantum (or Quantum-dot) Cellular Automata (QCA) based on ground state computing. This concept had been controversial from the very beginning [260, 270-272], because it leads to the problem of system trapping for exponentially large (read infinite) time in intermediate metastable states – for a detailed discussion, see the recent publications [282, 283]. Eventually the concept was repudiated by its authors – see the introduction section in Ref. 141. The second variety of logics based on similar cells ("adiabatic" or "clocked" QCA) [263] is very close to single-electron parametron circuits (that had been suggested earlier [139]), besides that the QCA cells have a more complex structure and as a result lower speed and narrower parameter margins [282, 283].

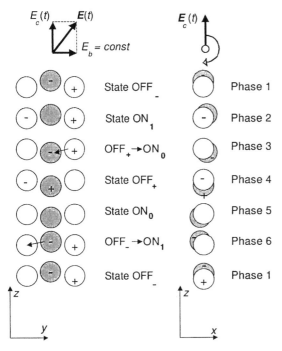

Fig. 21. A shift register based on single-electron parametrons.

the potential insensitivity of its cells to the random background charges, if a resistor with quasi-continuous charge transfer is used for the insertion of dc current I. Unfortunately, as was discussed in Sec. 3.7 above, the known implementations lead to very large size of such resistors, making the implementation of practical (room-temperature) phase-mode logic circuits hardly feasible.

Unfortunately, *all* the single-electron logic circuits discussed above face two crucial problems:

(i) In order to ensure an acceptable bit error rate, they require single-electron islands with very high electron addition energy, of the order of 100 $k_B T$. For room temperature operation, this means the island size below 1 nm (Fig. 13). At the same time, parameter margins of these devices are not very high ($\sim$20% or lower). This means that the critical dimension control (on 3σ level) should be of the order of 0.1 nm , i.e., even more tough than for nanoFETs.

(ii) Traditional logic circuit architectures (e.g., microprocessors, digital signal processors, etc.) do not allow a straightforward determination and ex-

clusion of faulty logic gates. This is why the effect of random background charge (Sec. 3.7 above) cannot be circumvented as simply as in memories (Sec. 3.9).

4. CMOL: Devices, circuits and architectures

4.1. *The concept*

The hard challenges faced by both sub-10-nm field-effect transistors and room-temperature single-electron logic circuits make me believe that the only chance of breakthrough beyond the 10-nm frontier is the complete change of the integrated circuit paradigm. The approach I am advocating is the transfer to "CMOL" hybrid circuits combining CMOS components and molecular (e.g., single-electron) devices, interconnected by nanowires (Fig. 22) [292]. [11]

The lower level is occupied by a CMOS stack. The transistor density of this sub-system cells may be relatively low, of the order of 10^8 cm^2. (The implementation of this density would require the 45-nm-node technology, which should be commercially available by the end of this decade [1].) Vertical plugs connect the CMOS circuit to I/O pins (including those providing power supply for the whole circuit), CMOS wiring, and the next circuit level, occupied by nanowiring.

In order to sustain the necessary density of molecular devices, these wires should be extremely narrow (a few-nm half-pitch). Simultaneously, the wire resistance per unit length should be not too high, below $\sim$10 MΩ/μm, to sustain relatively high operation speed. Because of this, as well as the necessary chemical compatibility with molecular devices of the top level, I prefer to think about explicitly patterned gold wires rather than molecular-wire options. Of course, patterning with this resolution is very challenging; however, a CMOL circuit would use just a few (say, two) layers of uniform, parallel, nanowire sets in two

---

[11] The hybrid approach makes CMOL rather different from several earlier concepts of molecular electronics circuits - see, e.g., Ref. 289.

56

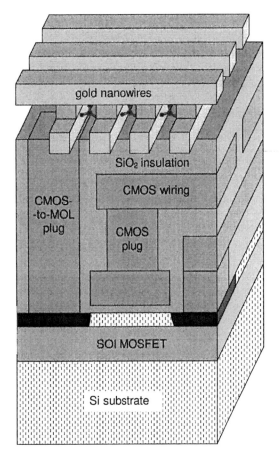

Fig. 22. General concept of CMOL (CMOS/nan-owire/MOLecular hybrid) circuit.

mutually perpendicular directions. Such a simple line pattern does not require nanometer-scale level alignment, and may be formed, e.g., by nanoimprinting [290]. (See also Chapter 3 of this collection.)

Finally, the nanowire layers are connected by self-assembling molecular devices having the smallest individual footprint and hence, the highest possibly density. With the 3-nm nanowire half-pitch (close to the limit imposed by wire-to-wire tunneling, see Sec. 2.4 above), the density of active molecular components would exceed 3×10^{12} functions per cm^2. (Just for comparison, Encyclopedia Britannica contains about 10^{10} bits of information.)

I believe that the advantage of the CMOL approach is that if allows an optimum combination

of strengths of its components: robust and universal CMOS circuits may take care of the functions requiring high reliability, high voltage gain and high ON currents, in particular signaling over long (1-cm-scale) distances, as well as I/O functions. On the other hand the molecular devices will sustain simple functions requiring highest integration scale. But most importantly, self-assembly of molecular devices may allow to keep the total fabrication costs of CMOL chip comparable with today's industrial level, thus avoiding the largest threat to the Moore-Law progress.

4.2. *Defect-tolerant architectures*

Even if/when VLSI CMOL circuits are successfully implemented, the yield of self-assembled molecular devices will hardly approach 100%. Moreover, as was discussed in previous sections, a-few-nm-scale devices will always be sensitive to random charged impurities, regardless whether they use FET-like or SET-like physics for electron transport control. Hence the CMOL approach requires defect-tolerant architectures that would allow to either tolerate or exclude bad devices. The most evident opportunity here is to use the molecular level as massive embedded memory for CMOS-based circuits [291, 308]. (Several industrial groups are already working in this direction – see, e.g., Ref. 245.) For more advanced circuits, there are several opportunities, [12] and I will describe just one of them, which is the focus of the recent work of our Stony Brook group [292-294, 316].

It is well known that neuromorphic networks [295-302] that mimic the basic functions of the cerebral cortex [303, 304] may be used for extremely efficient information processing. For example, the human brain can carry out an almost perfect recognition of a visual image in approximately 100 milliseconds, i.e., in just ~30 elementary "ticks" (neural spike durations) of the cortical circuitry. For comparison, the best modern microprocessor, using the best software available,

---

[12] For example, an interesting idea of defect-tolerant digital computing based on look-up-tables (essentially, large memories) is discussed in Ref. 310.

can perform a less reliable recognition on the scale of minutes, i.e. in $\sim 10^{12}$ of its 1-ns-scale "ticks" (clock periods).

The implementation of this advantage in artificial systems may require, however, VLSI circuits with the number of active components comparable with that in the cerebral cortex. The number of neural cells in mammal's cortex ($\sim 10^{10}$ [303, 304]) is not overwhelming, but they feature a very large *connectivity*: each cell is directly connected, on the average, to $M \sim 10^4$ other cells. Each connection is served by an active device, the synapse. Thus the number of active functions in the cerebral cortex is of the order of 10^{14}, and only molecular nanoelectronics gives the first hope for the fabrication of comparable structures on acceptable scale of circuit size, power dissipation and (yes, Virginia!) fabrication costs. Within the CMOL approach (Fig. 22), CMOS may take care of functions of neural cell bodies ("somas"), but synapses should be implemented using more dense molecular devices.

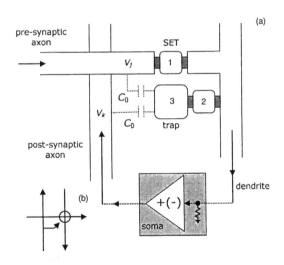

Fig. 23. (a) The simplest three-terminal single-electron latching switch and (b) its circuit notation.

Figure 23a shows the three-terminal version [294] of earlier suggested [292] device that is a simple combination of the single-electron transistor and the trap, working together as a "latching

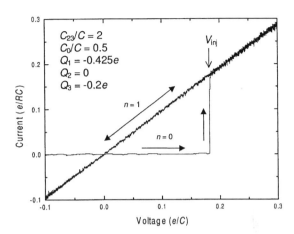

Fig. 24. DC $I - V$ curve of the latching switch, calculated using the "orthodox" theory [292]. For the three-terminal switch (Fig. 23), the effective applied voltage is $V_j + V_k$.

switch".[13] The device consists of three small islands connected by four tunnel junctions. Island 1, together with input and output wires serving as source and drain, forms a single-electron transistor. Islands 2 and 3 form a single-electron trap (cf. Fig. 11a), with trapping island 3 capacitively coupled to the SET island 1, thus playing the role of a single-electron floating gate. If the effective voltage $V \equiv V_j + V_k$ applied to the device is low, the trap in equilibrium has no extra electrons and its total electric charge is zero. As a result, the transistor remains in the Coulomb blockade state, and input and output wires are essentially disconnected. If V is increased beyond a certain threshold V_{inj} (which should be lower than the Coulomb blockade threshold voltage V_t of the transistor), one electron is injected into the trap. In this charge state the Coulomb blockade in the transistor is lifted, keeping the wires connected at any V. However, if the node activity (voltage V) is low for a long time, either thermal fluctuations or

---

[13] Notice also that island 2 is not really necessary and may be replaced by a thicker tunnel barrier – see the discussion in the end of Section 3.4. The device so modified is essentially a three-terminal version of a four-terminal device that had been discussed qualitatively in Ref. 310. Earlier suggestions to use single-electron devices in neuromorphic networks [305, 309–311] focused on the implementations of the somatic functions that should be, in my opinion, left for more robust CMOS circuits.

58

co-tunneling eventually kick the trapped electron out of the trap and the transistor closes, disconnecting the wires. Figure 24 shows typical results of numerical simulation of the latching switch dynamics with a perfect background charge of the SET island (it may be adjusted by voltage applied to an additional global gate). [14]

Figure 25 shows the possible molecular implementation of the three-terminal latching switch [294]. The role of single-electron island is played by diimide groups well known for their acceptor properties (see, e.g., Ref. 306); OPE bridges [307] are used as tunnel junctions, while thiol groups play the role of "alligator clips" that should allow the molecule to self-assemble from solution on gold nanowires [308].

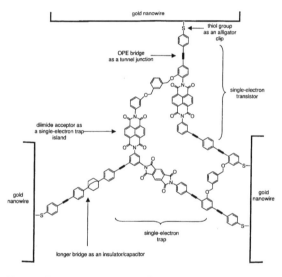

Fig. 25. Possible molecular implementation of the latching switch shown in Fig. 23 [294].

We have shown [294] that a group of four devices shown in Fig. 23a may implement a "Hebbian" synaptic function: the net synaptic weight is

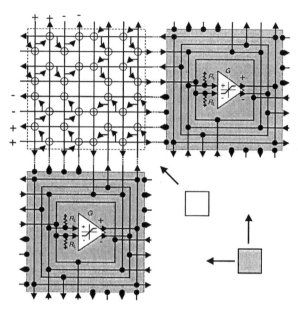

Fig. 26. Structure of a distributed crossbar network (Cross-Net), with the synaptic and somatic (gray) plaquettes.

strengthened when the post-synaptic activity immediately follows the pre-synaptic activity [295-302].

Figures 26 and 27 show the general structure, and two most promising species of the so-called distributed crossbar arrays for neuromorphic networks ("CrossNets") based on such synapses [293]. In each CrossNet, somatic cell interfaces (gray cells) are embedded sparsely into a 2D array of synaptic plaquettes, each containing 8 synapses (each synapse is a group of 4 latching switches). [15] As a result, each somatic cell is hard-wired to a large number, $4M$, of other somas, with the binary synaptic weights controlling which of these connections are currently active. Vice versa, the signal activity of the network determines whether the synapses are open or closed, though the state of any particular synapse is also affected by the underlying randomness of single-electron tunneling.

In InBar (which currently looks like the most promising CrossNet option), the gray cells sit on a square lattice inclined (hence the name) relative

[14] These simulations were based on the "orthodox" theory of single-electron tunneling [115]. Though this theory gives a good qualitative guide to properties of even molecular-size single-electron devices, their serious quantitative theory should be based on *ab initio* calculations of molecular orbitals – see, e.g., Refs. [181-186] and Chapter 6 of this collection and the transport theory taking into account the electron state discreteness [142-144].

[15] The CMOS-implemented somatic cell itself may be much larger than the synaptic plaquette, occupying all the chip area between two adjacent gray cells.

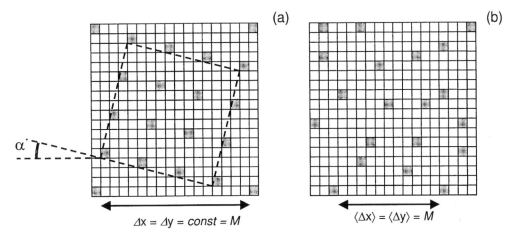

(a)   (b)

$\Delta x = \Delta y = const = M$ $\langle \Delta x \rangle = \langle \Delta y \rangle = M$

Fig. 27. Two CrossNet architectures: (a) InBar and (b) RandBar. M is the connectivity parameter. For the plaquette notation, see Fig. 26.

to the synaptic plaquette array. The incline angle α determines the network connectivity parameter: $M = 1/\tan^2 \alpha$. As can be seen from Figs. 26 and 27, the total "Manhattan" (dendritic plus axonic) distance between each pair of coupled cells in In-Bar, measured in synaptic plaquettes, is the same and equals to M. On the other hand, in RandBar the gray cell terminals are distributed randomly. This creates the Poissonian distribution of intercell distances, so that there is a small amount of very long interconnects. We see that this property creates some difference in network dynamics, but are still not sure whether this difference is substantial for information processing.

Our collaboration is currently working on the molecular implementation of CrossNet synapses and, in parallel, on numerical simulation of limited fragments of these networks on usual supercomputers, trying to train them to perform various functions. The important rule of this game is that, like in the future hardware implementations, the external tutor system has access only to (sparse) somatic cells, rather that to individual synaptic weights. This restriction, as well as deeply-recurrent nature of CrossNets and the statistical character of single-electron synaptic weights, does not allow for the straightforward use of the well-known methods of neuromorphic network training [295-302]. However, we have already succeeded to demonstrate [316] that in spite of these restric-

tions, as well as quasi-local coupling of somatic cells (finite M), CrossNets with Hebbian synapses may operate rather well as Hopfield networks. (Figure 28 shows the example of black-and-white image recognition in this mode.)

For more advanced applications, such as pattern classification and feature detection, the continuous-mode training is necessary. Very recently, we suggested [316] a way for such training, using chaotic self-excitation of CrossNets at large values of somatic amplifier gain. If these expectations are confirmed, CrossNets may work as pattern classifiers with very impressive characteristics. Estimates show [292] that at the neural cell connectivity $4M = 10^4$ (comparable with that of the human cerebral cortex) and a nanowiring half-pitch of 3 nm, the neural cell density may be as high as 10^7 cm^{-2}. The estimated time of signal propagation between the neural cells of the order of 20 ns, at high but acceptable power dissipation (100W/cm^2). This speed is approximately 6 orders of magnitude higher than that in biological neural networks. It is expected that the introduction of such circuits will create a new electronic market comparable, if not larger than the PC market.

This success would pave the way toward much more ambitious goals. It seems completely plausible that a cerebral-cortex-scale CrossNet-based system (with $\sim 10^{10}$ neurons and $\sim 10^{14}$ synapses, that would require $\sim 30 \times 30$ cm^2 silicon sub-

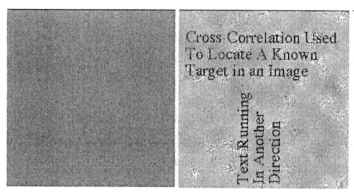

Fig. 28. The process of recall of one of three trained black-and-white images by an InBar-type CrossNet with 256×256 neural cells and connectivity parameter $M = 64$. The initial image (left panel) was obtained from the trained image (identical to the one shown in the right panel) by flipping 40% of randomly selected pixels. $\tau_0 = MR_LC_0 \lesssim RC_0$ is the effective time constant of intercell interaction. (C_0 is the dendrite wire capacitance per one synaptic plaquette.) [294].

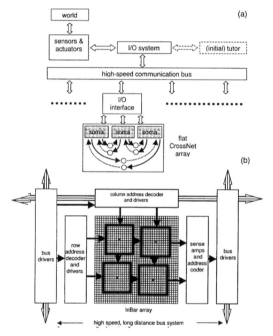

Fig. 29. (a) Possible hierarchical organization of a large-scale, CrossBar-based information system and (b) a simple scheme for incorporation of InBar matrices into the system fast (though relatively rare) communications of distant neural cells [294].

strate) [16] would be able, after a period of initial

---

[16] Such large-scale system would require a hierarchical organization (Fig. 29a) involving at least the means for fast signal transfer over long distances. Fortunately, for the InBar-type CrossNet with its regular location of somatic cell interfaces, such communication is easy to organize (Fig. 29b).

training by a dedicated external tutor (Fig. 29a), to learn directly from its interaction with environment. In this case one can speak of a "self-evolving" system.

If these expectations are confirmed, we may be able to revisit the initial dream of the neural network science of providing hardware means for reproducing the natural evolution of the neocortex on a much faster time scale. Such evolution may lead to self-development of such advanced features as system self-awareness (consciousness) and reasoning. If a substantial success along these lines materializes, it will have a strong impact not on the information technology, but also on the society as a whole.

5. Conclusions

I believe that the following main conclusions may be drawn from the materials presented in this chapter:

(i) Physics allows scaling of the silicon FETs, in their advanced (ultra-thin-channel, double-gate) form, to approximately 10 nm gate length, without an essential loss of performance. Further scaling, all way down to $\sim$ 5-nm-long channels, is also physically possible, but leads to an extremely high sensitivity of transistor characteristics (in particular its gate threshold voltage V_t) to minute variations of geometric dimensions. This sensitivity will probably lead to unacceptable cost of fabrication facilities and, as a result, to the necessity of transfer to alternative electron devices in order to continue the Moore-Law-type exponential progress.

(ii) So far there is no single universal electronic device capable of replacing silicon FETs when their scaling down runs out of steam. The most natural candidates, single electron transistors, have the advantage of being remarkably material-insensitive, but suffer (as essentially all nanoscale devices) from low voltage gain and high sensitivity to single charge impurities in the dielectric environment of their islands. Nevertheless, there is a hope that advanced lithography may reach the level necessary for the fabrication of hybrid SET/FET memories. The major competition to this concept may come from NOVORAM memories based on crested tunnel barriers and 2T-cell memories, both using nanoscale FETs with ultra-thin channels.

(iii) For more advanced applications, the introduction of "CMOL" circuits, combining a CMOS stack, a few nanowiring layers, and a subsystem of molecular devices self-assembled on the wires seems unavoidable. If this concept is developed and implemented within the following 10 to 15 years, it will prevent the impending crisis of the exponential, Moore-Law-type progress of microelectronics, and extend this progress to virtually atomic dimensions.

(iv) The first application of the CMOL concept could be in the dedicated ultra-dense memory chips, as well as digital microprocessors with the molecular sub-system working as an embedded memory. However, CMOL circuits are the very natural hardware for the implementation of much more sophisticated architectures, in particular neuromorphic networks performing such advanced functions as pattern classification and feature detection, and possibly other human brain functions, at much higher speed. If such systems are eventually developed, the technological and social consequences of their practical introduction may be extremely significant.

Acknowledgments

The author is grateful to many colleagues for numerous discussions of the issues mentioned in this chapter. T. J. Walls and V. A. Sverdlov, have generously permitted the use of some results of our joint works [59, 61] prior to their publication, while A. Mayr, I. Muckra and O. Turel have granted the similar permission for work [294]. Yu. Pashkin has kindly provided original graphics for Fig. 10b. M. Macucci has brought to my attention some important recent publications and provided preprint [283]. Research at Stony Brook on certain topics discussed in this paper was supported in part by AFOSR, ARDA via ONR, DOE, NSF, and SRC. This paper is a substantially updated and extended version of the earlier review [99].

References

[1] International Technology Roadmap for Semiconductors, 2001 Edition, 2002 Update; available on the Web at http://public.itrs.net/.

[2] S. Sze, Physics of Semiconductor Devices, 2nd ed. (Wiley, New York, 1981).

[3] Y. P. Tsividis, Operation and Modeling of the MOS Transistor (McGraw Hill, New York, 1987).

[4] R. F. Pierret, Semiconductor Device Fundamentals (Addison-Wesley, Reading, MA, 1996).

[5] S. Dmitrijev, Understanding Semiconductor Devices (Oxford Univ. Press, New York, 2000).

[6] D. J. Frank, R. H. Dennard, E. Nowak, P. M. Solomon, Y. Taur, and H.-S. P. Wong, Proc. IEEE 89 (2001) 259-288.

[7] H. Kawaura, T. Sakamoto, T. Baba, Y. Ochiai, J. Fujita, and J. Sone, IEEE Trans. on Electron Dev. 47 (2000) 856-859.

[8] H. Kawaura, T. Sakamoto, and T. Baba, Appl. Phys. Lett. 76 (2000) 3810-3812.

[9] R. Chau et al., in: Techn. Dig. IEDM (IEEE Press, Piscataway, NJ, 2000) 45-48.

[10] H. Wakabayashi et al., in: Techn. Dig. IEDM (IEEE Press, Piscataway, NJ, 2000) 49-52.

[11] M. Takahashi, T. Ohno, Y. Sakakibara, and K. Takayama, IEEE Trans. on Electron Dev. 48 (2001) 1380-1385.

[12] R. Chau, in: Proc. of Silicon Nanoelectronics Workshop (Honolulu, Hawaii, 2001) 57-60.

62

[13] B. Yu, H. Wang., A. Joshi, Q. Xiang, E. Ibok, and M.-R. Lin, in: Techn. Dig. IEDM (IEEE Press, Piscataway, NJ, 2001) 937-939.

[14] A. Hokazono et al., in: Techn. Dig. IEDM (IEEE Press, Piscataway, NJ, 2002) 639-642.

[15] J. Appenzeller et al., Appl. Phys. Lett. 77 (2000) 298-300.

[16] B. Doris et al., in: Techn. Dig. IEDM (IEEE Press, Piscataway, NJ, 2002) 267-270.

[17] F. Balestra, S. Cristoloveanu, M. Benachir, and T. Elewa, IEEE Electron Dev. Lett. 8 (1987) 410-412.

[18] T. Tanaka, K. Suzuki, H. Horie, and T. Sugii, IEEE Electron Dev. Lett. 15 (1994) 386-388.

[19] I. Y. Yang, C. Vieri, A. Chandrakasan, and D. A. Antoniadis, in: Techn. Dig. IEDM (IEEE Press, Piscataway, NJ, 1995) 877-880.

[20] H.-S. P. Wong, K. K. Chan, and Y. Taur, in: Techn. Dig. IEDM (IEEE Press, Piscataway, NJ, 1997) 427-429.

[21] S. Tiwari, A. Kumar, and J. J. Welser, Int. J. High Speed Electron. Syst. 10 (2000) 231-245.

[22] V. W. C. Chan and P. C. H. Chan, IEEE Electron Dev. Lett. 22 (2001) 80-82.

[23] K. W. Guarini et al., in: Techn. Dig. IEDM (IEEE Press, Piscataway, NJ, 2001) 425-428.

[24] M. Ieong et al., in: Techn. Dig. IEDM (IEEE Press, Piscataway, NJ, 2001) 441-445.

[25] D. Hisamoto, T. Kaga, and E. Takeda, IEEE Trans. on Electron. Dev. 38 (1991)1419-1424.

[26] L. Guo, E. Leobandung, and S. Y. Chou, Science 275 (1997) 649-651.

[27] D. Hisamoto et al., IEEE Trans. on Electron Dev. 47 (2000) 2320-2325.

[28] J. Kedzierski, P. Xuan, E. H. Anderson, J. Bokor, T.-J. King, and C. Hu, in: Techn. Dig. IEDM (IEEE Press, Piscataway, NJ, 2000) 57-60.

[29] Y.-K. Choi et al., in: Techn. Dig. IEDM (IEEE Press, Piscataway, NJ, 2000) 421-424.

[30] X. Huang et al., IEEE Trans. on Electron. Dev. 48 (2001) 880-884.

[31] T. Hiramoto, in: Proc. 2001 IEEE Int. SOI Conf. (IEEE Press, Piscataway, NJ, 2001) 8-10.

[32] J. Kedzierski et al., in: Techn. Dig. IEDM (IEEE Press, Piscataway, NJ, 2001) 437-440.

[33] B. Yu et al., in: Techn. Dig. IEDM (IEEE Press, Piscataway, NJ, 2002) 251-255.

[34] H. Takato et al., IEEE Trans. on Electron Dev. 38 (1991) 573-577.

[35] M. Yang, C. L. Chang, M. Carroll and J. C. Sturm, IEEE Electron Dev. Lett. 20 (1999) 301-303.

[36] J. M. Hergenrother et al., in: Techn. Dig. IEDM (IEEE Press, Piscataway, NJ, 1999) 75-78.

[37] J. M. Hergenrother et al., in: Techn. Dig. IEDM (IEEE Press, Piscataway, NJ, 2001) 51-54.

[38] D. J. Frank, S. E. Laux, and M. V. Fischetti, in: Techn. Dig. IEDM (IEEE Press, Piscataway, NJ, 1992) 553-556.

[39] K. Natori, J. Appl. Phys. 76 (1994) 4879-4890.

[40] M. S. Lundstrom, IEEE Electron Dev. Lett. 18 (1997) 361-363.

[41] F. G. Pikus and K. K. Likharev, Appl. Phys. Lett. 71 (1997) 3661-3663.

[42] S. Datta, F. Assad, and M. S. Lundstrom, Superlattices and Microstructures 23 (1998) 771-780.

[43] D. J. Frank, Y. Taur, and H.-S. P. Wong, IEEE Electron Dev. Lett. 19 (1998) 385-387.

[44] H.-S. P. Wong, D. J. Frank, and P. M. Solomon, in: Techn. Dig. IEDM (IEEE Press, Piscataway, NJ, 1998) 407-410.

[45] Y. Naveh and K. K. Likharev, IEEE Electron Dev. Lett. 21 (2000) 242-244.

[46] S. Cristoloveanu, T. Ernst, D. Munteanu, and T. Ouisse, Int. J. of High Speed Electron. and Syst. 10 (2000) 217-230.

[47] F. Assad, Z. Ren, D. Vasileska, S. Datta, and M. Lundstrom, IEEE Trans. on Electron Dev. 47 (2000) 232-240.

[48] J. G. Fossum, Z. Ren, K. Kim, and M. Lundstrom, Superlattices and Microstructures 28, (2000) 525-530.

[49] M. S. Lundstrom, IEEE Electron Dev. Lett. 22 (2001) 293-295.

[50] V. Sverdlov, Y. Naveh, and K. Likharev, in: Proc. 2001 IEEE Int. SOI Conf. (IEEE Press, Piscataway, NJ, 2001) 151-152.

[51] P. M. Solomon, and S. E. Laux, in: Tech. Dig. of IEDM (IEEE Press, Piscataway, NJ, 2001) 95-98.

[52] Z. Ren, R. Venugopal, S. Datta, and M. Lundstrom, in: Tech. Dig. of IEDM (IEEE Press, Piscataway, NJ, 2001) 107-110.

[53] A. Svizhenko, M. P. Anantram, T. R. Govindan, and R. Venugopal, J. Appl. Phys. 91 (2002) 2343-2354.

[54] M. Lundstrom, and Z. B. Ren, IEEE Trans. on Electron Dev. (2002) 133-141 .

[55] R. Venugopal, Z. Ren, S. Datta, and M. S. Lundstrom, J. Appl. Phys. 92 (2002) 3730-3739.

[56] J. H. Rhew, Z. B. Ren, and M. S. Lundstrom, Solid-State Electronics 46 (2002) 1899-1906.

[57] J. Wang and M. Lundstrom, in: Tech. Dig. of IEDM (IEEE Press, Piscataway, NJ, 2002) 707-710.

[58] S. E. Laux, A. Kumar, and M. V. Fischetti, in: Tech. Dig. of IEDM (IEEE Press, Piscataway, NJ, 2002) 715-718.

[59] V. A. Sverdlov, T. J. Walls, and K. K. Likharev, Nanoscale silicon MOSFETs: A theoretical study, accepted for publication in IEEE Trans. on Electron Dev. (2003).

[60] A. Svizhenko and M. P. Anantram, Scattering in nanotransistors: An analytical study, preprint (2002).

[61] T. J. Walls, V. A. Sverdlov, and K. K. Likharev, MOSFETs below 10 nm: Quantum Theory, Report at NanoMES'03 (Tempe, AZ, Feb. 2003), to be published in Physica E (2003).

[62] D. Esseni et al., in: Techn. Dig. IEDM (IEEE Press, Piscataway, NJ, 2001) 445-449.

[63] S. Takagi, A. Torumi, M. Iwase, and H. Tango, IEEE Trans. on Electron Dev. 41 (1994) 2357-2362.

[64] M. Fischetti, S. E. Laux, and E. Crabbe, J. Appl. Phys. 78 (1995) 1058-1087.

[65] T. Sakamoto, H. Kawaura, T. Baba, and T. Iizuka, Appl. Phys. Lett. 76 (2000) 2618-2620.

[66] P. M. Solomon (solomon@watson.ibm.com), On contacts to small semiconductor devices, unpublished (1998).

[67] V. Sverdlov, X. Oriols, and K. Likharev, IEEE Trans. on Nanotechnology 2 (2002) 59-63.

[68] K. K. Young,. IEEE Trans. on Electron Dev. 36 (1989) 504-506.

[69] K. Suzuki, T. Tanaka, Y. Tosaka, H. Hoire, and Y. Arimoto, IEEE Trans. on Electron Dev. 40 (1993) 2326-2329.

[70] S. P. Tang, R. M. Wallace, A. Seabaugh, D. King-Smith, Appl. Surf. Sci. 135 (1998) 137-142.

[71] D. A. Muller, T. Sorsch, S. Moccio, F. H. Baumann, K. Evans-Lutterodt, and G. Timp, Nature, 399 (1999) 758-761.

[72] M. S. Krishnan, L. Chang, T.-J. King, J. Bokor, and C. Hu, in: Techn. Dig. IEDM (IEEE Press, Piscataway, NJ, 1999) 241-244.

[73] S. Song et al., in: Techn. Dig. IEDM (IEEE Press, Piscataway, NJ, 2001) 55-58.

[74] J. P. Colinge, Silicon-On-Insulator Technology (Kluwer, Dordrecht, 1991).

[75] S. Cristoloveanu, Int. J. of High Speed Electron. Syst. 10 (2000) 129-130.

[76] F. Wittman, H. Gossner, and I. Eisele, J. of Mat. Sci.: Materials in Electronics 6 (1995) 336-340.

[77] L. Wong, H. Qu, J.-M. Kuo, and K. K. Chin, IEEE Trans. on Electron Dev. 46 (1999) 865-870.

[78] K. Nakazato, K. Itoh, H. Mizuta, and H. Ahmed, Electron Lett. 35 (1999) 848-850.

[79] S. Xu, K. P. Gan, P. D. Foo, and Y. Liu, IEEE Electron Dev. Lett. 21 (2000) 176-178.

[80] F. Allibert, A. Zaslavsky, and S. Cristoloveanu, in: Proc. of 2001 IEEE Int. SOI Conf. (IEEE Press, Piscataway, NJ, 2001) 149-150.

[81] Q. Lu et al., IEEE Electron Dev. Lett. 22 (2001) 324-326.

[82] L. Kang et al., IEEE Electron Dev. Lett. 21 (2000) 181-183.

[83] W. J. Qi et al., Appl. Phys. Lett. 77 (2000) 1704-1706.

[84] E. P. Gusev et al., in: Tech. Dig. IEDM (IEEE Press, Piscataway, NJ, 2001) 451-454.

[85] L. D. Landau and E. M. Lifshitz, Quantum Mechanics (Non-Relativistic Theory), 3rd ed. (Pergamon Press, Oxford, 1977) 184-185.

[86] J. A. López-Villanueva, P. Cartujo-Castinello, F. Gámiz, J. Banqueri, and A. J. Palma, IEEE Trans. on Electron Dev. 47 (2000) 141-146.

[87] R. Sasajima, K. Fujimaru, and H. Matsumura, Appl. Phys. Lett. 74 (1999) 3215-3217.

[88] C.-K. Huang, W. E. Zhang, and C. H. Yang, IEEE Trans. on Electron Dev. 45 (1998) 842-848.

[89] C. Wang, J. P. Snyder, and J. R. Tucker, Appl. Phys. Lett. 65 (1999) 618-620.

[90] P. P. Gelsinger, in: ISSCC Tech. Dig. (IEEE Press, Piscataway, NJ, 2001) 22-25.

[91] D. J. Frank, IBM J. of Res. Devel. 46 (2002) 235-244.

[92] S. J. Tans, A. R. M. Verschueren, and C. Dekker, Nature 393 (1998) 49-52.

[93] C. Zhou, J. Kong, and H. Dai, Appl. Phys. Lett. 76 (2000) 1597-1599.

[94] R. Martel, H.-S. P. Wong, K. Chan, and P. Avouris, in: Techn. Dig. IEDM (IEEE Press, Piscataway, NJ, 2001) 159-161.

[95] S. Rosenblatt, Y. Yaish, J. Park, J. Gore, V. Sazonova, and P. L. McEuen, Nano Letters 2 (2002) 869-872.

[96] R. Martel, T. Schmidt, H. R. Shea, T. Hertel, and P. Avouris, Appl. Phys. Lett. 73 (1998) 2447-2449.

[97] W. Roush, IEEE Spectrum 38 (2001) 25-26.

[98] R. F. Service, Science 293 (2001) 785-786.

[99] K. K. Likharev, in: H. Morkoç, ed., Advanced Semiconductor and Organic Nano-Techniques, Part I (Elsevier, San Diego, 2003) 239-302.

[100] B. Altshuler, P. Lee, and R. Webb, eds., Mesoscopic Phenomena in Solids (Elsevier, Amsterdam, 1991).

[101] Y. Imry, Introduction to Mesoscopic Physics (Oxford Univ. Press, Oxford, 1997).

[102] S. Datta, Electronic Transport in Mesoscopic Systems (Cambridge Univ. Press. Cambridge, MA, 1997).

[103] S. Datta, Superlattices and Microstructures 6 (1989) 83-93.

[104] F. Capasso, ed., Physics of Quantum Electron Devices (Springer, Berlin, 1990).

[105] A. J. Holden, in: K. E. Singer, ed., Gallium Arsenide and Related Compounds (IOP Publishing, Bristol, UK, 1990) 1-78.

[106] T. C. L. G. Sollner, E. R. Brown, W. D. Goodhue, and H. Q. Lee, in: F. Capasso, ed., Physics of Quantum Electron Devices (Springer, Berlin, 1990) 147-158.

[107] L. L. Chang, E. E. Mendez, and C. Tejedor, eds., Resonant Tunneling in Semiconductors (Plenum, New York, 1992).

[108] P. E. Thompson et al., Thin Solid Films 380 (2000) 145-150.

[109] M. Tinkham, Introduction to Superconductivity, 2nd ed. (McGraw-Hill, New York, 1996).

[110] K. Likharev, Dynamics of Josephson Junctions and Circuits (Gordon and Breach, New York, 1986).

[111] D. K. Brock, E. K. Track, and J. M. Rowell, IEEE Spectrum 37 (2000) 40-46.

[112] P. Bunyk, K. Likharev, and D. Zinoviev, Int. J. on High Speed Electronics and Systems 11 (2001) 257-305.

[113] M. A. Nielsen and I. L. Chuang, Quantum Computation and Quantum Information (Cambridge Univ. Press, Cambridge, UK, 2000).

[114] K. K. Likharev, IBM J. Res. Devel. 32 (1988) 144-158.

[115] D. V. Averin and K. K. Likharev, in: B. Altshuler, P. Lee, and R. Webb, eds., Mesoscopic Phenomena in Solids (Elsevier, Amsterdam, 1991) 173-271.

[116] M. H. Devoret and H. Grabert, eds., Single Charge Tunneling (Plenum, New York, 1992).

[117] K. K. Likharev, Proc. IEEE 87 (1999) 606-632.

[118] J. Lambe and R. C. Jaklevic, Phys. Rev. Lett. 22 (1969) 1371-1375.

[119] I. O. Kulik, and R. I. Shekhter, Sov. Phys. - JETP 41 (1975) 308-316.

[120] P. Lafarge et al., Z. Phys. B 85 (1991) 327-332.

[121] Sh. Farhangfar et al. J. Low Temp. Phys. 108 (1997) 191-215.

[122] J. P. Pekola et al. J. Low Temp. Phys. 128 (2002) 263-269.

[123] D. V. Averin and K. K. Likharev, J. Low Temp. Phys. 62 (1986) 345-372.

[124] K. K. Likharev, IEEE Trans. on Magn. 23 (1987) 1142-1145.

[125] T. A. Fulton, and G. D. Dolan, Phys. Rev. Lett. 59 (1987) 109-112.

[126] Yu. A. Pashkin, Y. Nakamura, and J. S. Tsai, Appl. Phys. Lett. 76 (2000) 2256-2258.

[127] G. Zimmerli, R. L. Kautz, and J. M. Martinis, Appl. Phys. Lett. 61 (1992) 2616-2618.

[128] R. A. Smith, and H. Ahmed, Appl. Phys. Lett. 71 (1997) 3838-3840.

[129] Y. Ono, Y. Takahashi, K. Yamazaki, M. Nagase, H. Namatsu, K. Kurihara, and K. Murase, Appl. Phys. Lett. 76 (2000) 3121-3123.

[130] T. A. Fulton, P. L. Gammel, and L. N. Dunkleberger, Phys. Rev. Lett. 67 (1991) 3148-3151.

[131] D. V. Averin and K. K. Likharev, in: M. H. Devoret and H. Grabert, eds., Single Charge Tunneling (Plenum, New York, 1992) 311-322.

[132] L. J. Geerligs et al. Phys. Rev. Lett. 64 (1990) 2691-2694.

[133] K. K. Likharev, N. S. Bakhvalov, G. S. Kazacha, and S. I. Serdyukova, IEEE Trans. on Magn. 25 (1989) 1436-1439.

[134] K. K. Likharev and K. A. Matsuoka, Appl. Phys. Lett. 67 (1995) 3037-3039.

[135] D. V. Averin and A. A. Odintsov, Phys. Lett. A 140 (1989) 251-255.

[136] P. D. Dresselhaus, L. Ji, S. Y. Han, J. E. Lukens, and K. K. Likharev, Phys. Rev. Lett. 72 (1994) 3226-3229.

[137] L. Ji, P. D. Dresselhaus, S. Y. Han, K. Lin, W. Zheng, and J. E. Lukens, J. Vac. Sci. Technol. B 12 (1994) 3619-3622.

[138] K. A. Matsuoka, K. K. Likharev, P. D., Dresselhaus, L. Ji, S. Y. Han and J. E. Lukens, J. Appl. Phys. 81 (1997) 2269-2281.

[139] K. K. Likharev and A. N. Korotkov, Science 273 (1996) 763-765.

[140] A. N. Korotkov and K. K. Likharev, J. Appl. Phys. 84 (1998) 6114-6126.

[141] A. O. Orlov, R. K. Kummamuru, R. Ramasubramaniam, T. Toth, C. S. Lent, G. H. Bernstein, and G. L. Snider, Appl. Phys. Lett. 78 (2001) 1625-1627.

[142] D. V. Averin and A. N. Korotkov, J. Low Temp. Phys. 80 (1991) 173-185.

[143] D. V. Averin, A. N. Korotkov and K.K. Likharev, Phys. Rev. B 44 (1991) 6199-6211.

[144] L. P. Kouwenhoven et al., in: L. Sohn et al., eds., Mesoscopic Electron Transfer (Kluwer, Dordrecht, 1997) 105-215.

[145] K. Matsumoto, M. Ishii, K. Segawa, Y. Oka, B. J. Vartanian, and J. S. Harris, Appl. Phys. Lett. 68 (1996) 34-36.

[146] K. Matsumoto, Y. Gotoh, T. Maeda, J. A. Dagata, and J. S. Harris, Appl. Phys. Lett. 76 (2000) 239-241.

[147] H. W. Ch. Postma, T. Teepen, Z. Yao, M. Grifoni, and C. Dekker, Science 293 (2001) 76-79.

[148] C. Thelander et al., Appl. Phys. Lett. 79 (2001) 2106-2108.

[149] J. B. Cui, M. Burghard, and M. Kern, Nano Letters 2 (2002) 117-120.

[150] Y. Takahashi et al., Electron. Lett. 31 (1995) 136-138.

[151] Y. Ono, Y. Takahashi, K. Yamazaki, M. Nagase, H. Namatsu, K. Kurihara, and K. Murase, IEEE Trans on Electron. Dev. 47 (2000) 147-153.

[152] D. H. Kim, J. D. Lee, and B.-G. Park, Jpn. J. Appl. Phys. (pt. 1) 39 (2000) 2329-2333.

[153] K. Uchida, J. Koga, R. Ohba, and A. Torumi, in: Techn. Dig. IEDM (IEEE Press, Piscataway, NJ, 2000) 863-865.

[154] M. Saitoh, N. Takahashi, H. Ishikuro, and T. Hiramoto, Jpn. J. Appl. Phys. (pt. 1) 40 (2001) 2010-2012.

[155] N. J. Tao, Phys. Rev. Lett. 76 (1996) 4066-4069.

[156] E. S. Soldatov et al., JETP Lett. 64 (1996) 556-558.

[157] M. A. Reed, C. Zhou, C. J. Miller, T. P. Burgin, and J. M. Tour, Science 278 (1997) 252-254.

[158] W. Han et al., J. Phys. Chem. B 101 (1997) 10719-10725.

[159] C. P. Collier et al., Science 285 (1999) 391-394.

[160] C. Kergueris et al., Phys. Rev. B 59 (1999) 12505-12513.

[161] J. Chen, M. A. Reed, A. M. Rawlett, and J. M. Tour, Science 286 (1999) 1550-1552.

[162] D. Porath, A. Bezryadin, S. de Vries, and C. Dekker, Nature 403 (2000) 635-638.

[163] C. P. Collier et al., Science 289 (2000) 1172-1175.

[164] D. I. Gittins, D. Bethell, D. J. Schiffrin, and R. J. Nichols, Nature 408 (2000) 67-69.

[165] H. Watanabe, C. Manabe, T. Shigematsu, and M. Shimuzu, Appl. Phys. Lett. 78 (2001) 2928-2930.

[166] Z. J. Donhauser et al., Science 292 (2001) 2303-2306.

[167] H. Xe et al., J. Am. Chem. Soc. 123 (2001) 7730-7731.

[168] X. D. Cui et al., Science 294 (2002) 571-574.

[169] S. P. Gubin et al., Nanotechnology 13 (2002) 185-194.

[170] J. Reichert, R. Ochs, D. Beckman, H. B. Weber, M. Mayor, and H. v. Löhneysen, Phys. Rev. Lett. 88 (2002) 176804 1-4.

[171] N. B. Zhitenev, H. Meng, and Z. Bao, Phys. Rev. Lett. 88 (2002) 226801 1-4.

[172] J. Park et al., Nature 417 (2002) 722-725.

[173] W. Liang, M. P. Shores, M. Bockrath, J. R. Long, and H. Park, Nature 417 (2002) 725-729.

[174] H. B. Weber et al., Chem. Physics 281 (2002) 113-125.

[175] J. Chen and M. Reed, Chem. Physics 281 (2002) 127-145.

[176] R. H. M. Smit et al., Nature 419 (2002) 906-909.

[177] D. B. Janes, H. Halimun, J. Choi, and S. Burns, Metal-molecule-metal; structures with pre-fabricated contacts, Report at the 6th Conf. on Molecular-scale Electronics (Key West, FL, Dec. 2002), to be published by New York Acad. Sci. (2003).

[178] M. Reed, Proc. IEEE 87 (1999) 652-705.

[179] M. Di Ventra, S. T. Pantelides, and N. D. Lang, Phys. Rev. Lett. 84 (2000) 979-982.

[180] V. V. Shorokhov, P. Johanson, and E. S. Soldatov, J. Appl. Phys. 91 (2002) 3049-3053.

[181] G. Cuniberti, L. Craco, D. Porath, and C. Dekker, Phys. Rev. B 65 (2002) 241314 (R) 1-4.

[182] G. Cuniberti, J. Yi, and M. Porto, Appl. Phys. Lett. 81 (2002) 850-852.

[183] N. Zimbovskaya and G. Gumbs, Appl. Phys. Lett. 81 (2002) 1518-1520.

[184] P. E. Kornilovitch, A. M. Bratkovsky, and R. S. Williams, Phys. Rev. B 66 (2002) 165436 1-12.

[185] D. L. Allara et al., Ann. of New York Acad. Sci. 852 (1998) 349-370.

[186] E. H. Nicollian and J. R. Brews, MOS physics and technology (Wiley, New York, 1982).

[187] A. N. Korotkov, Appl. Phys. Lett. 72 (1998) 3226-3228.

[188] Yu. A. Pashkin, Y. Nakamura, and J. S. Tsai, Appl. Phys. Lett. 74 (1999) 132-134.

[189] A. M. Ionescu, M. J. Declercq, K. Banerjee, and J. Gautier, in: Proc. of ACM IEEE Design Automation Conference (ACM, New York, 2002) 88-93.

[190] W. Zheng, J. R. Friedman, D. V. Averin, S. Han, and J. E. Lukens, Solid State Commun. 108 (1998) 839-843.

66

[191] K. E. Nagaev, Phys. Lett. A 169 (1992) 103-107.

[192] K. E. Nagaev, Phys. Rev. B 52 (1995) 4740-4743.

[193] Y. Naveh, D. V. Averin, and K. K. Likharev, Phys. Rev. B 58 (1998) 15371-15374.

[194] V. K. Sverdlov, A. N. Korotkov, and K. K. Likharev, Phys. Rev. B 63 (2000) R081302 1-4.

[195] Y. A. Kinkhabwala, V. A. Sverdlov, A. N. Korotkov, and K. Likharev, Shot noise at hopping: A numerical study, ArXiv preprint cond-mat/0302445, submitted to Phys. Rev. B (2003).

[196] D. M. Kaplan, V. A. Sverdlov, and K. K. Likharev, Coulomb gap, Coulomb blockade, and dynamic activation energy in frustrated single-electron arrays, ArXiv preprint cond-mat/0303439, submitted to Phys. Rev. B (2003).

[197] Y. Pekola, in: H. Morkoç, ed., Advanced Semiconductor and Organic Nano-Techniques, (Academic Press, San Diego, 2003), pt. 1.

[198] K. Likharev, Single-electron devices, to be published in: Encyclopedia of Nanoscience and Nanotechnology (American Scientific Publishers, Stevenson Ranch, CA, 2003).

[199] A. N. Korotkov, D. V. Averin, K. K. Likharev, and S. A. Vasenko, in: H. Koch and H. Lübbig, eds., Single Electron Tunneling and Mesoscopic Devices (Springer, Berlin, 1992) 45-60.

[200] A. N. Korotkov, Phys. Rev. B 49 (1994) 10381-10392.

[201] V. A. Krupenin, D. E. Presnov, A. B. Zorin, and J. Niemeyer, Physica B 284 (2000) 1800-1801.

[202] R. J. Schoelkopf, P. Wahlgren, A. A. Kozhevnikov, P. Delsing, and D. E. Prober, Science 280 (1998) 1238-1242.

[203] A. Aassime, G. Johansson, G. Wendin, R. J. Schoelkopf, and P. Delsing, Phys. Rev. Lett. 86 (2001) 3376-3379.

[204] A. N. Korotkov and M. A. Paalanen, Appl. Phys. Lett. 74 (1999) 4052-4054.

[205] B. Prince, Semiconductor Memories, 2nd ed. (Wiley, Chichester, UK, 1991).

[206] W. D. Brown and J. E. Brewer, Nonvolatile Semiconductor Memory Technology (IEEE Press, Piscataway, NJ, 1998).

[207] S. Tiwari, F. Rana, H. Hanafi, A. Hartstein, E. F. Crabbé, and K. Chan, Appl. Phys. Lett. 68 (1996) 1377-1379.

[208] H. I. Hanafi, S. Tiwari, and I. Khan, IEEE Trans. on Electron. Dev. 43 (1996) 1553-1558.

[209] Y. Shi, K. Saito, H. Ishikuro, and T. Hiramoto, J. Appl. Phys. 84 (1998) 2358-2360.

[210] Y. C. King, T. J. King, and C. M. Hu, IEEE Electron. Device Lett. 20 (1999) 409-411.

[211] I. Kim, S. Han, K. Han, J. Lee, and H. Shin, IEEE Electron. Device Lett. 20 (1999) 630-631.

[212] K. Han, I. Kim, and H. Shin, IEEE Trans. on Electron Devices 48 (2001) 874-879.

[213] R. Ohba, N. Sugiyama, J. Koga, K. Uchida, and A. Torumi, Jpn. J. Appl. Phys. (pt. 1) 39 (2000) 989-993.

[214] A. Fernandes et al., in: Techn. Dig. IEDM (IEEE Press, Piscataway, NJ, 2001) 155-158.

[215] T. Ishii, T. Osabe, T. Mine, F. Murai, and K. Yano, in: Techn. Dig. IEDM (IEEE Press, Piscataway, NJ, 2001) 305-308.

[216] A. Nakajima, T. Futatsugi, K. Kosemura, T. Fukano, and N. Yokoyama, Appl. Phys. Lett. 70 (1997) 1742-1744.

[217] J. J. Welser, S. Tiwari, S. Rishton, K. Y. Lee, and Y. Lee, IEEE Electron Device Lett. 18 (1997) 278-280.

[218] T. Tsutsumi, K. Ishii, E. Suzuki, H. Hiroshima, M. Yamanaka, I. Sakata, S. Kanemaru, S. Hazra, T. Maeda, and K. Tomizawa, Electron. Lett. 36 (2000) 1322-1323.

[219] K. Yano, T. Ishii, T. Sano, T. Mine, F. Murai, T. Hashimoto, T. Kobayashi, T. Kure, and K. Seki, Proc. IEEE 87 (1999) 633-651.

[220] K. Yano, T. Ishii, T. Sano, T. Mine, F. Murai, T. Kure, and K. Seki, in:ISSCC'98 Digest of Technical Papers (IEEE Press, New York, 1998) 344-345.

[221] K. K. Likharev, and A. N. Korotkov, in: Proc. of the 1995 Int. Semicond. Device Res. Symp. (Univ. of Virginia, Charlottesville, VA, 1995) 355-358.

[222] C. D. Chen, Y. Nakamura, and J. S. Tsai, Appl. Phys. Lett. 71 (1997) 2038-2040.

[223] A. N. Korotkov, J. Appl. Phys. 92 (2002) 7291-7295.

[224] K. K. Likharev, Appl. Phys. Lett. 73 (1998) 2137-2139.

[225] A. Korotkov and K. Likharev, in: Techn. Dig. IEDM (IEEE Press, Piscataway, NJ, 1999) 223-226.

[226] K. K. Likharev, IEEE Circuits and Devices Mag. 16 (2000) 16-21.

[227] H. Shichijyo et al., in: Ext. Abstr. of the 16th Int. Conf. on Solid State Devices and Materials (Business Center for Academic Societies, Tokyo, 1984) 265-267.

[228] N. J. Stone and H. Ahmed, Appl. Phys. Lett. 73 (1998) 2134-2136.

[229] K. Nakazato, and H. Ahmed, Appl. Phys. Lett. 66 (1995) 3170-3172.

[230] Z. A. K. Durrani, A. C. Irvine, and H. Ahmed, Appl. Phys. Lett. 74 (1999) 1293-1295.

[231] Z. A. K. Durrani, A. C. Irvine, and H. Ahmed, IEEE Trans. on Electron. Devices 47 (2000) 2334-2339.

[232] A. C. Irvine, Z. A. K. Durrani, and H. Ahmed, J. Appl. Phys. 87, 8594-8603 (2000).

[233] H.-O. Müller, K. Katayama, and H. Mizuta, J. Appl. Phys. 84 (1998) 5603-5609.

[234] T. Osabe, T. Ishii, T. Mine, F. Murai, and K. Yano, in: Techn. Digest IEDM (IEEE Press, Piscataway, NJ, 2000) 301-304.

[235] K. Nakazato, K. Itoh, H. Mizuta, and H. Ahmed, Electron Lett. 35 (1999) 848-850.

[236] K. Nakazato, K. Itoh, H. Ahmed, H. Mizuta, T. Kisu, M. Kato, and T. Sakata, in: Dig. of Techn. Papers of ISSCC (IEEE Press, Piscataway, NJ, 2000) 132-133.

[237] J. H. Yi, W. S. Kim, S. Song, Y. Khang, H.-I. Kim, J. H. Choi, H. H. Lim, N. I. Lee, K. Fujihara, H.-K. Kang, J. T. Moon, and M. Y. Lee, in: Techn. Dig. IEDM (IEEE Press, Piscataway, NJ, 2001) 787-790.

[238] J. F. Scott, Ferroelectric Memories (Springer, New York, 2000).

[239] S. Tehrani, B. Engel, J. M. Slaughter, E. Chen, M. DeHerrera, M. Durlam, P. Naji, P. Whig, J. Janesky, and J. Calder, IEEE Trans. on Magn. 36 (2000) 2752-2757.

[240] K. Inomata, IEICE Trans. on Electronics E84C (2001) 740-746.

[241] S. Lai and T. Lowrey, in: Techn. Dig. IEDM (IEEE Press, Piscataway, NJ, 2001), 803-806.

[242] Ovonic unified memory, available on the Web at http://www.ovonyx.com/ovonyxtech.html.

[243] R. K. Kummamuru et al., Appl. Phys. Lett. 81 (2002) 1332-1334.

[244] C. Kothandaramann, S. K. Iyer, and S. S. Iyer, IEEE Electron Dev. Lett. 23 (2002) 523-525.

[245] HP announces breakthroughs in molecular electronics, press release, available on the Web at http://www.hp.com/hpinfo/newsroom/press/09sep02a.htm (2002).

[246] M. J. Yoo et al., Science 276 (1997) 579-582.

[247] J. R. Tucker, J. Appl. Phys. 72 (1992) 4399-4413.

[248] A. N. Korotkov, R. H. Chen, and K. K. Likharev, J. Appl. Phys. 78 (1995) 2520-2530.

[249] R. H. Chen, A. N. Korotkov, and K. K. Likharev, Appl. Phys. Lett. 68 (1996) 1954-1956.

[250] H. Iwamura, M. Akazawa, and Y. Amemiya, IEICE Trans. on Electron. E81-C (1998) 42-48.

[251] R. H. Chen and K. K. Likharev, Appl. Phys. Lett. 72 (1998) 61-63.

[252] K. K. Likharev, and V. K. Semenov, in: Ext. Abstr. of Int. Supercond. Electronics Conf., (Tokyo, 1987) 128-131.

[253] S. V. Vyshenskii, S. P. Polonsky, and K. K. Likharev, Single electron logic circuits, unpublished (1990), described in detail in Ref. 131.

[254] P. Bakshi, D. A. Broido, and K. Kempa, J. Appl. Phys. 70 (1991) 5150-5152.

[255] C. S. Lent, P. D. Tougaw, W. Porod, and G. Bernstein, Nanotechnology 4 (1993) 49-57.

[256] K. Nomoto, R. Ugajin, T. Suzuki and I. Hase, Electron. Lett. 29, (1993) 1380-1381.

[257] D. V. Averin, L. F. Register, K. K. Likharev, and K. Hess, Appl. Phys. Lett. 64 (1994) 126-128.

[258] P. D. Tougaw and C. S. Lent, J. Appl. Phys. 75 (1994) 1818-1825.

[259] S. Bandyopadhyay, B. Das, and A. E. Miller, Nanotechnology 4 (1994) 113-133.

[260] A. N. Korotkov, Appl. Phys. Lett. 67 (1995) 2412-2414.

[261] M. G. Ancona, J. Appl. Phys. 78 (1995) 3311-3314.

[262] K. Nomoto, R. Ugajin, T. Suzuki and I. Hase, J. Appl. Phys. 79 (1996) 291-300.

[263] C. S. Lent and P. D. Tougaw, J. Appl. Phys. 80 (1996) 4722-4736.

[264] S. Bandyopadhyay and V. Roychowdhury, Jpn. J. Appl. Phys. (pt. 1) 35 (1996) 3350-3362.

[265] A. V. Krasheninnikov and L. A. Openov, JETP Lett. 64 (1996) 231-218.

[266] S. N. Molotkov and S. S. Nazin, JETP 83 (1996) 794-802.

[267] C. S. Lent and P. D. Tougaw, Proc. IEEE 85 (1997) 541-557.

[268] N.-J. Wu, N. Asahi, and Y. Amemiya, Jpn. J. Appl. Phys. 36 (1997) 2621-2627.

[269] N. Asahi, M. Akazawa, and Y. Amemiya, IEEE Trans. on Electron. Dev. 44 (1997) 1109-1116.

[270] A. M. Bychkov, L. A. Openov, and I. A. Semenihin, JETP Lett. 66 (1997) 298-303.

[271] Y. Fu and M. Willander, J. Appl. Phys. 83 (1998) 3186-3191.

[272] M. P. Anantram and V. P. Roychowdhury, J. Appl. Phys. 85 (1999) 1622-1625.

[273] M. Governale, M. Macucci, G. Iannaccone, and C. Ungarelli, J. Appl. Phys. 85 (1999) 2962-2971.

[274] A. Gin, S. Williams, H. Y. Meng, and P. D. Tougaw, J. Appl. Phys. 85 (1999) 3713-3720.

[275] A. Gin, P. D. Tougaw, S. Willams, J. Appl. Phys. 85 (1999) 8281-8286.

[276] M. Girlanda, M. Governale, M. Macucci, and G. Iannaccone, Appl. Phys. Lett. 75 (1999) 3198-3200.

[277] C. Ungarelli, S. Francaviglia, and M. Macucci, and G. Iannaccone, J. Appl. Phys. 87 (2000) 7320-7325.

[278] J. R. Pasky, L. Henry, and P. D. Tougaw, J. Appl. Phys. 87 (2000) 8604-8609.

[279] M. T. Niemier and P. M. Kogge, Int. J. of Circuit Theory and Appl. 29 (2001) 49-62.

[280] L. Bonci, G. Iannaccone, and M. Macucci, J. Appl. Phys. 89 (2001) 6435-6443.

[281] J. Timler and C. S. Lent, J. Appl. Phys. 91 (2002) 823-831.

[282] L. Bonci, M. Gattobigio, G. Iannaccone, and M. Macucci, J. Appl. Phys. 92 (3169) 3169-3178.

[283] M. Macucci et al., Critical assessment of the QCA architecture as a viable alternative to large scale integration, preprint (2003).

[284] R. A. Kiehl and T. Ohshima, Appl. Phys. Lett. 67 (1995) 2494-2496.

[285] T. Ohshima, and R. A. Kiehl, J. Appl. Phys. 80 (1996) 912-923.

[286] T. Ohshima, Appl. Phys. Lett. 69 (1996) 4059-4061

[287] F. Y. Liu, F. T. An, and R. A. Kiehl, Appl. Phys. Lett. 74 (1999) 4040-4042.

[288] P. Delsing, L. S. Kuzmin, K. K. Likharev, and T. Claeson, Phys. Rev. Lett. 63 (1989) 1861-1865.

[289] Y. Wada, Ann. New York Acad. Sci. 960 (2002) 39-61.

[290] S. Zankovych, T. Hoffmann, J. Seekamp, J. U. Bruch, and C. M. S. Torres, Nanotechnology 12 (2001) 91-95.

[291] J. M. Tour, M. Kozaki, and J. M. Seminario, J. Am. Phys. Soc. 120 (1998) 8486-8493.

[292] S. Fölling, Ö. Türel, and K. K. Likharev, in: Proc. of the 2001 Int. Joint Conf. on Neural Networks (Int. Neural Network Society, Mount Royal, NJ, 2001) 216-221.

[293] Ö. Türel, and K. K. Likharev, Int. J. of Circuit Theory and Appl. 31 (2003) 37-54.

[294] K. Likharev, A. Mayr, I. Muckra, Ö. Türel, CrossNets: Fault-tolerant architectures for molecular electronic circuits, report at the 6th Conf. On Molecular-Scale Electronics (Key West, Dec. 2002), to be published by the New York Acad. Sci. (2003).

[295] D. J. Amit, Modeling Brain Function (Cambridge U. Press, Cambridge, UK, 1989).

[296] J. Hertz, A. Krogh, R. G. Palmer, Introduction to the Theory of Neural Computation. (Perseus, Cambridge, MA, 1991).

[297] P. S. Churchland and T. J. Sejnowski, The Computational Brain (MIT Press, Cambridge, MA, 1992).

[298] L. Fausett, Fundamentals of Neural Networks (Prentice Hall, Upper Saddle River, NJ, 1994).

[299] B. Müller, J. Reinhard, M. T. Strickland, Neural Networks, 2nd ed. (Springer, Berlin, 1995).

[300] M. H. Hassoun, Fundamentals of Artificial Neural Networks (MIT Press, Cambridge, MA, 1995).

[301] S. Haykin, Neural networks (Prentice Hall, Upper Saddle River, NJ, 1999).

[302] W. S. Sarle, Neural networks: FAQ, papers, and various other things, available for ftp download from ftp.sas.com/pub/neural/FAQ.html.zip, 2002.

[303] V. B. Mountcastle, Perceptual Neuroscience. The cerebral cortex (Harvard U. Press, Cambridge, MA, 1998).

[304] V. Braitenberg and A. Schüz, Cortex: Statistics and Geometry of Neuronal Connectivity, 2nd ed. (Springer, Berlin, 1998).

[305] M. Akazawa and Y. Ameniya, Appl. Phys. Lett. 70 (1997) 670-673.

[306] D. Gosztola, M. P. Niemczyk, W. Svec, A. S. Lukas, and M. R. Wasielewski, J. Phys. Chem. A 104 (2000) 6545-6551.

[307] H. D. Sikes et al., Science 291 (2001) 1519-1523.

[308] H. McNally, D. B. Jones, B. Kasibhatla, C. P. Kubiak, Superlattices and Microstructures 31 (2002) 239-245.

[309] E. G. Emiroglu, Z. A. K. Duranni, D. C. Hasko, D. A. Williams, J. Vac. Sci. Technol. B 20 (2002) 2806-2809.

[310] J. R. Heath, P. K. Kuekes, G. S. Snider, R. S. Williams, Science 280 (1998) 1716-1721.

[311] M. Kirihara and K. Taniguchi, Jpn. J. Appl. Phys. (pt. 1) 36 (1997) 4172-4175.

[312] C. Gerousis, S. M. Goodnick, W. Porod, Int. J. of Circuit Theory and Appl. 28 (2000) 523-535.

[313] J. C. Da Costa et al., Analog Integrated Circuits and Signal Processing 24 (2000) 59-71.

[314] H. Park et al., Nature 407 (2000) 57-60.

[315] J. Chen et al., in: H. Morkoç, ed. Advanced Semiconductor and Organic Nano-Techniques, Part III (Elsevier, Amsterdam, 2003) 43-187.

[316] Ö. Türel, I. Muckra, and K. Likharev, Possible Nanoelectronic Implementation of Neuromorphic Networks, accepted for presentation at the Int. Joint Conference on Neural Networks (Portland, OR, July 2003); preprint available on the Web at http://rsfq1.physics.sunysb.edu/~likharev/ nano/Portland.pdf.

Nano and Giga Challenges in Microelectronics
Greer at al (Editors)
© *2003 Elsevier Science B.V. All rights reserved*

Lithography: Concepts, Challenges and Prospects

Kevin Lucas [a], Sergei Postnikov [a] Cliff Henderson [b], and Scott Hector [a].

[a] *Motorola Advanced Products Research and Development Laboratory, DigitalDNA* [TM] *Labs,*
3501 Ed Bluestein Blvd., Mail Drop: K-10, Austin, Texas 78721 USA
[b] *School of Chemical Engineering, Georgia Institute of Technology, 778 Atlantic Dr., Atlanta, GA 30332-0100 USA*

Abstract

Lithography is a key technology enabling the continued miniaturization of integrated circuits. In this chapter, an overview is presented of all important aspects of lithography. Some basic scaling equations governing the theory of optical lithography are presented, but the chapter focuses on describing the practical challenges in applying lithography for integrated circuit manufacturing. Practical aspects of pattern layout, exposure tool design, masks, lens design, CD control, overlay control, resists, substrate control, simulation, optical proximity correction, phase shift masks, cost of ownership and emerging lithographic techniques are described in detail.

Key words: lithography, photolithography, masks, resists, steppers, scanners.

1. Background

Semiconductor integrated circuits have become ubiquitous in modern tools and toys. As costs have fallen quickly and functionality has risen dramatically, powerful circuits are now available at low cost for a wide range of uses. Improvements in optical lithography have been the driving force behind the extraordinary advances in integrated circuit cost and performance for 30 years [1] (Fig. 1). The number of elements in a circuit has been doubling every eighteen months (Moore's law), and computational power has been increasing even faster [1]. Semiconductor industry roadmaps show this trend continuing [2] into the foreseeable future, although extending optical lithography is becoming progressively more difficult. This chapter aims to provide insight into the past, present and future of optical lithography by explaining basic concepts, develop-ment techniques, and possible successor technologies.

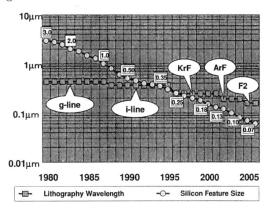

Fig. 1. Progress of optical lithography resolution over the last 20+ years and the expected progress in the next few years. Resolution is shown to be shrinking much faster than the lithographic exposure wavelength.

2. Introduction

The transfer of a pattern from one medium onto another medium is called lithography. This name comes from the Greek words 'lithos' (rock) and 'graphia' (writing). Lithography is typically a binary process; namely, the features in the pattern and in its copy are either opaque or clear (black or white). In integrated circuit fabrication, a circuit designer uses a computer database to specify patterns, which will be transferred into thin film layers or in bulk substrates necessary to fabricate electronic devices on a substrate material such as silicon. Lithography is the process of fabricating the designer's pattern as a relief structure in a sacrificial thin film, called photoresist (or resist), on the surface of a wafer substrate such as silicon (Fig. 2). The resist material relief structure then provides a selective window for further processing of thin films on the substrate beneath the resist or the substrate directly. Processes such as etching (Fig. 3) or ion implantation (Fig. 4) are typically performed with the resist as a selective mask of the underlying layers or substrate. Additive processes such as electroplating or deposition on top of patterned resist are often performed for research applications that use lithography for fabricating devices besides commercial integrated circuits (ICs).

The IC designer's pattern for each of the many layers needed to selectively process the device films or substrate is encoded on a master mask used to replicate the circuit pattern repeatedly in resist. The mask for complex circuit patterns is usually fabricated using a pattern-generating exposure tool where an electron or laser beam traces out the desired pattern in a resist that is sensitive to the beam. Because writing the mask is a serial process, it takes considerable time, making it generally impractical to write the pattern thousands of times on the wafer substrate directly. The pattern in resist on the mask is then transferred into a film that absorbs or reflects photons for optical lithography (Fig. 2). For present optical lithography the mask is a flat, square plate of highly transmissive quartz that supports a patterned layer of chromium. Other types of masks are also used and will be discussed later in the chap-

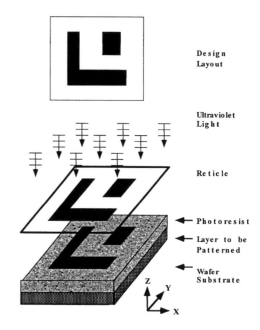

Fig. 2. Idealized schematic of optical lithography exposure process. The pattern on the reticle is transferred to a photosensitive polymer film, "photoresist", via ultraviolet radiation.

ter. The mask is illuminated with visible or ultraviolet radiation, and the pattern on the mask is projected or shadow-cast onto the wafer substrate with the device being fabricated. This process is called photolithography.

3. History

At the inception of integrated circuit fabrication, the mask was placed in intimate contact with the substrate, and the mask pattern was shadow cast into resist using visible light. This technique is known as contact printing. Intimate contact is difficult to achieve using a mask and substrate that are not perfectly flat; thus, 0.2-mm thick glass masks were developed so that the mask could conform to the substrate. Particulates lodged between the mask and substrate during exposure introduced pattern distortions or blocked pattern transfer. These distortions or missing pattern regions are called defects. The particulates also damaged the mask and the substrate, [3] resulting

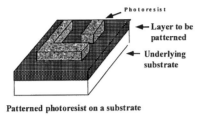

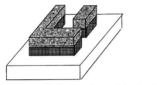

Fig. 3. Idealized schematic showing the role of photoresist as a hardmask in the etch process.

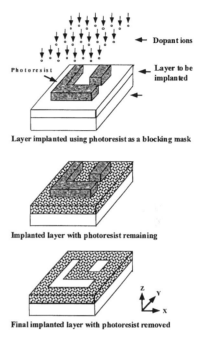

Fig. 4. Idealized schematic showing the role of photoresist in the implant process as a blocking mask.

in defects on the mask and possibly making the device being fabricated inoperable. The lifetime of these flexible masks was limited to about 15 exposures. [4] To overcome this limit, the industry shifted to proximity printing in which the mask was held in close proximity to the wafer but not in contact. The pattern on the mask is shadow cast into resist. Particles on the wafer can then no longer abrade the mask and substrate as long as the gap between them is larger than the particles. The size of the gap, however, is limited by diffraction and by the finite size of the illuminating source [5,6].

To overcome diffraction and practical limits with proximity printing, optical projection lithography has been used for commercial fabrication of integrated circuits since the mid-seventies [7]. In projection lithography an optical system forms a real image of the mask on the wafer. Originally, unity magnification systems were employed and the entire wafer was exposed. As the size of the features desired on the wafer decreased, imaging systems capable of resolving fine features while simultane-

ously imaging the entire mask accurately onto the wafer became impractical [3]. Scanning and step-and repeat exposure schemes, which in most cases used demagnification of the mask pattern by a factor of four or five, were introduced (Fig. 5). In the step-and-repeat scheme, the mask is broken into smaller fields that are individually imaged during several exposures. In the scanning approach, the image of the mask is built up by scanning the mask through a narrow illuminating beam and imaging the illuminated strip onto the substrate. The scanning approach requires a smaller field of view of the optical system simplifying its design enough to allow for highest resolution imaging. Scanning is predominantly used for advanced IC manufacturing today. Ultraviolet light has been used in these projection printers.

Optical projection lithography has been the principal vehicle of semiconductor manufacturing for more than 20 years. Resolution of optical projection lithography is proportional to the wavelength of the radiation used for exposure and inversely proportional to the numerical aperture

72

of the projection lens. To reduce minimum resolution, wavelength can be reduced, and numerical aperture can be increased. The main wavelengths used in optical projection lithography for exposure are 436, 365, 248, 193 or 157 nm. The wavelength has been continually reduced in these steps to increase resolution from tens of microns down to roughly 50 nanometers pattern width with 100-nm pattern pitch in the near future. Introducing a shorter wavelength requires new lens materials, new sources of radiation, new sensors in the exposure tool and new resist materials. Increasing numerical aperture increases the lens volume as roughly the fourth power of numerical aperture. The lens design and fabrication also become more sensitive to aberrations. To further enhance resolution, the constant of proportionality between wavelength divided by numerical aperture and minimum resolution can be reduced. Reducing this constant requires modified illumination schemes such as off-axis illumination, use of phase shifting masks, pupil filters, optical proximity correction of the mask to compensate for diffraction, and or better resists.

The complexity of optical lithography, especially in making the masks, continues to increase as wavelength reduction continues to 157nm. The International Technology Roadmap for Semiconductors (ITRS)[2] predicts that a new lithography paradigm will probably be required for manufacturing 45-nm half-pitch patterns. Leading candidates to succeed optical projection lithography are extreme ultraviolet lithography (EUVL) and electron projection lithography (EPL). The ITRS is a roadmap developed by representatives of the semiconductor industry, and it focuses on complementary metal oxide semiconductor (CMOS) device requirements [8]. The timing of various device generations is debated at various international forums. A committee decides the timing of the device generations, which are called nodes. The rule of thumb for the nodes is based on doubling the density of the integrated circuit features every node [9]. Therefore, the minimum pitch of the patterns is the fundamental dimension. Lithographic resolution is also governed by pattern pitch. The roadmap document contains detailed requirements for lithography, including requirements for mask and resists. These requirements are developed by working groups in each major semiconductor-producing region of the world. The working group members represent integrated device manufacturers (IDMs), academia, research consortia, and suppliers of exposure tools, masks and resist.

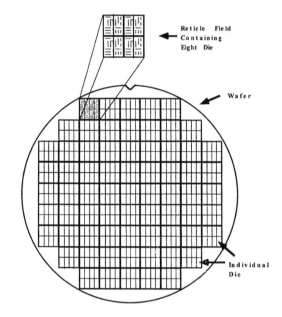

Fig. 5. Semiconductor wafer being patterned lithographically by a multiple die reticle. The notch at the top of the wafer provides orientation to the manufacturing equipment during processing.

The main requirements of lithography for integrated circuit manufacturing are control of pattern feature placement, feature size, and feature fidelity with a rapid process that has high yield. In the vocabulary of lithography, the requirement is high precision critical dimension (CD) control and accurate overlay that is free of printed or process defects in a high throughput process with minimal cost of ownership. For a lithography technology to successfully meet the needs of economically producing integrated circuit devices, these elements are essential. Fabrication of the mask is a crucial element in all of these requirements, and will be discussed briefly in this chapter. However, for a more in depth discussion of mask fabrication, which is also critical to cost effective manufacturing with lithography, see Levinson et al. [10].

In this chapter, the fundamentals of mainstream lithography used for integrated circuit manufacturing will be discussed. The basic theory of optical projection lithography resolution and pattern width or linewidth control will be given. Next, the fundamentals of resist chemistry will be introduced. To develop robust processes for manufacturing with lithography, process optimization is required. To be most cost effective, this optimization requires accurate simulation of the optical image formation in the exposure tool. Modeling of optical lithography will therefore be discussed. A discussion of applying this modeling in conjunction with experimental data for process optimization will follow. After focusing primarily on optical projection lithography, the chapter will conclude with a survey of next generation lithography approaches.

4. Optical Lithography Background

The optical lithography process can be defined as the method by which light, tools, materials, photoresists and environment are combined to create a 3D pattern in the photoresist film on silicon wafers. The optical lithography process is composed of multiple complex sub-processes. These can be classified as wafer substrate preparation, alignment, exposure, development and metrology. Each sub-process has unique tools, materials and process steps. The sub-processes are tailored to comprehend design, reticle, optical lithography and etch /implant process interactions to produce optimum final (etched or implanted) patterning results. In this overview section, we introduce the lithography process elements in greater detail.

In substrate preparation, the wafer surface is prepared for photoresist, commonly referred to as simply resist, application. The substrate is defined as the stack of material layers, including the semiconductor wafer itself, on which the resist is applied. The subsequent process will use the resist as a mask to etching of the top layer(s) of material or as a mask to ion implantation of dopants. The wafer is first carefully cleaned to remove any

potential contaminants. The wafer surface is then treated with an adhesion promoter before a resist solution is spin coated onto the wafer [11]. The newly coated wafer is heated, or baked, to remove solvents in the resist and create a physically stable polymer film [11]. This is known as the pre-bake, post-apply bake or soft-bake step. These operations are performed on a coat and develop tool, also known as a coater track or a track. After the substrate is prepared, the wafer is then transferred to an exposure tool, known as a stepper or a scanner, which is essentially an imaging camera [12]. The stepper contains the reticle, an exposure source, imaging optics and an alignment system. In alignment, the reticle pattern is aligned to the pattern of a previously patterned layer at each wafer exposure field (Fig. 6). Dedicated features on the reticle and on a previously patterned layer are used by the alignment system to ensure accurate positioning before the image transfer occurs.

After alignment, the stepper is used to expose each field on the wafer to transfer the reticle pattern to the resist. During exposure, light from the illumination source is shaped and directed onto the reticle by the illuminator optics, or condenser. The light transferred through the reticle is focused onto the wafer by the projection optics [13] (Fig. 7). The wafer stage on the stepper moves the wafer into the correct alignment position for each field exposure. Shutters, or blades, restrict the exposure light to the correct wafer field. Sensors on the stepper ensure that the desired exposure energy and optimum wafer focus position are achieved. The clear and opaque patterns on the reticle create exposed and unexposed regions of photoresist. The photochemistry of the resist creates large solubility differences between these exposed and unexposed regions to an aqueous developing solution known simply as developer [14,15].

After exposure, the wafer is transferred back to the track to be baked again in a post-exposure bake [11,14,15,16]. The post-bake, as it is called, reduces the undesirable effects of thin film interference during exposure. In chemically amplified resists the post-bake also acts to complete the chemical reactions initiated by exposure. The resist is then introduced to the developer [11,14,15]. During development, the highly soluble resist areas are

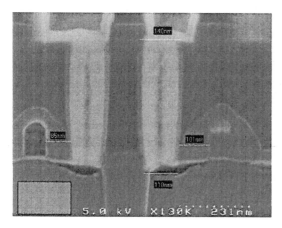

Ideal overlay

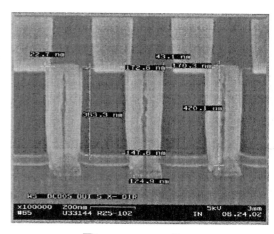

Poor overlay

Fig. 6. Xsection view of ideal and non-ideal overlay between features on adjacent layers of a device. The patterns at left are making good electrical connection between layers. The patterns at right are making very poor electrical connection between layers.

selectively dissolved away to produce the final resist pattern. The wafer is then rinsed in water to remove the developer solution completely. The resist pattern may be subjected to UV light and/or an additional bake, called UV bake or hard bake, to increase the resistance to the etchant used during the etch process or resistance to high energy incident ions in the implant process [17].

After the resist pattern is formed on the wafer, the wafer may be inspected and measured to ensure correct processing [18]. To increase throughput, only a small percentage of the wafers and the die on these wafers are typically analyzed. If a problem is identified, the resist pattern on the wafer can be removed and reworked. Once the inspection and metrology tests have been completed with satisfactory results, the wafer is allowed to continue on to the next process, which is either etch or implant. Additional inspections and metrology are done after the etch process to ensure that the final pattern was formed correctly. If correct, the patterning step is completed.

5. Introduction to Patterning Issues

Many optical lithography issues impact whether a layer will be correctly patterned. These issues include the designed feature size and shape; the accuracy of the reticle pattern; the imaging capability of the stepper; alignment between patterning layers; capability of the photoresist; interactions between the resist and the etch or implant process; control of the films on the wafer substrate; the ability of metrology to measure and inspect the pattern; and the addition of pattern altering particles to the wafer at any step in the process. The patterning of a die is correct if all dimensions of every feature are within specified tolerances of the designed dimension and also the overlay of every feature to features on the underlying layer(s) within specified bounds. These specifications are given by the ability of the circuit functionality or subsequent processing steps to tolerate patterning deviations. Meeting the specifications is a difficult task as circuit layers can contain more than 100 million microscopic features each of which is allowed only a small deviation. Process latitude is the ability of the patterning process to tolerate manufacturing deviations and produce size and overlay outputs within the allowed specifications. This section will

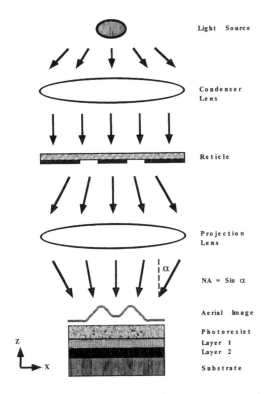

Light Source

Condenser Lens

Reticle

Projection Lens

NA = Sin α

Aerial Image

Photoresist
Layer 1
Layer 2
Substrate

z

x

Fig. 7. Idealized schematic of a stepper exposing a photoresist coated film stack. The condenser and projection optics are each approximated by a single lens. The projection lens numerical aperture (NA) is the sin of the largest angle imaged by the lens onto the wafer.

further explain the main steps and elements leading to functionally correct optical lithography patterning. We will also describe resolution enhancement techniques, software which improves patterning results, simulation methods for optimizing patterning performance, and optical lithography factors critical to financial success in semiconductor manufacturing.

6. Design

An integrated circuit design contains, in a computer database, the description of the desired pattern features, which will make up the circuit. The design describes the ideal size, shape and relative position of all features on all patterning layers [19,20,21,22] (Fig. 8). The features are described

by the placement of their vertices on a uniform grid of allowed vertex locations. The size of the grid and the design rules determine what features and feature locations are allowed on each layer [22,23]. Design rules incorporate knowledge of processing and device capability so that circuits produced using these rules are manufacturable and will function properly [24]. For optical lithography, the design rules specify minimum and/or maximum feature dimensions on each layer. The rules also specify minimum and/or maximum spacings between features on each layer and required overlaps or spacings between features on neighboring layers. New design rules typically need to be created for each device technology due to increasing process complexity. However, to maximize reuse in the design stage, successive generations of a device technology family will attempt to scale down, or shrink, versions of the original design rules with only minor modifications [22]. New or altered design rules can have a significant impact upon the chip area, the power dissipation, and the design effort. As chip designers are very sensitive to schedules, proposed rule changes which cause significant design rework or delays are extremely difficult to get implemented. Derbyshire described the design flow and emphasized the need to increase collaboration among designers and lithographers [25].

There are two main classes of designs, which are logic (Fig. 9) and memory [20] (Fig. 10); however, actual circuits generally contain both types. These classes traditionally have distinct design characteristics and patterning requirements. Logic, or random logic, designs contain a large number of hand-designed groups of features. Logic designs typically contain a wide array of feature types such as long lines, short bars and small squares. They also contain a large variety of local environments for features: dense arrays of features, isolated features and seemingly randomly placed neighboring features. Memory designs contain one main feature set, namely the memory cell or bit cell. The cell is repeated in a regular array across the design. Therefore, the designed feature types and local environment are extremely limited in comparison to a logic design. These limitations allow the features in the memory cell to be highly optimized for patternability, yield, area consumed and circuit performance [26]. Thus, for a given patterning or manufacturing capability, design rules for a memory design may be more aggressive than for a logic design. Recent trends show logic designs to contain an increasing amount of memory, as circuits migrate to system on a chip (SOC) design styles. Memory cells now often consume greater than 50% of the area in a so-called logic chip. Therefore, a blend of logic and memory design styles and patterning techniques are now being used.

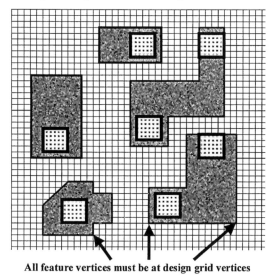

All feature vertices must be at design grid vertices

Fig. 8. Example of a layout design showing metal (solid pattern) and via (dotted pattern) features and the design grid of allowed feature vertices.

Regardless of the design type, the design is stored in computer memory in a format optimized for the design process. Often this format is GDSII (also called GDS, Calma format or Stream format) [27]. GDSII has been the dominant data format for more than a decade, but new data formats may soon be adopted to address the burgeoning data volume. Before a design can be transferred onto a reticle, it must be translated into a format for controlling the reticle patterning tool [28]. Typically this is the Mebes format [29], although other formats including GDS are becoming prevalent. The process of translating the design into the reticle patterning format is called fracture. Fracture also includes any scaling of the design (sizing of the features and the grid together), resizing of partic-

ular design features (sizing of the features only), addition of alignment features and any calibration structures [30]. The calibration structures are used by the metrology to determine the deviation from the ideal alignment and sizing of the features on each patterned wafer layer [18]. Once fracture is performed, the design data is ready to be used in reticle manufacturing.

Fig. 9. Example portion of a random logic design layout showing two patterning layers. Note the wide range of feature geometries.

7. Reticle Patterning

A reticle is a master optical mask containing the design features [31]. It is used to pattern light in the optical lithography exposure of a wafer field. A reticle is composed of clear and opaque (typically chrome) regions on a quartz substrate. In addition to the design pattern, the reticle contains alignment features, or marks, for the stepper to align the reticle to the wafer [18]. The size of the reticle is typically 5 or 6 inches square [31]. The reticle is patterned with the design at a reticle manufacturing factory, often referred to as the mask shop. In the reticle lithography process, chrome is selectively removed to define the pattern. Reticles generally fall into two categories: bright field - where chrome features exist in a mainly clear background, and dark field - where clear features exist in a mainly chrome background [18].

Reticles are also patterned with a resist exposure and a substrate (chrome) etch process. However, the pattern to be transferred is contained only in the fractured design data. The reticle exposure is typically performed by a scanning electron beam (E-beam) or optical laser tool, where the photoresist will be optimized for the exposure tool type. A round exposure spot approximates the square fractured design grid. In a typical reticle exposure process that uses positive resist, each grid point that is designed to be clear on the reticle is exposed by a beam spot. The reticle exposure tools may have one or many scanning beams, but the reticle grid points are exposed individually. Therefore, the exposure time of a reticle (~1-12 hours) is much slower than the relatively instantaneous (<1 second) exposure of a wafer field. Additionally, on many reticle writing tools the exposure time is linearly dependent upon the number of grid points in the design to be exposed. Thus, fine reticle write grids and complicated patterns are often more expensive to pattern.

Fig. 10. Example portion of one layer in a memory design layout. A single memory cell (bitcell) is repeated a large number of times throughout the layout.

Reticles are typically patterned at a larger size than the desired wafer pattern size [12] (Fig. 7). Reticle magnification factors in use are 1X, 2X, 2.5X, 4X, 5X and 10X, but 4X and 5X are the most commonly used magnifications. The reticle magnification factor is equal to the lens reduction factor in the exposure tool. The reticle pattern is reduced by this magnification factor when imaged onto the wafer. For advanced devices, accurately writing the features and placing them in correct relationship to each other is extremely difficult at 1X reticle magnification. Errors in feature dimension or placement on the reticle will affect the wafer pattern, but these errors are also reduced by the magnification factor. A large magnification factor makes the reticle features easier to manufacture but the design then takes up more area on the reticle and limits the wafer die size. However, control of the critical feature dimensions, or CDs, is still crucial. Reticle CD errors often use much of the overall wafer CD budget. Global CD sizing errors, also known as mean to target error , can generally be compensated for by the wafer exposure dose and are of secondary importance [32].

Random reticle CD errors can be classified as width, length or corner rounding errors [31,33]. Corner rounding results from the inability of the reticle write process to reproduce the sharp edges of features in the design. Severe corner rounding causes line shortening or length error, typically on narrow rectangular features. Reticle CD errors can be caused by variations in exposure energy, neighboring feature environment (proximity), feature size, reticle substrate, reticle resist variations and chrome etch [31,33,34]. Defects, typically extra or missing chrome, can also cause CD errors, depending upon size and placement.

To ensure usability in wafer patterning, the reticle is inspected for particles and to verify pattern correctness. The pattern verification is done either with die-to-die (for multi-die reticles) or die-to-database inspection [31]. If a defect is found, and the majority of reticles contain these initially, an attempt will be made to repair the defect [31]. Missing chrome spots can be filled in with vapor deposited metal. Unwanted chrome pieces can be evaporated with a laser or a focused ion beam. The accuracy of the repair procedures is limited and may not be successful in all cases. The correct placement of features on the reticle is also critical. Most systematic placement errors, e.g., all features offset by the same amount, can be corrected for in the stepper during wafer exposure. However, random reticle feature placement errors are a considerable problem for overlay control [31]. Larger reticle field sizes increase the difficulty of placement control.

Measurement of feature placement accuracy or registration and of CD errors are other important reticle manufacturing steps [31,35]. Due to throughput restrictions, only a minute fraction of the features on each reticle can be measured accurately. Reticle CD measurements are often made with an optical microscope using visible light although the use of SEM metrology is increasing. Once the inspection and metrology determine that the reticle was correctly patterned, a pellicle is applied [31]. The pellicle is a thin transparent polymer film mounted above the chromium side of the reticle, which keeps particles away from the reticle surface and out of the focal plane of the light image during wafer exposure. Therefore, the particles that land on the pellicle cannot be printed on the wafers. After the pellicle has been correctly installed the reticle is sent to the stepper in the wafer fab.

8. Optical Performance

A wafer stepper is a tool, similar to a large camera, which illuminates a reticle with light to optically transfer the reticle pattern to the resist film on the wafer substrate [12] (Fig. 7). During exposure, diffraction limitations of the stepper optics transform the discrete, also called binary, reticle pattern into a smoothly varying light intensity pattern, or aerial image. It is the photo-chemical interaction of the aerial image with the resist components, which allows the creation of 3D resist patterns (Fig. 11). The creation of the aerial image on the wafer by the stepper can be decomposed into three separate steps: illumination of light upon the reticle, reticle diffraction of light, and imaging of the diffracted light onto the wafer. The illumination of the reticle is described by the spatial coherence function of the stepper's illuminator (Fig. 12). This function essentially explains the amplitude and direction of the light incident upon the reti-

cle. The diffracted angular light pattern at the reticle from a non-normally incident angle is generally identical to the diffracted light pattern from normally incident light but is shifted at an angle equal to the incident angle. Light incident from a particular direction is assumed to be incoherent in phase with light incident from any other direction. This means that each individual illumination direction creates its own intensity image on the wafer, and the total wafer image is the sum of all the individual intensity images.

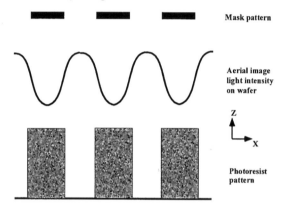

Fig. 11. Xsection view showing a typical relationship between the binary mask pattern, the smooth light intensity aerial image and the resulting vertical photoresist pattern.

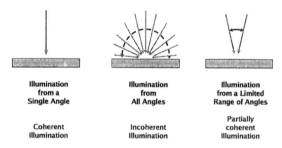

Fig. 12. Figure explaining the differenced between coherent, incoherent and partially coherent illumination.

The diffraction of the incident light by the reticle is, in general, well described by Fraunhofer diffraction [37,38] (Fig. 13). The Fraunhofer conditions of a thin opaque mask material, mask opening widths >> the wavelength of light, and the imaging point being optically far from the mask are generally well met in lithography systems. Therefore, the light image captured by the stepper projection lens is

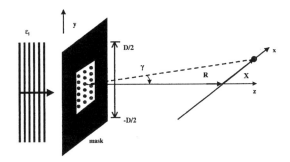

Fig. 13. Diagram showing Frauenhofer diffraction for a two-dimensional mask opening. The diffracted light amplitude can be calculated as if it is the summation of the amplitudes created by a mask opening full of individual light source points.

$K_x = kX/R = k\sin\gamma, K_y = kY/R = k\sin\gamma$
$E = A(x,y)\sum\sum e^{i(k_y X = K_y Y)}dxdy.$

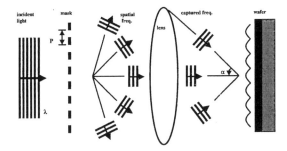

Fig. 14. Diagram showing how a periodic mask pattern diffracts incident light into discrete traveling frequencies. The diffracted light with traveling angle less than the projection lens NA is imaged onto the wafer to create the aerial image. Lens is a spatial filter, only passes lower spatial freq. (angles). If largest freq. captured is α, define $\sin\alpha =$ Numerical Aperture (NA).

the Fourier transform of the mask pattern, and the frequency information of the mask pattern is transformed into physically traveling light directions or spatial frequency values. Lower frequency information travels at low angles to the normal direction (the optical axis), while high frequency information travels at larger angles to the normal direction. For periodic mask patterns we can consider these traveling directions to be discrete and we will label them as traveling light orders where each order has a unique propagation angle. The lowest order (0th order) travels along the normal direction and its amplitude corresponds to the average mask am-

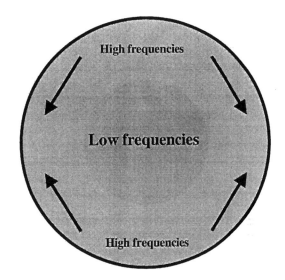

Fig. 15. Diagram of the projection lens showing that high frequency information is imaged onto the wafer by the outer parts of the lens while low frequency information is imaged by the inner parts of the lens.

plitude transmission. The higher (+/-1 or higher) orders travel at angles, θ_n, given by the diffraction grating equation [38]:

$$sin(\theta_n) = n * \lambda/P \qquad (1)$$

where n is the order number and P is the grating periodicity, or mask pitch. The amplitude of these orders relates to the mask pattern pitch, its duty cycle, the complex transmittance of the mask pattern and 2D shape information for corners and line-ends.

The imaging of the diffracted mask pattern by the projection optics on to the wafer is well described by Fourier optical theory, also known as Abbe's theory [36,37,38] (Fig. 14). The projection optics acts as both a low pass spatial frequency filter and an inverse Fourier transform 3 (Fig. 15). The projection optics have a maximum light diffraction angle, θ_{max}, which can be imaged onto the wafer. The spatial frequency corresponding to this angle times the wavelength is the numerical aperture (NA) of the system [38].

$$NA = n_{medium}sin(\theta_{max}) \qquad (2)$$

where n_{medium} is the real part of the refractive index of the medium where the image is formed.

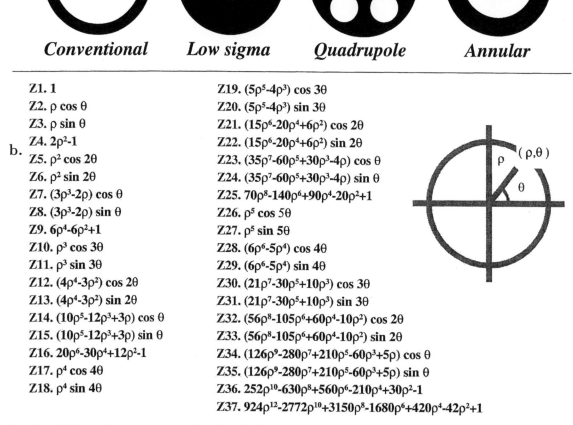

a.

Conventional **Low sigma** **Quadrupole** **Annular**

b.

Z1. 1
Z2. $\rho \cos \theta$
Z3. $\rho \sin \theta$
Z4. $2\rho^2-1$
Z5. $\rho^2 \cos 2\theta$
Z6. $\rho^2 \sin 2\theta$
Z7. $(3\rho^3-2\rho) \cos \theta$
Z8. $(3\rho^3-2\rho) \sin \theta$
Z9. $6\rho^4-6\rho^2+1$
Z10. $\rho^3 \cos 3\theta$
Z11. $\rho^3 \sin 3\theta$
Z12. $(4\rho^4-3\rho^2) \cos 2\theta$
Z13. $(4\rho^4-3\rho^2) \sin 2\theta$
Z14. $(10\rho^5-12\rho^3+3\rho) \cos \theta$
Z15. $(10\rho^5-12\rho^3+3\rho) \sin \theta$
Z16. $20\rho^6-30\rho^4+12\rho^2-1$
Z17. $\rho^4 \cos 4\theta$
Z18. $\rho^4 \sin 4\theta$

Z19. $(5\rho^5-4\rho^3) \cos 3\theta$
Z20. $(5\rho^5-4\rho^3) \sin 3\theta$
Z21. $(15\rho^6-20\rho^4+6\rho^2) \cos 2\theta$
Z22. $(15\rho^6-20\rho^4+6\rho^2) \sin 2\theta$
Z23. $(35\rho^7-60\rho^5+30\rho^3-4\rho) \cos \theta$
Z24. $(35\rho^7-60\rho^5+30\rho^3-4\rho) \sin \theta$
Z25. $70\rho^8-140\rho^6+90\rho^4-20\rho^2+1$
Z26. $\rho^5 \cos 5\theta$
Z27. $\rho^5 \sin 5\theta$
Z28. $(6\rho^6-5\rho^4) \cos 4\theta$
Z29. $(6\rho^6-5\rho^4) \sin 4\theta$
Z30. $(21\rho^7-30\rho^5+10\rho^3) \cos 3\theta$
Z31. $(21\rho^7-30\rho^5+10\rho^3) \sin 3\theta$
Z32. $(56\rho^8-105\rho^6+60\rho^4-10\rho^2) \cos 2\theta$
Z33. $(56\rho^8-105\rho^6+60\rho^4-10\rho^2) \sin 2\theta$
Z34. $(126\rho^9-280\rho^7+210\rho^5-60\rho^3+5\rho) \cos \theta$
Z35. $(126\rho^9-280\rho^7+210\rho^5-60\rho^3+5\rho) \sin \theta$
Z36. $252\rho^{10}-630\rho^8+560\rho^6-210\rho^4+30\rho^2-1$
Z37. $924\rho^{12}-2772\rho^{10}+3150\rho^8-1680\rho^6+420\rho^4-42\rho^2+1$

Fig. 16. **a.** Different illumination source functions. Light is incident upon the reticle from the white colored regions of the source. **b.** List of the first 37 Zernike polynomials and a figure showing polar coordinates used in the polynomials.

Diffracted light orders from the mask whose angles are greater than the stepper NA are not captured by the projection optics and do not get imaged onto the wafer. Diffracted light orders from the mask whose angles are less than the stepper NA are captured by the projection optics and are imaged onto the wafer. The projection optics create an inverse Fourier transform of these captured light frequencies when it forms the aerial image on the wafer. Thus, the wafer aerial image for each illumination angle incident upon the reticle is a low-pass spatial frequency filtered version of the mask

image.

The goal of the stepper is to optimize the contrast and positioning of this light image on the wafer. To correctly image and align these exposures, the stepper requires many different optical and mechanical elements. The illumination source provides the light for the exposure. It is typically a mercury vapor lamp or a laser. Mercury vapor sources are often wavelength filtered to one or more atomic transition lines at the 436nm (G-line), 365nm (I-line), or 248nm (deep ultraviolet or DUV) [12] wavelengths. Laser wavelengths are generally tuned to one of the I-line, DUV, 193nm or 157nm wavelengths.

The condenser optics, or illumination optics, image the light from the source uniformly onto the reticle. These optics control the spatial coherence, or sigma, of the incident light [40,41]. As the coherence and the spatial distribution of the light affects the resolution and process latitude of the patterning process, special optics or filters can be used to shape the condenser image seen by the reticle. The most common condenser shapes are annular or quadrupole illumination [41] (Fig. 16). These off-axis illumination (OAI) methods direct most of the light onto the reticle from angles not on the optical axis. Increasing the angle of incident light in the illumination system improves the spatial resolution of the imaging system, but also tends to increase the pitch dependence of the imaging results. The reticle is held by the reticle stage in the proper position for exposure where the projection optics then image the light diffracted by the reticle onto the wafer field. As described above, the NA of the projection optics greatly impacts the resolution and process latitude of the patterning process [41,42]. The projection optics also reduce the reticle pattern size upon exposure by the reticle magnification factor. The goal of the projection optics is to deliver a nearly ideal (diffraction limited) light image to the wafer (Fig. 16B).

The design and fabrication of the projection lens is perhaps the most critical element in the exposure tool. Tibbets and Wilczynski [43], Braat [44], Williamson [45], Glatzel [46], Philips and Buzawa [47], and Stover [48] describe the design criteria and design methods for projection lenses. The projection lens design is governed by several main requirements, which include:

- The requirement of high resolution drives the design to require large numerical aperture (NA).
- A large field of view is required, which makes correction of aberrations more difficult especially at high NA. The large field of view also necessitates low field curvature so that the image is focused in a single plane over the exposure field.
- Low distortion is required for overlay accuracy.
- The lens should have a large reduction ratio to make the exposure less sensitive to CD errors, placement errors and defects on the mask.

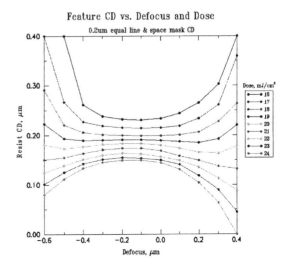

Fig. 17. Bossung style plot of photoresist feature CD vs. DUV stepper defocus as a function of exposure dose for a 0.2 μm line/space pattern. The lines on the plot are in the same vertical order as the legend. For instance, the CD versus focus curve for a dose of 16 mJ/cm2 has the largest CD values while a dose of 24 mJ/cm2 has the smallest CD values.

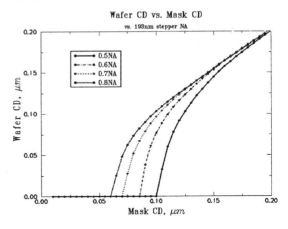

Fig. 18. Plot of photoresist feature CD on the wafer vs. the 1X chrome feature CD on the reticle as a function of projection lens NA for a 193 nm stepper.

- A real image of the mask must be formed in the plane where resist is exposed.
- The lens should have a long working distance to keep the wafer as far as possible from the final lens element to avoid defect and contamination problems.
- The chromatic aberration must be minimized,

which is complicated by the lack of glass choices at wavelengths <365 nm. The index of refraction of available materials is smaller at shorter wavelengths, forcing them to have larger radii of curvature and be more subject to aberrations. More recently at 157-nm wavelength, the birefringence of CaF_2, the only practical lens material at this wavelength complicates lens design.

- Glass volume should be minimized to fit the lens in the exposure tool and to minimize the cost of the raw materials for the lens.
- The lens should have minimal track length to fit in the exposure tool.

The double Gauss lens form, adapted from photographic lenses, was used to develop initial projection lenses for steppers. Scanner lenses were influenced by a design form developed by Offner [49]. The design process goal is to meet the requirements above and achieve less than 20 milliwaves root mean square (rms) aberration [50]. The rms specification may also be misleading since aberrations of opposite sign may cancel. The 3rd order Seidel aberrations must be driven to nearly zero to prevent deleterious effects on imaging. A CAD program for lens design that uses ray tracing is typically used to optimize the lens, but the designer must use prior experience to choose the surfaces as input to the optimization routine.

To increase numerical aperture and to lower aberrations with limited glass types at shorter wavelengths, modern exposure tools have progressed to a scanner design. The field of view that is well corrected for aberrations is made smaller, and the wafer and mask are scanned through this field of view to expose the field. This scanner design could be implemented with all refractive lenses or catadioptric designs, where beamsplitters are used to allow for a mirror to be placed in the design. The advantage of a mirror is that it is achromatic. Radiation with all wavelengths reflects from it in the same direction. The mirror may also be aspheric, providing further aberration correction. Aspheric surfaces are now more routinely manufactured as transmissive lenses, and they are being incorporated in refractive portions of the design as well.

Manufacturing the lens is also challenging. Interferometers that operate at the exposure wavelength are typically required. They must have higher accuracy than the 20 milliwaves to reach the design goal. The surfaces must be manufactured and the entire lens assembled to a typical goal of <50 milliwaves of aberrations. The manufacturing and assembly process is subject to many errors, including:

- The lenses or mirrors may be tilted in their mount.
- The axial thickness of the lens may not be correct. Lenses must be moved along the axis to correct for these errors.
- The lenses or mirrors might be decentered. Decentering may lead to trapezoid distortion.
- The curvature may not be correct. The lens spacing is adjusted to compensate for these errors.
- The index of refraction of the lens material may have change due to internal stress or compositional inhomogeneity. This must be compensated with lens spacing adjustments or refiguring.
- Surface asphericity must be compensated by lens spacing adjustments.
- Astigmatism must be adjusted with clocking, which is a process where individual lenses are rotated about their optical axis to empirically minimize astigmatism and birefringence.

Chapman has described a rigorous orthogonal decomposition algorithm to calculate the appropriate degrees of freedom in the alignment of an off-axis aspherical mirror system [51]. These techniques may be applied to adjust the assembled lens to meet wavefront quality goals.

The wafer rests on a wafer stage which moves, or steps, the wafer so each field can be exposed sequentially. Alignment optics are used to accurately ensure the correct reticle and wafer stage horizontal positions for proper alignment of each exposure to a previous patterned layer on the wafer. Focus sensors ensure that the wafer and reticle stages are at the proper vertical locations, without tilt, for best pattern transfer [12]. Dose sensors ensure that the correct amount of light energy is incident upon each wafer field. The exposure dose is used to optimize the size of the resist features [14,15,41] (Fig. 17). Software programs control the workings of the different stepper mechanical subsystems and the interaction between the tool and the user. The software is used to create groups of instructions, or exposure recipes, to control and automate the tool's functions. The stepper generally connects directly to the coater track for automatic wafer transfers.

Half of the minimum pitch, R, resolvable for a diffraction limited (ideal lens) optical system is given by: [12,39,41]

$$R = K_1 * \lambda/NA \qquad (3)$$

which can be derived from the diffraction grating equation. K_1 is a process dependent factor determined mainly by the resist capability, the tool control, reticle pattern adjustments and the process control. Typical minimum values of K_1 are between 0.4 and 0.6 in advanced manufacturing. Therefore, the trends in optical lithography are towards lower wavelength, higher NA imaging systems and smaller K_1 values to allow the printing of smaller features and denser patterns. The sensitivity of the patterning process to expected variations determines the manufacturability of the process. A patterning process can be characterized by its sensitivity to two main process control parameters, focus and exposure energy, or exposure dose [41] (Fig. 17). The usable focus and exposure latitude budgets actually incorporate a number of process variations, many unrelated to the stepper performance. Exposure errors alter resist CDs and limit focus latitude. The parameters which cause effective exposure errors include: substrate reflectivity variations, nonuniform illumination intensity, reticle CD errors, resist sensitivity variations, developer variations, post exposure bake temperature variations and feature proximity effects. Differences in designed feature size can also reduce the exposure latitude of a process as different exposure doses are required to correctly pattern large and small features. Reticle CD variations of smaller features must also be controlled more tightly than those of larger features [52] (Fig. 18).

Focusing errors, or defocus, lower the definition and contrast of the aerial image, alter the resist CDs and limit exposure latitude (Fig. 19). The focus latitude, or depth of focus (DOF), expected at a single point in a stepper field is: [12,41]

$$DOF = K_2 * \lambda/NA^2 \qquad (4)$$

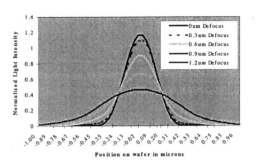

Fig. 19. Plot of aerial image light intensity vs. position on the wafer as a function of 0.6NA i-line stepper defocus for a 0.5μm opening on the reticle (1X). Note the iso-focal point at a relative intensity of approximately 0.4 in which the image width is independent of focus.

K_2 is a tool, process and pattern size dependent parameter, where small features have lower focus latitude. A typical value of K_2 is 1.0 for a minimum dimension feature. Therefore, the trends in lambda, NA and feature size require patterning with lower overall focus margin. Parameters which cause effective focus errors between exposures or across the exposure field include: wafer flatness, lens aberrations, substrate topography, stepper focusing errors, stage tilt, stage flatness and resist thickness variations. When NA is large or imaging is carried out in a medium that has index of fraction, n, different than unity, the DOF becomes: [54]

$$DOF = k_3 \frac{\lambda}{n \sin^2 \left[\frac{1}{2} \sin^{-1} \left(\frac{1}{n} \sin \alpha_o \right) \right]} \qquad (5)$$

where α_0 is the maximum angle of the ray imaged by the lens. This angle is limited by the numerical aperture of the lens. Lens aberrations cause imperfections in the image transfer of patterns. The manufacture of lenses with high NA, wide field size and low aberrations is extremely difficult, especially for shorter illumination wavelengths [53].

Tuning the coherence and spatially filtering the illumination have been shown to improve process margin for certain feature patterns. Focus latitude can also be increased by optimizing the mask feature size together with the stepper exposure dose. This allows the feature width to be determined by an aerial image intensity level, which is less

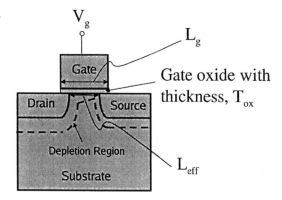

Fig. 20. Schematic cross section of a metal oxide semiconductor field effect transistor (MOSFET), showing various dimensions that relate to CD control requirements for lithography.

sensitive to focus variations, the iso-focal point (Fig. 19). To improve process control, a new type of stepper has become dominant, the step-and-scan exposure tool, or scanner [55]. A scanner exposes only a strip of the reticle (and wafer field) at any time, scanning the exposure across the reticle to complete the image transfer. The stage then steps the next field into position to be exposed by scanning. This method allows finer lens optimization and lower aberration imaging. However, image transfer errors due to scattered light, or flare, and mechanical movement are increased.

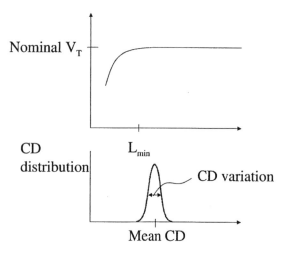

Fig. 21. Illustration of Vt roll off and its consequences for CD control at the gate layer.

One way to decrease the minimum resolvable half pitch is to decrease the exposure wavelength as shown in Equation 3. Decreasing wavelength typically requires: new lens materials, new resist materials, new sensors and new sources. The 0.09μm and 0.065μm lithography generations will use 193nm exposure light. There are a number of difficult challenges in developing exposure tools at this shorter wavelength [10,56-58]. The absorption coefficient of most materials increases as the wavelength of the radiation decreases. At 193nm, the quartz material used for making the lens elements and mask blanks begins to absorb at levels that can cause imaging problems and damage to the optical system. Irradiating quartz lens elements in experimental exposure tools with high intensity 193-nm radiation can damage the lens both by the formation of color centers within the material and by induced compaction of the material. As the lens elements absorb energy, lens heating also becomes a larger problem and compensation for its effect on imaging performance becomes more crucial. The end result is that lens lifetimes in these new exposure tools may be shorter than those of past tool sets. One potential solution to some of these problems is the use of calcium fluorite (CaF_2) lens elements at critical locations. CaF_2 has lower absorbance than similar quartz elements. [12,59] To reduce lens manufacturing difficulties, only scanning exposure tools are being considered for 193nm tools. The exposure sources for the 193-nm exposure tools are ArF excimer lasers.

Another way to increase resolution is to increase NA. To increase imaging resolution, very high projection NAs of 0.8 or greater are being manufactured for 248nm and 193nm illumination step-and-scan systems [60]. Shifting to higher NA tools provides an increase in patterning capability without the high costs and infrastructure changes required to switch to a new lithography wavelength [61,62]. For patterning of small features, especially with dense pitches and darkfield patterns, higher NA illumination offers improved process latitude over lower NA illumination. Exposure latitude is especially improved. However, very tight tool and process focus control will be required to work with the extremely small focus budgets of these very high NA lenses. Another major challenge for this extension is in the manufacture of wide field, low aberration very high NA projection lenses. Lens volume and lens cost, scale roughly as NA^4. Furthermore, the general effect of aberrations increase in proportion to the square root of (1 minus the NA). Initial results of these very high NA lenses suggest that the lens and tool manufacturers are indeed able to meet this challenge.

9. Critical Dimension Control

The success of IC manufacturing depends on the projection of the mask image onto a silicon wafer with high fidelity and repeatability. However, in practice the pattern printed on the wafer is not an exact replica of the mask or design pattern. Variations in printed features on the wafer result largely from three sources. The first source is the fundamental diffraction limit of the projection optics. Second, the mask pattern itself does not look like the design due to limitations of the mask fabrication process (i.e. mask errors). The third source is random and systematic variations of the multitude of lithographic process parameters, such as focus, exposure, etc.

The modern IC consists of tens to hundreds of millions of transistors. Ideally all transistors should have identical effective gate length across the entire chip, across all chips printed on the wafer, across all wafers in a lot, and finally across all lots allocated for production. The effective gate length, also known as effective channel length or L_{eff}, is defined in Figure 20. The effective channel length depends on the poly silicon line width of the device gate level, L_G, and on the dopant level and dopant locations in the source, drain and channel. Typically, $L_{eff} < L_G$. However, the variations in the manufacturing process parameters inevitably cause variations of poly silicon line width and consequently L_{eff}. In a classical long channel MOSFET current entering the drain when the maximum bias is applied between the source and drain is given by [63]:

$$I_{DSAT} = \frac{W}{2L_{eff}} \frac{\varepsilon_{ox}}{T_{ox}\mu_{eff}} (V_G - V_T)^2 \quad (6)$$

where W is the channel width, ε_{ox} is the permittivity of the gate dielectric, T_{ox} is the thickness of the gate dielectric, $g\mu_{eff}$ is the effective mobility of minority carriers in the channel, V_G is the gate voltage and V_T is the saturation threshold voltage. For digital circuits, the transistor is either operated in saturation when "on" or in the subthreshold region when "off." The control of the dimensions of the lithographic pattern directly affects W and L_g. The relationship between L_g and L_{eff} depends on the device fabrication process. However, the dopants in the source and drain regions are implanted using the gate as a mask, so the gate width controls their location with respect to the edge of the gate to a large degree. The speed of the circuit is determined in part by gate delay, τs which is given by:

$$\tau = 0.25 C_{load} V_{DD} \left(I_{DSAT\_N}^{-1} + I_{DSAT\_p}^{-1} \right) \quad (7)$$

The capacitive load on the transistor, C_{load}, depends on the circuit design and might be dominated by parasitic capacitance of the MOSFET structure or by the capacitance of the interconnects between transistors. The power supply voltage, VDD, has to be reduced as circuit complexity increases since static power is given by:

$$P_{static} = W I_{leak} V_{DD} \quad (8)$$

Therefore, I_{DSAT} should be maximized. Two main methods of modifying I_{DSAT} are decreasing T_{ox}, which imposes significant challenges to materials development, or to decrease L_{eff} and hence L_G. Increasing W increases circuit area ($A_{circuit}$), which makes static power increase and makes it more difficult to fabricate circuits with high yield since the yield is related to the random process defect density, D_0, as [65, 66]:

$$Y = exp(-D_0 * k * A_{circuit}) \quad (9)$$

The kill ratio, k, is a term that describes what fraction ($k < 1$) of the circuit is sensitive to defects. There are many relations between yield and D_0 that depend on the distribution of D_0 values in the process, but the one above is the most conservative. Because of these basic relations, the lithography process is pushed to reduce L_G as much as possible.

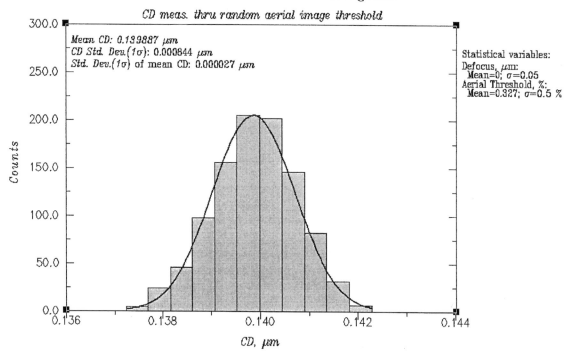

ST–LITH Resist Width CD Histogram

CD meas. thru random aerial image threshold

Mean CD: 0.139887 μm
CD Std. Dev.(1σ): 0.000844 μm
Std. Dev.(1σ) of mean CD: 0.000027 μm

Statistical variables:
Defocus, μm:
  Mean=0; σ=0.05
Aerial Threshold, %:
  Mean=0.327; σ=0.5 %

Fig. 22. Example of simulated statistical CD control distribution.

The overall speed of the circuit is also related to the speed of the individual transistors. If some are faster than others, at some value of the speed difference, the circuit will not operate properly. If some are only slightly faster, the speed of the circuit will only be as fast as the slowest devices. Therefore, the distribution of L_G must be made as small as possible [67-71]. This is the primary challenge for CD control. On the isolation layer, lithography must be optimized to minimize variations in W. On the gate layer, lithography must be optimized to minimize mean L_G and to minimize the variation in L_G. As the channel length of MOSFETs becomes small and the electric field between the source and drain becomes large, the MOSFET threshold voltage no longer remains constant as gate length changes. This effect is known as a "short channel effect," and one of the main short channel effects is "Vt roll-off" [63]. Figure 21 shows an example plot of threshold voltage as a function of channel length. Below a gate length value, denoted L_{min}, the threshold

voltage drops. For these smaller channel lengths, the gate no longer can fully control the flow of current between the source and drain when a voltage is applied between the source and drain. The transistor becomes overly leaky and fails to operate as designed. Therefore, the mean value of gate length becomes the critical dimension (CD) at the gate layer and must be tightly controlled.

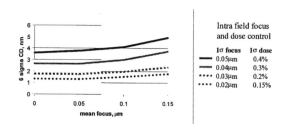

Fig. 23. Predicted CD control as a function of mean focus at varying pairs of intra field focus/dose control. Nominal line width is 90nm on 180 pitch, patterned with 193nm exposure wave length, 0.85NA lens.

The distribution of CDs over the exposure field is known as across chip linewidth variation (ACLV). In optical lithography, the width of the distribution is characterized by 3σ, where σ is the standard deviation in the case of a normal or close to normal distribution. In practice, the statistical sampling is often too small to judge its normality, and the distribution width can be simply described by its total indicated range (TIR), i.e. maximum CD minus minimum CD measured in a statistical sample. The values of 3σ or TIR are termed CD control. An example of a CD distribution is shown in Figure 22. To maximize the speed of the circuit, the mean of the gate CD must be pushed as close as possible to L_{min}. How close the mean may be to L_{min} depends on the variation of CD. As Figure 21 illustrates, the smallest CD in the distribution of CD must be equal to L_{min}. Minimizing the variation and controlling the mean is therefore the primary driver of lithographic control of CD at the gate layer.

For interconnect or contact layers, CD control is required to maintain the desired resistance of the contact or wire. Resistance is proportional to cross sectional area, so that area must be controlled. The area of the contact is directly affected by lithography. The width of metal lines is affected by lithography, but the area is also influenced by the thickness of the line. The most demanding requirement on CD control is at the gate layer, and therefore the most advanced lithography technology is typically used to pattern that layer.

Modern ICs normally require 3σ CD control to be 10% to 15% of the nominal linewidth after pattern transfer (etch). The final physical dimension of the poly silicon linewidth is rapidly decreasing with every new generation of ICs. Consequently, the total CD error budget becomes exceedingly small. For instance, 50 nm gates should have 3σ CD control of about 5 nm. It is very important to understand how the allocated CD error budget is consumed during the patterning process. Understanding the origin and the impact of various error sources on CD error helps determine what sources of error should be minimized first in order to improve CD control. There are two types of CD errors: systematic and random. Systematic errors repeat in the exact same way from one exposure field

to another. For example, mask CD error, total focal plane deviation, illumination uniformity, dose drift, zero focus drift, and wafer topography resulting from underlying device layer are some of the systematic CD error sources. On the other hand, random CD errors do not reproduce from die to die. Some examples of random CD error sources are random dose variation within field, dose and focus repeatability from field to field, scan synchronization error effects, chuck and wafer flatness, post-exposure bake non-uniformity. The contributions from random errors are statistically independent from one another and can be added in quadrature in order to obtain the overall CD error induced by random process variables. Final CD error can be found by linear addition of the total random CD error component and 80% to 100% of the total systematic CD error component. [72]

The Monte Carlo method can be employed for estimating CD variations stemming from the exposure dose and focus variation. Figure 22 shows an example of a simulated CD distribution using the Monte Carlo method. The methodology described below can be used for proper assignment of input variances for the Monte Carlo simulation. The focus error budget is broken down into three categories:

(i) Within-chip: field tilt, level sensor precision, total focal plane deviation, lens aberrations, reticle non-flatness, short-range wafer non-flatness, and topography of the underlying device layer σ_{f1}

(ii) Chip-to-Chip: auto focus precision, wafer stage non-flatness, and clamping σ_{f2}

(iii) Calibration: daily best focus determination and adjustment, tool stability, tool-to-tool matching, operator error in choosing the best focus.

The components within each of these categories are added in quadrature and denoted as σ_{f1}, σ_{f2} and σ_{f3}, respectively. Similarly, dose variations are classified as follows:

(i) Within-chip: across slit dose non-uniformity denoted as σ_{e1}

(ii) Chip-Chip: dose repeatability denoted as σ_{e2}

(iii) Dose accuracy: long term dose drift and tool-to-tool matching and operator error and measurement in determining the best

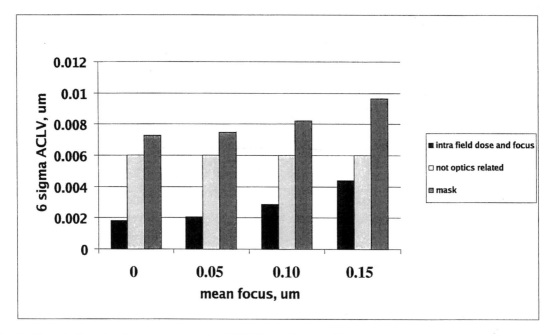

Fig. 24. Contribution of various error sources to ACLV. Nominal line width is 90nm on 180 pitch, patterned with 193nm exposure wave length, 0.85NA lens. Non-optics related errors, e.g. PEB temperature non-uniformity, are assumed, and shown for illustrative purpose only.

focus denoted as σ_{e3}.

Finding the magnitude of each component is a separate and quite complex task. The error values can be inferred from experiments and/or vendor specifications of the exposure tool. The errors are usually expressed in terms of 3σ or a range. Typically, whenever the root sum of squares is used to add different error components, 3σ is equal to half TIR. Although the exact distributions of focus and dose across the field are not known, they are usually considered to be Gaussian with a certain mean value and standard deviation, σ_{f1} and σ_{e1}respectively, for intra-field focus and dose errors. The variation of the mean values of focus and dose from chip-to-chip on a wafer are described by σ_{f2} and σ_{e2}, respectively. The overall maximum error in mean focus and dose that might occur on any given chip on a wafer is a linear sum of the corresponding errors, which assumes no cancellation of the terms. Thus, ($3\sigma_{f2}+3\sigma_{f3}$) and ($3\sigma_{e2}+3\sigma_{e3}$) are the maximum error in mean defocus and dose. Others have used weighting factors [72] or uniform random variables [73] to account for systematic er-

rors. Linear superposition of the error terms represents a worst case.

The mask CD control is usually measured in terms of a CD range on the mask. The magnitude of the range depends on the mask making process. A more comprehensive mask error budget has been discussed previously [72]. Typically, the mask CD range is less than 5% of the nominal mask linewidth. Simulations can be used effectively to analyze the major contributors to CD variation. The patterned image in the photoresist film depends upon hundreds of input parameters of the lithographic process. However, variations in resist CD can accurately be simulated using simple aerial images of the light incident upon the wafer plane. The resist CD is evaluated at a threshold level of the aerial image intensity. This method is called a constant threshold resist (CTR) model. In order to estimate ACLV, a Monte Carlo analysis can be performed in which resist CD is quickly evaluated 1000 times or more. Focus and dose are the only two statistical input parameters being varied around their corresponding means F and E with standard deviations σ_{f1} and σ_{e1}. The CD errors resulting from photo resist processing (post exposure bake non-uniformity), substrate reflectivity variations, etch process, metrology are combined into one CD error contributor, herein referred to as non-optical CD error. The result of the Monte Carlo analysis is a CD distribution on a chip exposed with a certain mean focus F and mean exposure dose E.

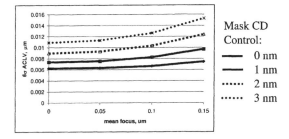

Fig. 25. CD control as a function of mean focus at varying mask CD control . Nominal line width is 90nm on 180 pitch, patterned with 193nm exposure wave length, 0.85NA lens.

The mask error is not included in a statistical run. The mask error contribution is simulated as mask error factor (MEF) multiplied by mask CD range, where MEF equals $\Delta CD_{wafer}/\Delta CD_{mask}$. The resulting ACLV is a sum of MEF times mask CD range and the 6σ CD variation obtained from the Monte Carlo analysis. The reason for not including mask CD distribution in a Monte Carlo analysis is uncertainty in assigning a standard deviation to CD mask distribution. In order to address the focus and dose repeatability from chip to chip, the mean focus and dose are varied in a loop from F_0-$(3\sigma_{f2}+3\sigma_{f3})$ to F_0+$(3\sigma_{f2}+3\sigma_{f3})$ for focus and E_0-$(3\sigma_{e2}+3\sigma_{e3})$ to E_0+$(3\sigma_{e2}+3\sigma_{e3})$ for dose. The Monte Carlo analysis is performed for each value of the mean focus, dose and mask. The results of a series of Monte Carlo runs are mean CD and 1σ CD as a function of mean focus and dose for the same statistical input of intrafield variances of focus and dose. Using these results, the factors that affect the CD most strongly can be identified [74].

The approach described above was applied to ACLV simulations for 90-nm lines on a 180-nm pitch printed with an exposure tool at 193-nm wavelength, a 0.85 NA lens and an annular illumination scheme with a binary mask. This optical

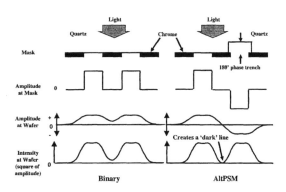

Fig. 26. Cartoon showing how an alternating PSM improves wafer image quality compared to a binary mask.

condition corresponds to K_1=0.36. The simulation technique is used to estimate exposure tool control of focus and dose necessary to achieve the required 10% CD control for the 90 nm technology generation. Chip-to-chip mean focus was varied from 0 to 0.15μm emulating possible error in mean focus from chip-to-chip on the wafer. Figure 23 shows the expected CD control resulting from different intrafield focus and dose variations. Figure 24 summarizes the contributions of various sources to resist CD. Mask CD control range was assumed to be 3 nm at the wafer scale (after stepper reduction). Within chip variation of focus and dose seem to be relatively small contributors to ACLV, which is a result of very tight intrafield focus and dose control values of $1\sigma_{f1}$=0.03μm and $1\sigma_{e1}$=0.3%, respectively.

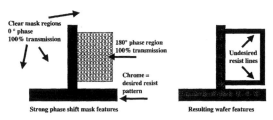

Fig. 27. Top down view of strong phase shift design and subsequent printed wafer resist pattern. Undesired small resist lines appear on the wafer from the junction of phase shifted and non-phase shifted regions. The thick resist lines directly below the chrome are the only desired lines.

The results shown in Figure 24 suggest that the mask contribution to ACLV is the largest contribution, and it quickly increases with defocus error as a result of very large MEF at such low K_1

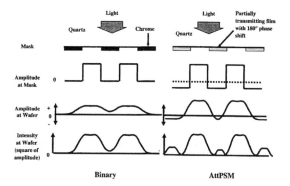

Fig. 28. Cartoon showing how an attenuating PSM improves wafer image quality compared to a binary mask.

values, e.g. MEF=2.4 at best focus and MEF=3.2 at 0.15μm defocus. Even though we do not have any valid *a priori* reason to assign 6 nm to the non-optical contributions to 6σ ACLV, it is shown in Figure 24 purely for comparison with the other sources that we can simulate. The final ACLV is the root-sum square of the error components as shown in Figure 25 with varying values of mask CD control. Thus, in order to benefit from tight specifications imposed on the mask and the exposure tool, the magnitude of the non-optical error component should be close to the mask component. Otherwise, the non-optical errors will dominate the final ACLV and further improvements in mask CD control, focus and dose control will not effectively lower ALCV. In this example, the simulation suggests that the non-optical contribution to ACLV should be around 8 nm 6σ. The results also highlight the impact of defocus on CD control. Calibration errors are large contributors to the overall error in mean focus.

10. Phase Shifting Masks

A new class of reticles known as phase shifting masks (PSMs) has provided improvements in imaging performance over traditional chrome on glass, or binary, reticles. The largest improvement is provided by the strong or alternating PSM (Alt. PSM) [75] (Fig. 26). During exposure, Alt. PSMs create low light intensity areas in the image by varying the phase of the transmitted light between

0 degrees and 180 degrees. During coherent (low sigma) illumination of a PSM, this phase transition causes the light amplitude to transition from a positive to a negative value. As the amplitude must cross zero during the transition, the light intensity (being the square of the amplitude) must also reach zero. This phase shifting effect increases the contrast of the image. Alt. PSMs can theoretically image considerably smaller features dimensions and spacings compared to those imaged with a binary reticle. Actual production factors limit the improvement somewhat, but aerial images from Alt. PSMs are still better than from binary masks [76].

The alternating phase shift effect is typically created by selectively etching into the quartz reticle substrate to provide a 180 degree optical path difference between shifted and unshifted regions [77]. Alt. PSMs have many issues which must be solved before they can be used effectively. In the manufacture of Alt. PSMs, additional challenging reticle patterning steps are performed, phase control is difficult, and defectivity is increased [77]. Serious design difficulties also exist [78,79]. Upon exposure, narrow low intensity areas appear at the edges of every phase regions. In a positive resist process, these areas will appear as resist lines (Fig. 27). Many design patterns exist which cannot be implemented with Alt. PSMs without creating undesired lines. Techniques have been developed to remove these unwanted phase transition effects known as "phase conflicts". The 0 degree to 180 degree phase junction can be performed in 0 degree, 60 degree, 120 degree, and 180 degree steps to smooth the transition and prevent unwanted lines from printing. However, defocus effects require considerable space in the design for this solution to be effective [80,81]. Unwanted lines can also be prevented in a positive resist process by using a second exposure of the phase conflict areas [82]. This method decreases throughput, requires tight exposure1 to exposure2 overlay control, and also has design difficulties.

Other PSM types have been developed to reduce the difficulties encountered with Alt. PSMs. The most accepted is the attenuating PSM (Att. PSM) [83,84] (Fig. 28). In an Att. PSM the chrome layer on a binary reticle is replaced with an attenuat-

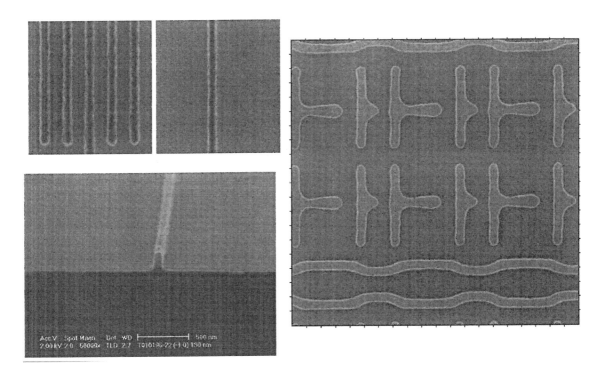

Fig. 29. SEM images of photoresist line patterns on the wafer. Top left image is of 193nm photoresist lines, Right image is of a 248nm photoresist bitcell pattern. Bottom left image is a X-section of a 193nm photoresist line.

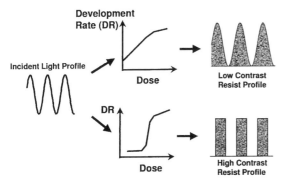

Fig. 30. Cartoon highlighting the need for a non-linear development rate response (high contrast) to dose changes. The higher contrast photoresist in the bottom row provides more vertical final profiles and improves lithography process latitude in many cases.

ing phase shifting layer. The attenuating layer allows light to be transmitted with a uniform intensity, often between 4 - 11% of the clear reticle area light transmission although much higher transmission values are being investigated. The light trans-

mitted through the attenuating layer has an ideal phase shift of 180° relative to the clear area light transmission. Att. PSMs have found considerable use improving the image contrast of both dark field and clear field patterns. The imaging benefits of the Att. PSM are less than those of the Alt. PSM, but phase conflicts are eliminated. However, Att. PSM technology also has challenges. Development and process control of attenuating materials for deep ultraviolet illumination has proven difficult [85]. Light diffraction effects at the edges of features can also cause secondary features, or sidelobes, to resolve in the resist [86]. Finally, small variations in phase or transmission across the reticle field can have a negative impact upon the lithographic processing margins.

A more recently developed PSM type is the chromeless phase lithography (CPL) mask [87]. A CPL mask contains etched quartz features of phase 180 degrees. The mask features used to pattern larger feature sizes are covered by chrome and are effectively traditional binary features. Smaller

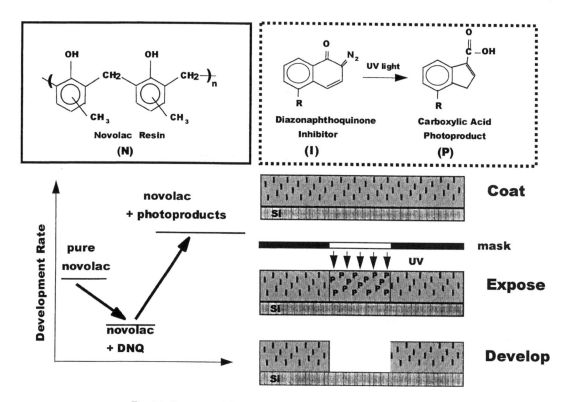

Fig. 31. Overview of Diazonaphthoquinone-novolac resist behavior.

feature sizes are purely quartz on the mask. CPL masks are used in combination with single exposure off-axis illumination and do not suffer from phase conflict problems. The combination of the two mask phase edges for the small features in close proximity causes a dark line to appear in the aerial image. These masks have shown excellent feature size and pitch resolution capability. CPL technology is still immature, but the known drawbacks include mask manufacturing and inspection, moderate design optimization issues, and the need for additional techniques to improve the depth of focus of small isolated features.

11. Photoresists

The goal of the photoresist is to translate the smoothly varying aerial image produced by the exposure tool into a vertical profile relief image in a protective barrier material (Fig. 29). To do this, the

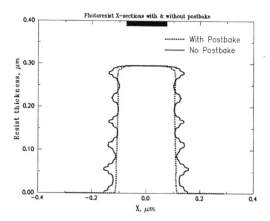

Fig. 32. Xsection view of simulated photoresist profiles patterned on a reflective substrate with and without post-exposure bake (postbake). The scalloped photoresist profile in the non-postbake case is due to PAC variations caused by thin film interference. The postbake step diffuses the PAC locally to produce the desired smooth final profile.

photoresist must undergo some physical or chemi-

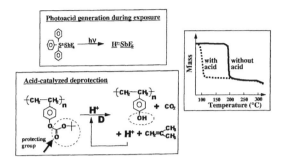

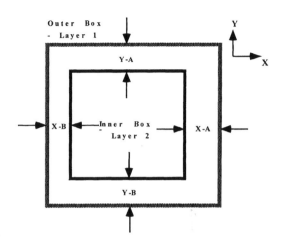

Fig. 33. Example of chemically amplified resist function. Exposure of the resist generates acid from the photoacid generator molecules. Presence of the acid in the insoluble "protected" or "blocked" polymer matrix catalyzes the deprotection of the polymer to yield hydroxyl functional groups that increase the solubility of the resist in developer. This mechanism is responsible for switching the solubility of the exposed areas of the resist and permits relief image formation.

Fig. 35. Determination of overlay error between layers 1 and 2 by measuring offset between box in box structures in X and Y directions. The offset x,y = (XB-XA)/2, (YB-YA)/2.

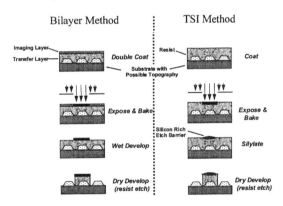

Fig. 34. Example of two alternative resist processing methods, bilayer and top surface imaging (TSI).

cal change upon exposure to light that can result in the generation of the relief image through further processing, i.e. it must be "photo" sensitive. The photoresist must also "resist" or withstand further processing such as ion implantation or plasma etching in order to protect the regions covered by the resist. In general, there are a number of requirements that a photoresist must satisfy in order to be useful for integrated circuit manufacturing. First, the photoresist must be able to be spin-coated into defect free thin films that will adhere to a variety of underlying substrates. The resist should have a relatively long shelf life (i.e. not change chemically or physically over the span of at least months) and should give repeatable coating thickness with good uniformity. The coated resist films should be relatively chemically and physically stable, and the resist material must have good physical and mechanical properties in order to withstand elevated temperatures and harsh environments such as corrosive etches without losing pattern shape or adhesion to the substrate.

In order to avoid stepper lens contamination, the resist must not outgas excessively during exposure. The resist must also possess a high sensitivity to light energy to allow for the desired wafer throughput in the production line. High quality 3D resist patterns should have vertical or close to vertical sidewalls. However, the aerial image from small features on the mask has a significant gradient in the X direction, see Figure 11. Therefore, the resist dissolution rate dependence must be strongly non-linear or in other words, threshold like. Resists with such dissolution rate dependence upon exposure dose are referred to as high contrast resists. Only high contrast resists can convert sloped aerial images into vertical 3D pattern. Figure 30 illustrates how would the resist print if the dissolution rate was not a strong non-linear function of absorbed energy. Finally, the resist should have a resolution capability, which exceeds the CDs of the desired patterns [14].

Resists typically have two main components, a

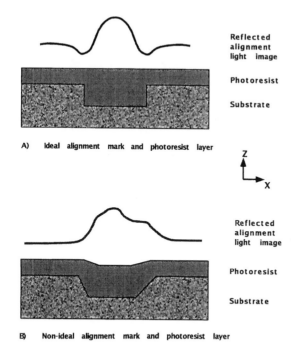

Reflected
alignment
light image

Photoresist

Substrate

Z

X

Reflected
alignment
light image

Photoresist

Substrate

B)   Non-ideal alignment mark and photoresist layer

Fig. 36. Reflected light intensity of alignment image from (A) ideal and (B) non-ideal alignment marks on the wafer substrate.

base polymer resin and a photosensitive additive or sensitizer. The polymer resin gives the resist its good film forming and physical property characteristics while the sensitizer makes the material respond to radiation exposure and allows the material to be imaged. Resists can be either positive or negative tone depending on their response to radiation exposure and development. In positive tone resists, the areas which are exposed to radiation are dissolved away during the development step leaving behind resist in the unexposed areas (and vice-versa for negative resists). Resists can also be classified by their general design as either non-chemically amplified or chemically amplified. Both of these schemes will be explained in more detail.

The previous dominant "workhorse" resists in the microelectronics industry were the diazonaphthoquinone-novolac materials [15]. These resists are used by exposing them to the G, H, or I-line wavelengths. Figure 31 shows an overview of how these resists function. The polymer resin

in these resists is novolac, a copolymer of phenol and formaldehyde. Novolac is soluble in many common organic solvents and can be coated from solution to form high quality thin films. Novolac is also soluble in aqueous base solutions by virtue of the acidic nature of the phenolic groups on the polymer, giving rise to the common basic developers for these resists. The photoactive compound (PAC) in these resists are substituted diazonaphthoquinones (DNQ). Upon exposure to UV radiation, the DNQ is converted into a carboxylic acid photoproduct which is itself soluble in basic developers. The presence of the DNQ photoactive compounds in the novolac resin serves to drastically reduce the dissolution rate of the novolac polymer in aqueous base developers. Once the DNQ has been exposed and converted to the carboxylic acid photoproduct, the presence of the carboxylic acid photoproduct dramatically increases the dissolution rate of the polymer in basic developers. Thus, by converting the DNQ in the resist film using exposure to UV radiation, it is possible to cause a dramatic change in the development rate of the resist and thus form relief images in the material. An important property of this type of resists is the phenomenon known as bleaching. Bleaching refers to the fact that the optical absorbance of the photoproduct is significantly lower than its parent DNQ molecule, which results in the resist becoming more transparent in the ultraviolet as it is exposed. This bleaching phenomenon permits more light to propagate through to the bottom of the resist film as the exposure process is carried out.

When resist is exposed over reflective substrates, the formation of standing waves due to interference of the various reflections within the thin resist film can lead to scalloped looking PAC profiles in the resist at the edge of the feature, that without further processing would be transferred into the final resist relief image [15,41] (Fig. 32). During a post-exposure bake, the PAC can diffuse from areas of high concentration in the anti-nodes of the standing waves into areas of low concentration thus smoothing out the concentration gradients at the edge of the feature. In this manner, the sidewalls of the relief image can be returned to a smooth vertical profile.

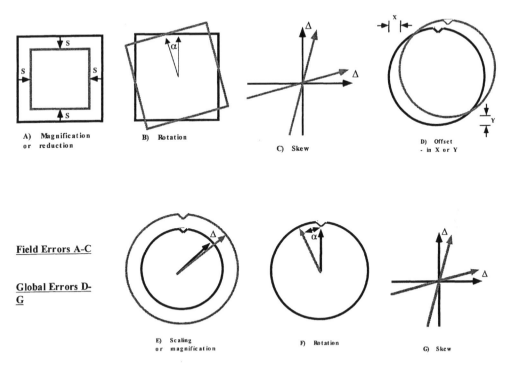

A) Magnification or reduction

B) Rotation

C) Skew

D) Offset - in X or Y

Field Errors A-C

Global Errors D-G

E) Scaling or magnification

F) Rotation

G) Skew

Fig. 37. Description of field and global overlay errors.

Traditional DNQ-novolac resists were found to be unsuitable for 248nm lithography because of their high absorbance and low sensitivity to the limited output of mercury vapor light sources at 248nm [14]. These problems were the basis for the invention of the second major class of resists, the chemically amplified resists (CARs). Figure 33 illustrates how CARs function. Exposure to light generates a catalyst in the resist, typically an acid, that acts on the surrounding matrix polymer in the presence of heat to catalyze a series of reactions which modify the matrix polymer's dissolution behavior. Again, it is this change in the dissolution or development behavior of the photoresist that allows for the generation of a relief image in the resist. The catalytic action of the photogenerated acid serves to dramatically increase the sensitivity of CAR resists since a single acid molecule can perform hundreds or thousands of other chemical events in the resist that modify its solubility. This catalytic action is what is referred to as the "chemical amplification" in the resist. However, CARs also have their share of problems. For ex-

ample, the photogenerated acid migrates from exposed regions into unexposed regions during post exposure bake thus contributing to CD bias and line end shortening [89,90,91,92]. During the early implementation of these systems, environmental contamination and acid neutralization of the resist by airborne basic species were shown to be major problems. A number of solutions have been proposed to solve this problem including: independent filtration of the air in the exposure tool systems, special top coat layers to protect the resist, and development of new chemically amplified resist systems that have lower sensitivities to contamination.

Improvements to resist chemistries and the development of new resist processing schemes have been and will continue to be a key ingredient to the success and extension of current optical lithography technologies [14]. For example, increased contrast and surface inhibition improvements to i-line resist chemistries, in conjunction with the development of high-NA i-line exposure tools, have allowed i-line lithography to be extended to $0.30\mu m$

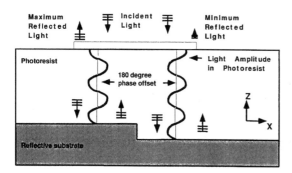

Fig. 38. Figure showing the cause of the swing effect in which CD varies sinusoidally with resist thickness. Thin film interference causes the absorbed light energy to vary with resist thickness often leading to large CD variations.

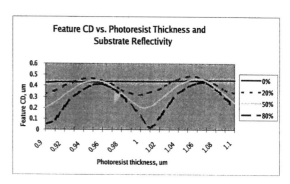

Fig. 39. Plot of a swing effect for a nominal $0.4\mu m$ photoresist feature showing CD vs. photoresist film thickness as a function of % light reflectivity intensity at the photoresist/substrate interface.

resolution or below [88]. Further improvements to resist technology are under development. Resists with higher surface inhibition are useful for attenuated phase-shift mask exposure to help eliminate the common problem of side-lobe formation. Improvements in photospeed and etch resistance will help extend the capability and economic advantages of current optical technologies. For CARs, developing resists with lower sensitivity to environmental contamination will also be of great benefit. Advanced resists often perform better at imaging specific type of features, i.e. either lines or spaces and either dense or isolated features. Resist properties can thus be specifically tailored to the needs of particular patterning layers. This specialization and selection of layer specific photoresists expands patterning capabilities but adds to the complexity of the overall lithographic process.

In addition to improving the resists used for typical single layer resist (SLR) processes, the possibility of using hardmask, bilayer or top-surface imaging (TSI) processes offers the potential to extend the capability of current optical technologies [14]. Figure 34 shows examples of the bilayer and TSI resist processing schemes. In the hardmask and bilayer approaches, a thin resist imaging layer is coated upon a transfer layer. In the hardmask approach, this transfer layer is an inorganic film designed for etch selectivity to the underlying film(s) to be etched and for good ARC properties. In the bilayer approach, the transfer layer is a thick organic planarizing layer [93]. The initial pattern formation is performed in this top imaging layer using

conventional lithographic techniques, and then the pattern is replicated in the transfer layer using an anisotropic etch process. In TSI, the exposure process creates a chemical change in the top surface layers of a thick resist which then selectively prevents or allows silylation of this thin exposed region using a subsequent chemical treatment [94]. The resulting silylated areas of the resist are resistant to an oxygen plasma etch and, thus, the pattern can be transferred through the entire thickness of the resist using an anisotropic etch process. The advantage of all these processing schemes is that initial image formation takes place in a very thin layer at the top surface of the resist or patterning layer, thus increasing focus latitude and reducing resist opacity and substrate reflection issues. The drawback for these processes are the added complexity and additional steps required to complete the patterning process. Nonetheless, these advanced processing schemes offer the opportunity to image smaller features with existing optical lithography technologies.

193nm and future 157nm lithographic technologies require the development of entirely new resist chemistries for these shorter wavelength [95-99]. The main problem for resist designers is the lack of transparent matrix polymers. The polyhydroxystyrene (PHS) polymers used in 248nm resists absorb too strongly to make them useful for 193nm and 157nm resists. There have been a number of polymer material families investigated as possible materials for 193nm resist design includ-

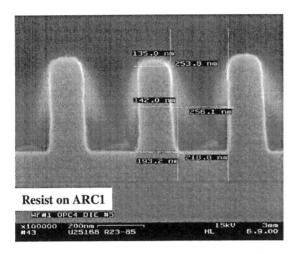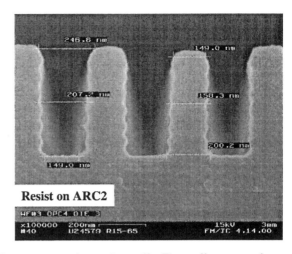

Fig. 40. X-section images showing the impact of substrate reflectivity upon photoresist profile. The profiles patterned on the lower reflectivity ARC1 are smooth. The profiles patterned on the higher reflectivity ARC2 show considerable standing waves.

ing acrylates, maleic anhydride copolymers, and cyclic olefin polymers. The key problem in development of these new resists is to satisfy all of the etch, imaging, and photospeed requirements in a single material. High photospeed resists will help reduce problems such as color center formation and lens compaction due to absorption of deep ultraviolet light by the lens system. Unfortunately, the move away from polymer resins containing aromatic rings that is dictated by transparency and imaging requirements has led to a reduction in the etch resistance of these new polymer resins as well. Thus, there is a difficult tradeoff that must be made between the imaging and etch characteristics in these new resist materials. In addition to the resists themselves, new ARC materials will also be required to enable the successful implementation of these advanced resist systems.

Once a wafer has been patterned with photoresist, it is then ready for further processing such as etching or ion implantation. It is at this point where the physical properties of the resist become very important in its resistance to the harsh environments and elevated temperatures possible in these processes. There are several criteria that must be met by both the etch process and masking resist to make the combination successful for a particular application. First, the resist must main-

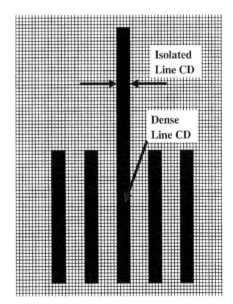

Fig. 41. Example of isolated and dense line patterning requirements for random logic applications.

tain good adhesion throughout the etch process to prevent etching in undesired areas. The resist must also maintain its profile during the process. The etch process must also show a high selectivity between the resist and the underlying material to be etched. This allows relatively thin resist layers (compared to underlying film thickness) to be used as etch masks without being completely consumed during the etch process. Resist lines may also need to be selectively shrunk in a so-called gate trim etch. This process step typically occurs just before the etching of the gate electrode substrate. The trim etch is tailored to controllably shrink the resist linewidth so a smaller resist CD is transferred into the transistor gate dimension. Finally, the resist layer must be relatively easy to remove after the etch process is completed [11].

In most cases there is a trade off between the selectivity and anisotropy of the etch process that must be balanced. Typically, etching is a plasma process. In plasma etching, or sputtering as it is sometimes called, ions generated in a plasma chamber are accelerated by a potential difference in the chamber toward the wafer and literally "chip off" atoms as they bombard the surface. Since sputtering is a physical process, this type of plasma etching tends to be non-selective but anisotropic. The high degree of anisotropy is achieved due to the fact that the ions can be accelerated along a single axis. One method used to improve the etch resistance of photoresist masking layers is to use a process known as UV hard baking. In this process, the patterned resist layer is heated and subjected to intense ultraviolet light that causes the resist polymer to crosslink and thus increase its etch resistance. One can also improve the selectivity of the process at some loss of anisotropy by forming chemically reactive species during the plasma etch that chemically react to etch the surface. This process is often referred to as reactive ion etching or RIE. There are also many other mechanisms which play roles in the performance of plasma etch processes that are beyond the scope of this work such as polymer redeposition and advanced plasma etch processes [19,100,101].

Photoresist is also used as a mask layer for ion implantation. In ion implantation, the masking layer must meet several requirements. Most importantly, the resist layer must be able to physically block the incoming ions. This typically means that the resist layers used for this process must be substantially thicker than resist layers in other steps of the manufacturing process. Due to this increased thickness, resists for this application require relatively high sensitivity to exposure energy

and nearly vertical sidewall angles for patterned features. A UV hard bake step can again be used to crosslink the resist and thereby increase its ability to stop incoming ions by increasing the density of the resist film. The implantation of ions into the resist layer can lead to charging on the surface of the resist [11]. This build up of charge can arc to the substrate damaging the device or can deflect other incoming ions leading to non-uniform doping profiles. Resists also can outgas substantially during ion implantation and can act as a source of secondary electrons in the implanter. During the implantation, the resist becomes more heavily cross-linked and thus can become difficult to remove. Extended exposure to an oxygen plasma, or plasma ashing, is often required to remove such layers.

12. Metrology and Inspection

In general, the purpose of a metrology step is either for process control or process analysis. Process control metrology and inspection steps are performed during production to determine if individual parts of the process are meeting their control requirements or whether a layer has been correctly patterned. In process analysis, metrology is used to develop, improve and test processes; find problem sources; and characterize tool, resist and material performance [18]. To provide results with a high degree of confidence, each metrology step should have adequate resolution, accuracy and repeatability. Automated metrology tools should also have high throughput, usable software interfaces and adequate sensitivity to process variations. Using an appropriate sampling plan, metrology for process control requires only moderate sensitivity to determine if the process is operating within allowed variations. Process analysis metrology requires higher sensitivity to better show the effects of process variations from the limited number of samples generated during optimization experiments or occasional process excursions.

Process control steps are consistently performed during the wafer patterning process, or in-line. A low resolution optical inspection looks for large

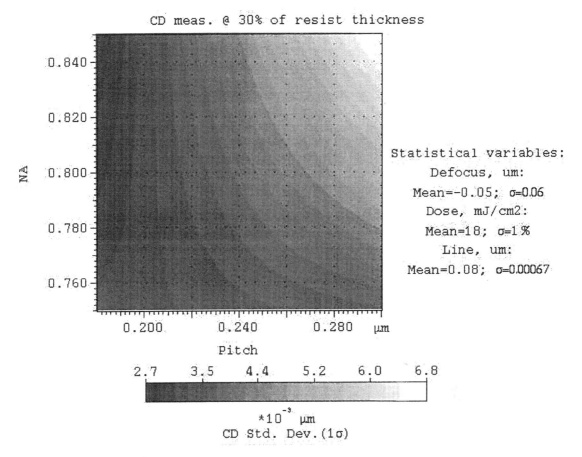

Fig. 42. Statistical CD control simulation optimization result showing the benefit of higher NA illumination for denser pitches and the benefit of lower NA illumination for more isolated pitches.

processing problems such as obvious process errors or large defects. A higher resolution optical field-to-field pattern comparison checks wafers for smaller random defects [18]. A top-down scanning electron microscope (SEM) measures with high accuracy the CDs of test features on a field to ensure that the printed pattern size is correct [18]. Extending the CD SEM capability for future generations poses many challenges [105]. Optical scatterometry quickly measures CDs and profiles of test feature gratings by analyzing the scattered reflected light diffraction order amplitudes and comparing to a data library [102]. Adequate CD control is often statistically defined to be when the 3sigma CD variation is within 10% of the targeted feature size. Additionally, optical measurements

determine the overlay error between different patterning layers [18] (Fig. 35). These measurements are done with a dedicated overlay metrology tool upon specially designed features which overlap layer to layer. Once the metrology and inspection tests have been passed, the wafer is assumed to be correctly patterned. In addition to in-line measurements of wafer patterns, tools (including metrology tools), resists and materials are periodically measured to ensure correct performance.

Process analysis may require additional metrology steps. A SEM is used for top-down viewing of printed feature shapes and CD measurements. Image analysis software is becoming popular for analyzing top-down pictures taken by SEMs [103]. X-section SEMs are used to measure and view feature

100

Optimum ARC thickness ~177A by vertical incidence model. Same optimum is predicted by rigorous statistical model.

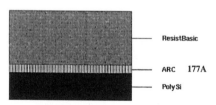

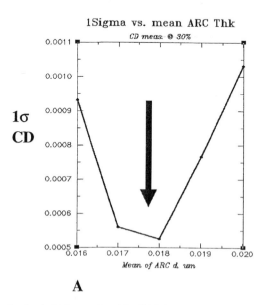

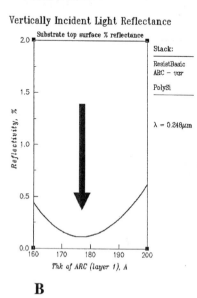

A   B

Fig. 43. Optimal thickness of a thin ARC layer predicted by the more accurate non-vertical model statistical simulations (A) matches the value predicted by the less accurate vertically incident model (B). Statistical simulations assume annular and high NA illumination.

profiles [104]. An atomic force microscope (AFM) can be used for extremely high accuracy profile measurements [15]. Electrical probing is used to quickly obtain many measurements of etched electrically conducting features or effective transistor gates [18,106-108]. Electrical probing is especially useful in lithography for lens performance characterization. Interferometric measurements are made of material and resist film thicknesses. Ellipsometry is used to determine thickness and optical properties of material and resist films; or to calibrate interferometry [109-111]. Significant metrology performance improvements in these areas are required to ensure continuing lithographic patterning progress.

Metrology is also a major consideration in areas other than wafer measurements. Reticles, stepper internal diagnostics (including lens aberrations),

photoresist properties, factory environmental control, track chemical delivery control, hotplate temperature control among many other factors must be measured and analyzed to ensure an optimum lithographic process. Metrology is also important in the use of resolution enhancement software [25], factory automation software, etc. Software metrology is also called software quality testing. Software quality testing analyzes whether the software output matches the specified requirements for a range of inputs, and also whether the procedures used to develop the software are in accordance with known good practices. Regardless of the specific application, metrology comprises the critical data collection and analysis to show whether each part of the complicated manufacturing process is behaving properly.

13. Overlay

Control of layer to layer pattern positioning is critical to the proper functioning of integrated circuits. The terms "overlay" and "alignment" are widely but inconsistently used across the semiconductor industry. In this paper, we will define overlay to be the layer-to-layer positioning of features, and alignment to be the determination of reticle and wafer field positions, which makes overlay possible. Improvements in overlay control are a necessary ingredient for increasing pattern density of designs [18]. The overlay control of features on one layer to features on another layer must be within specified tolerances (Fig. 6). These tolerances are determined by the ability of the circuit to function with less than ideal overlaps or spacings between features on adjoining layers. A typical value for the maximum allowed overlay error on a layer is one-third of the minimum feature size [2,18]. The first patterning step is performed on an unpatterned, or bare silicon, wafer which contains no alignment features. The stepper merely centers the exposed fields on the wafer. However, the first patterning layer is important to overlay control because it defines a reference pattern which other layers align to.

Overlay control is limited by errors in reticle patterning; alignment of the reticle to the stepper, also called reticle alignment; alignment of the wafer to the stepper, or wafer alignment; distortions in pattern transfer to the wafer; and processed induced distortions on the wafer. During reticle patterning, the reticle write tool may not print features at the correct positions relative to each other [31]. These errors are known as reticle registration errors. In both reticle and wafer alignment, each alignment system performs either a scanning or static illumination of alignment features, or alignment marks (Fig. 36). The reticle marks are typically chrome or clear lines on the reticle. The wafer marks are typically grouped or isolated lines on a previously patterned wafer layer. The alignment signal, the reflected or transmitted light profile of these marks, is analyzed to determine the locations of the alignment marks. Different light detection and signal processing schemes may be used to enhance the accuracy of this determination. Errors in reticle or wafer alignment can occur in the determination of the mark edge positions or from movement inaccuracies of the reticle or wafer stages to these positions. Lens aberrations create overlay errors by distorting the image placement of the reticle pattern during resist exposure [112]. These errors are often not measured by traditional overlay metrology structures, and thus, are not compensated for. Feature sizing errors strongly affect acceptable overlay errors. Dimensions which are either too large or too small limit the ability of features on different layers to connect properly with adequate overlay tolerance (Fig. 6).

As wafers distort during high temperature furnace or mechanical polish processing steps and as pattern transfer is never perfect, good overlay control requires matching the characteristics of previously patterned layers. The patterning errors affecting overlay are either intrafield or wafer based, although the actual error types are similar. The main correctable intrafield, or field errors are magnification, rotation and skew. Skew is also called orthogonality. Armitage and Kirk [113] described an orthonormal expansion of the overlay error , dx and dy, in terms of these errors:

$$dx = \delta x - \delta\alpha \cdot y + S_x x + \delta\Delta \cdot \frac{y}{2} + HOT \quad (10)$$

$$dy = \delta y + \delta\alpha \cdot x + S_y y + \delta\Delta \cdot \frac{x}{2} + HOT \quad (11)$$

where δx and δy are translation, $\delta\alpha$ is rotation, S_x and S_y are scaling or magnification in either direction, and $\delta\alpha$ is orthogonality or skew. The higher order terms (HOT) include barrel, pincushion, trapezoid distortion, cubic, fifth order, and asymmetry distortions. These terms are typically not orthonormal. The main correctable wafer errors, also known as global or grid errors, are offset, magnification, rotation and skew [12] (Fig. 37). These terms can be described mathematically in the same manner as the intrafield error terms. Before exposure, the stepper measures multiple alignment structures across the wafer and creates an internal model of the previous layer's field and wafer errors. The stepper then attempts to emulate these errors as closely as possible during resist exposure to minimize layer-to-layer overlay variations [12]. This is achieved by adjusting the stage stepping characteristics and lens reduction ratio. In scanning systems, additional corrections can be made for field skew and field magnification differences between the x and y axis. Special overlapping overlay calibration features from the two layers can be measured after patterning to analyze how well the layers were matched (Fig. 35).

Because individual steppers or types of steppers have characteristic image placement distortions, worse overlay matching is typically obtained between layers patterned on different tool types than between those patterned on the same tool or same tool type. Therefore, the common cost reduction strategy of using multiple exposure tool types for patterning different layers, called mix and match lithography, creates issues for overlay control [12,114]. The overlay error between layers exposed on two different exposure tools, called matching, also needs to be minimized. The stepper matching performance can be optimized based on the results of the analysis shown above [115]. Because the detection of the alignment signal is critical to overlay accuracy, the integrity of the alignment features is also important [12,116]. These features were created during previous patterning steps and have been subjected to all subsequent semiconductor manufacturing steps. These steps include film etching and resist application and may include film deposition, high temperature annealing and chemical mechanical polish, (CMP). These steps can introduce undesirable changes or nonuniformities to the alignment features which prevent the alignment signal from being accurately analyzed (Fig. 36). As each alignment analysis method has individual process sensitivities, steppers include multiple alignment options.

14. Substrate Control

Accurate control of substrate films on the wafer is a necessary ingredient for staying within the optical lithography CD and overlay budgets. Control is required because changes in substrate topography, film thickness, resist thickness, film optical properties or film chemical properties can cause CD and overlay errors. Many of these errors are at-

tributable to thin film interference effects, or swing effects, where the CD error magnitude is sinusoidally dependent upon film thickness [10,15,41] (Figs. 38 and 39). During resist exposure, light incident upon the wafer may be reflected at the air/resist interface, the resist/substrate interface and at any of the interfaces between substrate films. The interaction of incident and reflected light creates vertical standing waves of light intensity locally within the resist layer. (Fig. 40) These thin film interactions lead to variations in the energy absorbed by the resist with changes in resist film thickness, substrate topography or substrate reflectivity. Often the exposure error budget of the patterning process is dominated by these effects. Additionally, nonplanar features on reflective substrates can scatter a significant portion of light laterally causing undesired exposure of resist areas [10,15,117]. This effect is known as reflective notching.

As swing effects are due to variations in the energy coupled into the resist from incident and reflected light, they can be affected by a number of parameters [10,118]. Broadband mercury lamp illumination is less sensitive to swing effects than highly line narrowed laser illumination. Resists dyed with light absorbing additives allow less light to reach and reflect from the reflective substrate, although this light absorption also tends to limit resist imaging performance. Optimization of the resist thickness to an energy coupling minima or maxima will minimize CD variations due to small changes in resist thickness. An anti-reflective coating (ARC) on top of the resist can minimize variations in reflected light intensity, and therefore, absorbed light intensity [10,15]. The reflectivity of the resist/substrate interface can be lowered with the use of an ARC underneath the resist, a bottom ARC or BARC; or by optimizing substrate layer thickness [10,15,119] (Fig. 39 and Fig. 40). Bottom ARCs are the preferred strategy for improving substrate reflectivity effects as they can eliminate swing effects and reflective notching. Bottom ARCs can be either organic or inorganic films. They work either by absorbing incident light, phase cancellation of reflected light or a combination of both. Often process integration issues such as etch or planarization requirements will deter-

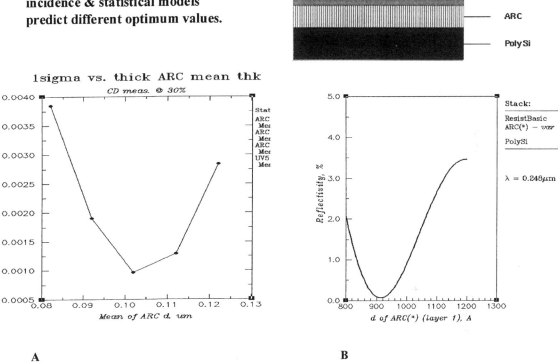

A

B

Fig. 44. Optimization of thick ARC layers by more accurate non-vertical simulations incorporating statistical variations (A) and by less accurate vertically incident model (B). Statistical simulations assume annular and high NA illumination.

mine an ARC strategy [119].

Due to light absorption in resists, large variations in resist thickness caused by substrate topography create CD errors even on non-reflective substrates, a so-called bulk absorption effect. The use of CMP on substrate layers will minimize these errors. However, typical variations in CMP polish depth across fields or wafers can cause substantial substrate thickness variations [120]. Although seldom considered, absorbed energy errors can lead directly to overlay errors, especially for narrow features [121]. Undersizing these features causes considerable line-end pullback, where generally the line ends are designed to connect to features on another layer (Fig. 8). Additionally, adequately reflected light contrast at alignment wavelengths is critical for the stepper to accurately determine the position of alignment features. Chemical control of the substrate is also important. Resist/substrate chemical interactions can prevent adequate resist/substrate adhesion or, alternatively, cause incomplete resist development at the substrate, known as resist scumming or resist poisoning [11].

15. Simulation

Lithography simulation has proven to be a useful tool for understanding, developing, improving, transferring and teaching optical lithography processes [122]. Many examples in this paper were generated using lithography simulation. Perform-

104

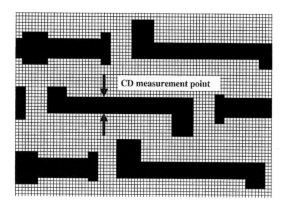

Fig. 45. Example of a bitcell pattern being optimized with CD measurement point highlighted.

ing optical lithography experiments via simulation reduces the number of experiments in the fab or research lab, saving time, effort and money. Many models and simulators exist for a large number of applications. These applications include:

– Optical element design and manufacture
– Stepper illumination and projection lens filtering optimization
– Reticle patterning and manufacture
– Design optimization
– Material thickness and property optimization
– Resist thickness, profile and performance optimization
– Yield prediction and optimization
– Defect printability analysis
– Overlay and metrology optimization
– Process throughput analysis
– ARC optimization
– Process cost
– CD control optimization
– Process latitude analysis.

The application will vary with the simulation tool as will accuracy, range of model validity, number of dimensions modeled, speed, scale, cost, software robustness and ease of use. However, shorter product development cycles, more complicated processes and larger equipment costs are increasing the need for simulated experiments. The most common uses of simulation are in optimization studies of design features, stepper illumination, ARCs, and in analysis of focus and exposure latitudes. Current research in simulation [123,124]

centers upon the speed of design optimization, the accuracy of resist photochemistry models, new data presentation methods, linkage to models for other semiconductor processes and development of models for non-optical lithography methods.

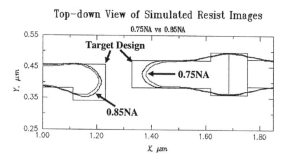

Fig. 46. Simulation example showing the benefit of higher NA imaging to reduce line-end pullback of patterned photoresist features.

A first example of the usages of lithography simulation is the optimization of the stepper illumination for a polysilicon gate layer of a 65nm generation circuit using a binary mask. The goal of the simulations is to optimize the CD control of dense and isolated feature patterns by analyzing the NA and sigma of the scanner (Fig. 41). In this example, the illumination source is chosen to be annular, so the inner sigma and outer sigma must both be optimized. The expected variations of mask CD, focus and exposure are known and used as statistical input parameters. Varying NA and annular conditions are simulated through a range of pitches to determine optimum values. The tradeoff between improved dense feature CD control with higher NA vs. improved isolated feature CD control at lower NA is shown in Figure 42. A similar tradeoff occurs between improved dense feature CD control with higher outer sigma vs. improved semidense feature CD control with lower outer sigma. A compromise illumination is finally chosen guided by the simulations.

A second example of lithography simulation application shows the optimization of ARC thickness using two different simulation methods. The first method is simple vertically incident light reflectivity calculations. These calculations have been shown to be accurate for thin (compared to the light wavelength) ARC layers on reflective sub-

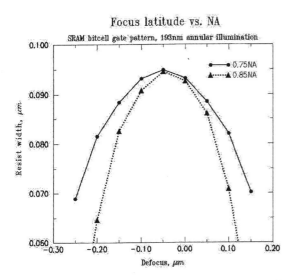

Fig. 47. Simulation result showing CD vs. focus response of the bitcell feature in Figure 45. The lower NA illumination is seen to have less CD sensitivity to defocus.

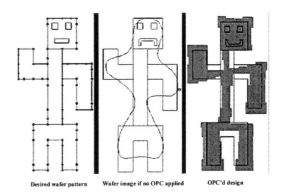

Fig. 48. Figure showing how OPC is applied to a design. The left picture shows the desired (designed) image which should be patterned on the wafer. Due to pattern transfer distortions, the wafer pattern will look like the smooth contour in the center if no OPC is applied. The right picture shows the OPC necessary to obtain a wafer image with a better match to the desired pattern shape.

strates (Fig. 43). However, the vertically incident model is not accurate for thicker ARC layers or dielectric substrate films, due to the path length difference between vertical and non-vertical propagating light through these thick films. In these situations, more accurate but slower full resist simulations utilizing non-vertically propagating light calculations should be used (Fig. 44). A final example of lithography simulation application is the optimization of a memory cell gate layer for two different stepper NAs with constant annular illumination settings (Fig. 45). The positions of the line-end and interior corner pattern edges are clearly seen to change as the stepper NA changes (Fig. 46). The higher NA setting shows higher pattern fidelity but at the cost of reduced CD control with focus variations (Fig. 47).

16. Resolution Enhancement Software

Starting approximately with the 0.35μm generation, lithography technology began to show significant mismatches between the desired features in the design database and the actual features obtained on the wafer. These differences were labeled as proximity effects (i.e., printing differ-

ences between isolated and dense features), corner rounding, line-end pullback, and/or pattern density loading effects (Fig. 48). As lithographers continue to push patterning processes into increasingly non-linear regimes (lower-K1 lithography), these effects increase [129,130]. These effects are often caused by optical diffraction limits, but other sources of error include reticle, substrate, resist and etch/implant properties [134]. New resolution enhancement software techniques were developed to reduce the systematic mismatches in pattern size, shape or spacing which were occurring. These software techniques became known as optical proximity correction (OPC) [135]. OPC software modifies the original design data to precompensate for these systematic errors in lithographic patterning and allows the final wafer features to more closely resemble the desired shape [131,132] (Fig. 48). The reduction of these systematic errors improves CD and overlay control. The correction of a design to improve its patterning is equivalent to expanding the lithographic design rules. Additional requirements for printability are added, causing the modified design to be more specific to a given patterning process.

New applications for design altering software have now become common in lithography [136,140,141,137]. The two main categories of use are: 1) OPC to reduce systematic differences between the designed and printed features, 2) implementation of reticle illumination or process techniques which reduce CD control sensitivity to random process variations. For category 1), the traditional OPC discussed earlier is still used but it has been extended. The earliest OPC work was done with manual, or hand drawn, changes to memory cell patterns to compensate for systematic differences observed visually. This method is still in use today but is obviously impractical for the large variety of random logic design features. The first automatic software methods typically used existing design layout or design rule checking (DRC) software tools to create extensions of line-ends, alter isolated-dense feature CD ratios, or add serifs to feature corners. These DRC-based methods used pattern analysis rules to define the design corrections and were successful at compensating for lithographic distortions on critical patterning layers through the $0.18\mu m$ device generation. However, it is generally accepted that the increased non-linearity and processing complexity of the 0.13g and later generations require higher accuracy OPC methods (Fig. 49).

The higher accuracy OPC methods that have been adopted by the industry are known as model-based OPC (MBOPC) methods [142]. These methods use extremely fast modeling techniques (based on multiple convolutions) to simulate the patterning of each feature in the design [143]. Design features are broken up into edges which can be independently simulated to determine where they will print on the wafer for a given patterning process. Neighboring edges are treated successively and iteratively to consider neighboring edge interactions. The goal of the model is to move each feature edge of the MBOPC'd design to a position at which the wafer feature edge will print exactly as on the original (desired) design. The simulations typically use an optical model to calculate the expected light intensity profile at each edge. They then use empirically tuned parameters to evaluate the intensity profile and predict the on-wafer edge location. The empirically fitted model enables the simulator to fit non-optical patterning behavior arising from reticle, resist or etch characteristics. The physically based optical model provides decent predictive capability in situations which were not used for model building.

The flexibility of the modeling methods allows the implementation of complex patterning schemes such as off-axis illumination (OAI) and novel post-exposure resist processing (e.g., trim etches to greatly shrink developed resist lines). Naturally, the final correction is limited by a number of restrictions including:
- The reticle exposure grid [28]
- Reticle manufacturing or inspection limits of small features/jogs [144]
- Output file size (typically 2-5X the original size)
- The need for symmetric final correction output for active device layers
- Runtime and memory usage of the MBOPC software
- Model accuracy

Until just recently MBOPC methods were considered nearly impossible to implement on large product designs, but the fast rate of improvements in speed, accuracy and flexibility from electronic design automation (EDA) software vendors has enabled MBOPC to become a standard lithographic technique.

The second category of resolution enhancement software application enables the use of techniques which improve process control. These techniques can be on the reticle, in the stepper illumination, or in the design. The software essentially coordinates the implementation of the technique with respect to the individual features in the design. The software will typically also perform traditional (first application category) OPC to reduce any systematic pattern transfer errors after the process improvement technique is applied. An example of a reticle process control improving technique is the use of sub-resolution assist features (SrAFs) [138]. To implement SrAFs the software modifies the original design by adding small (below the optical lithography resolution limit) features near isolated regular features. The SrAF features increase the pattern density of the design near these regular features and thereby alter the light diffraction characteristics. As dense pattern features have less sensitivity to focus variations than isolated features, the SrAFs improve focus control of the isolated regular features. The goal of the software is to place SrAFs in design locations where they are

WITHOUT OPC   **WITH OPC**

Fig. 49. Figures highlighting the importance of OPC for advanced patterning. The left figure shows large pattern distortions and photoresist line collapse with no OPC applied. The right figure shows a greatly improved pattern with OPC applied.

needed to improve regular feature process control. However, the software must also ensure that the SrAFs are not so large or so close to regular features that the SrAFs themselves print on the wafer, that is, they must remain sub-resolution. Other examples of reticle process improving techniques implemented by resolution enhancement software are alternating phase shift mask (AAPSM) use and high-transmission attenuated phase shift mask use.

Examples of software enabling the use of illumination techniques to improve process control are double exposure methods and quadrupole source methods. In double exposure techniques (including complementary AAPSM), the original design layer must be decomposed into two separate reticle databases which are then exposed sequentially to reconstruct the original design pattern on the wafer. These methods have the advantage of allowing different features on the design to be highly optimized during the separate exposures. For example, horizontally oriented features are imaged with the first exposure using a horizontally optimized illumination scheme, typically dipole [139], where vertically oriented features are imaged with the second exposure using a vertically optimized illumination scheme. These methods

require highly complex design decomposition software techniques, in combination with traditional MBOPC, in order to function properly.

An example of software enabling the use of a design technique to improve process control is the selective upsizing of isolated design features. Larger isolated features can be patterned lithographically with greater CD control than smaller isolated features. Therefore, in many interconnect layers, the original designed size of an isolated feature can be increased in order to improve process control. This change in isolated feature size is separate from any size changes which might be performed during traditional (first category) OPC. The goal of traditional OPC is to ensure that the final on-wafer feature dimension matches the original designed feature dimension. In order to achieve this goal, reticle feature dimensions may be upsized or downsized. In this new upsizing methodology, the final on-wafer dimension of the isolated feature is considerably larger than the original designed feature dimension. It is this upsizing of the on-wafer dimension which provides the process control benefit. Of course, the design groups which estimate resistance and capacitance of interconnect features need to be aware of this upsizing and incorporate

108

it into the design timing models.

Many challenges exist for successful design correction [140]. It is difficult to perform the alterations quickly and correctly on modern designs which contain hundreds of millions of features, multiple feature patterns and multiple design styles. Subresolution features and small jogs in altered features create enormous difficulties for reticle manufacture and inspection [134]. Additionally, the substantial increase in design file size can overload the capabilities of reticle patterning and inspection tools [28]. The usefulness of the rules and models may be limited by metrology accuracy, simulation accuracy and inherent process variability, both in reticle and wafer patterning. Finally, the traditionally separate roles of design, reticle, fracture, and process groups can make design correction projects difficult to successfully manage.

17. 157nm Lithography

Due to the reduced ability of fused silica lenses to work below the 193nm wavelength, any future optical lithography technologies will not use traditional quartz refractive optics technique. The next, and probably last, optical lithography technology will be at the 157nm wavelength and will use mainly CaF_2 refractive optics. An F_2 laser will be the illumination source and the reticle substrate will be made from fluorine-doped silica instead of the traditional quartz. Considerable technological innovation has occurred in the last few years to make this lithography technology appear possible [145]. Despite these recent advances, the difficult challenges in developing exposure tools for this shorter wavelength are many. These challenges include high quality CaF_2 material supply, lens purging with dry nitrogen to eliminate excessive light absorption, preventing contamination of the optical elements, optical element lifetimes, optical coating lifetimes, and 157nm wavelength metrology.

There are also many technological challenges to be solved before the 157nm lithography infrastructure is ready for scheduled production in approximately the year 2006. While 248nm and 193nm lithography had been under development for more than a decade before wide insertion, 157nm lithography will have had only about seven years of development, with many more fundamental issues to resolve. Entirely new photoresist chemistries need to be developed which will have low absorption of 157nm light. Current work in this area is focused on flourocarbon and siloxane-based polymer platforms [146]. A serious problem is the lack of a pellicle for use with 157nm radiation. As traditional thin polymer membrane pellicles are damaged by the high energy 157nm photons, thick quartz pellicles are being considered. The combination of these new pellicles with a new reticle substrate will require innovations in reticle handling and cleaning. Finally, a large obstacle to be overcome for 157nm lithography is the high cost of ownership, for device manufacturers and their suppliers. Because of the advances of very high NA 193nm scanners, probably only a single generation of 157nm scanners with NAs of approximately 0.80 to 0.85 will be used in production. Therefore, the cost of developing the infrastructure for 157nm materials, tools and supplies might not be amortized over many generations of products. Furthermore, the use of 157nm lithography is only expected to provide an approximately 20% reduction in feature size over very high NA 193nm lithography, considerably less than the typical 30% reduction. This reduces the financial rewards of circuit manufacturers who choose to implement 157nm lithography into production.

18. Immersion lithography

Interest is also growing in immersion lithography. The effective numerical aperture of the projection lens can be increased by imaging in a medium that has an index of refraction, n, greater than unity. As Lin has shown, minimum resolution and depth of focus each decrease by a factor of $1/n$ [54]. The possibility of introducing immersion lithography introduces many challenges, which include:
– The fluid must have no effect on resist performance, or new resists must be developed that have adequate performance when exposed in the fluid.

– No defects may be added due to residual fluid or precipitates remaining on the wafer after exposure.
– The index of refraction must be spatially uniform, so no large thermal gradients may exist since index of refraction is a function of temperature.
– Inserting wafers into fluid for exposure and removing the fluid after exposure is a significant engineering challenge for fabricating the exposure tool.
– Fluid must be nearly free of bubbles, which may scatter light. Bubbles might result from turbulence in the fluid or from outgassing from the resist during exposure.

The fluid must be transparent at the exposure wavelength and at the metrology wavelength. Absorption reduces throughput and could also cause thermal gradients.

The historical success of optical lithography is creating proponents of immersion lithography to extend the technology that they have come to trust for continued scaling of integrated circuits.

19. Cost of Ownership

As each of the many layers in a device requires patterning, the optical lithography process is a major part of the cost of manufacturing integrated circuits. Lithography and associated metrology currently comprises 30-40% of the entire cost of semiconductor manufacturing. This fraction depends strongly on the product mix and age of equipment in the factory, though [147]. The goal of cost of ownership (COO) analysis is to understand the cost components of lithography processing to minimize manufacturing cost. Gaining an understanding of the costs in this complex process is not easy. Many parameters outside of lithography must be included. However, the effort is worthwhile as large savings can be realized by even minor reductions in processing cost. The cost of a process is typically measured in cost per wafer, per process layer, or per die. Cost of ownership analysis of lithography is usually quantified in terms of cost per good wafer level exposed.

Much of the cost of lithography, or any process, comes from rapidly escalating equipment costs [12] or even more rapidly increasing reticle costs [149]. Other components important to COO and potential profitability include cleanroom space, facilities, materials and payroll. (Sadly for engineers, this component is typically minor). Arnold and coworkers set the standard for lithography cost of ownership [12,148]. The cost of ownership of lithography, expressed as cost per wafer level exposed (PWLE), can be quantified as:

$$C_{pwle} = \frac{C_e + C_l + C_f + C_c + C_r Q_{rw} N_c}{N_g} + \frac{C_m}{N_{wm}}$$

where:
C_{pwle} = cost per wafer level exposure
C_e = yearly cost of exposure, coating, and pattern transfer equipment (including depreciation, maintenance, and installation)
C_l = yearly cost of labor
C_f = yearly cost of cleanroom space
C_c = cost of other consumables (condenser, laser diodes)
C_r = cost of resist
Q_{rw} = quantity of resist used per wafer
N_c = number of wafers coated
T_{net} = net throughput = raw throughput * utilization
N_g = number of good wafers out = $\int T_{net} \times$ yieldoflithographydt
C_m = cost of mask
N_{wm} = number of wafers exposed per mask

C_e is determined from the price of the equipment including installation costs. This cost is allocated to each year using depreciation, which is typically assumed to be straight line for five years. In practice, the terms that usually have the greatest effect on cost of ownership are C_e, T_{net}, C_m and N_{wm}. The cost of the mask can be determined from cost of ownership models of the mask fabrication process [150, 151]. These models include the yield of various process steps, and models of yield and yield learning can be used to predict the possible cost of various options. Process performance tradeoffs can also be analyzed in terms of cost. Factors include tool capability, resist capability, tool throughput, tool size, tool utilization, field size, wafers exposed per reticle, wafer size (e.g., 200mm vs. 300mm), process development time, product volume, number of manufacturing steps, device size, device performance, yield, defectivity, engineering resources, intellectual property, and extendibility of processes to future device generations [147].

Additional, somewhat abstract, components are also needed for the financially successful production of semiconductor devices. These include adequate supplier support, design capability, marketing expertise, work force technical experience, work force motivation, organizational structure and management effectiveness. Customer, product and supplier roadmaps are also needed to provide long term cost analysis. By acknowledging the need for these abstract factors and building a model to analyze and optimize the more defined components, accurate COO comparisons can be made for different business scenarios [12]. Lithography technology (e.g., 193nm vs. 157nm technology), tool, material and process choices can be effectively guided by COO analysis. The importance of accurate planning would be difficult to overestimate. Improvements in optical lithography planning using detailed COO analysis methods can be shown to save a new semiconductor factory hundreds of millions of dollars per year [147].

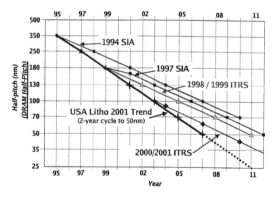

Fig. 50. Plot of the half pitch of lithography used in high volume manufacturing as predicted by various versions of the International Technology Roadmap for Semiconductors.

20. Introduction to emerging lithography options beyond optical lithography

Optical projection lithography has been the principal vehicle of semiconductor manufacturing for more than 20 years. It has long been predicted that optical projection lithography will eventually be incapable of meeting the demands of economically reducing feature size with adequate process control in integrated circuit manufacturing. The complexity of optical lithography, especially of making the masks, continues to increase as wavelength reduction continues to 157nm. The International Technology Roadmap for Semiconductors (ITRS) [2] predicts that a new lithography paradigm will probably be required for manufacturing 45-nm half-pitch patterns. Figure 50 shows a plot of the half pitch predicted with time for four versions of the ITRS. Each of these versions of the ITRS has predicted an acceleration of the reduction in half pitch. Leading candidates to succeed optical projection lithography are extreme ultraviolet lithography (EUVL) and electron projection lithography (EPL). Many other options also exist that are being actively developed. The list is organized in several broad categories, which include: next generation lithography (NGL), maskless lithography (ML2), interferometric lithography, and technologies with masks having unity magnification. The list includes:

Next generation lithography (use masks with reduction)
– Extreme ultraviolet lithography (EUVL) [152-154]
– E-beam projection lithography (EPL) [155-158]
– Ion-beam projection lithography (IPL) [159-160]
Maskless lithography (ML2)
– Multicolumn e-beam [161,162]
– Multisource e-beam [163]
– Low energy electron beam direct write (LEEBDW)
– Conversion of photon image with reduction optics to electrons for exposure [164]
– Spatial light modulators using UV light [165]
– EUV multimirror: spatial light modulators using EUV light [166]
– Arrayed zone plates with UV to 1-nm x-rays

[167]
- Arrayed AFM tips [168]
- Arrays of carbon nanotubes as electron emitters multicolumn ion beam [160,169]
Interferometric lithography
- Multiple laser beam interference [170,171]
1X masks
- Proximity x-ray lithography (PXL) [172,173]
- Proximity e-beam (LEEPL) [174]
- Step and flash imprint lithography (SFIL) [175]
- Nanoimprint [176-178]

Nearly all of the technologies in this list use one of three energetic particles, including photons, electrons or ions, to expose resist. The one exception is nanoimprint, which uses a master template to emboss resist using temperature and pressure without any exposure. Some technologies use a mask to expose large portions or all of the pattern in one exposure step. Other technologies use a serial exposure where the exposure apparatus is controlled by a computer and directed to "write" patterns on the wafer. Interferometric lithography uses a set of plane or spherical waves to form an interference pattern in the resist film. Without a master mask, the interference pattern can be as complex as can be synthesized with a set of plane or spherical waves. If a master mask is used, interferometric lithography is simply holography.

The key question that has lingered in the industry surrounding lithography has been: when will next generation lithography be used in high volume manufacturing? Others ask: will next generation lithography ever succeed optical lithography? Many individuals have predicted the end of optical lithography without success. Next generation lithography will succeed optical lithography in manufacturing when it meets all of the basic requirements of lithography in a more cost effective manner than optical lithography. A list of the basic requirements is:
- Critical dimension (CD) control: The size of many features in a design needs to be accurate and precise. CD control needs to be maintained within each exposure field, over each wafer and from wafer to wafer
- Overlay: The placement of the image with respect to underlying layers needs to be accurate on each integrated circuit in all locations.

- Defect control: The desired pattern must be present in all locations, and no additional patterns should be present. No particles should be added to the wafer during the lithography process.
- Lw cost: The cost of tools and masks needs to be as low as possible while still meeting the CD control, overlay and defect control requirements. The lithography step should be performed as quickly as possible. Masks should be used to expose as many wafers as possible. Equipment needs to be reliable and ready to expose wafers when needed.

More specifically, exposure tools, masks and resist must meet all lithography requirements; in addition, the new lithography operation must be carried out at lower cost than optical lithography. Lower cost might be achieved by having better capability so that yields for state-of-the-art products are better than can be obtained with optical lithography. Finally, the emerging lithography option must be capable of exposing the patterns needed to fabricate integrated circuits. Circuit designers are unlikely to significantly modify their designs to meet the requirements of the lithography unless there is no other option. Designers are making restrictions to their designs to achieve better performance, but the basic pattern shapes and ability to freely locate structures are not highly restricted. It also goes without saying that tools, masks and resists must be commercially available. This requirement actually drives the need for a large number of companies to adopt a technology so that suppliers of tools, masks and resist have a viable business plan to support their development of the technology.

Champions of emerging lithography have often espoused the elements of their technology that are superior to optical lithography while glossing over the attributes that do not meet some of the requirements listed above. For example, most proponents of emerging technologies stress resolution, but the difficulty of making the mask is often overlooked, especially masks with unity magnification or "1X masks." Overlay and defect control concerns are also not usually addressed in early work. The inability to adequately address many of the requirements of lithography at low cost have led to the delay of the introduction of emerging lithography into high volume manufacturing well beyond when its introduction had been predicted.

In the 1980s, debate in the lithography technical community focused on the timing when direct write e-beam lithography or proximity x-ray lithography might be introduced into high volume manufacturing. Since direct e-beam lithography is a serial process, tools that used a shaped beam to speed the delivery of the electrons to the wafer provided higher throughput. Although direct write e-beam lithography was used by IBM for some ICs that were highly customized in the late 1980s and early 1990s, direct write e-beam lithography has largely been relegated to the role of fabricating photomasks. In the early 1990s, the development of proximity x-ray lithography (PXRL) expanded with IBM installing a synchrotron as the source for exposure tools. The ITRS at one time predicted the introduction of proximity x-ray lithography at the 250-nm node in the late 1990s. Motorola, Lockheed Sanders and AT&T worked with IBM to develop the technology. The US government supported much of the effort at IBM, and it also supported programs at universities and at AT&T Bell Labs. Companies in Japan also actively developed PXRL under the support of the Japanese government. Mitsubishi installed a storage ring. By the second half of the 1990s, the difficulty of making and maintaining a defect free 1X mask with adequate mask CD control and pattern placement was one of the leading factors that led many companies to begin concentrating on other options. Present proponents of PXRL have now proposed the new name of collimated proximity lithography (CPL) [179].

In the early 1990s, AT&T Bell Labs developed an electron beam projection lithography technology they later named scattering with angular limitation project electron lithography (SCALPEL). IBM developed similar methods and later formed an alliance with Nikon to commercialize the technology. As Lucent Technologies, Bell Labs allied with ASML and Applied Materials to commercialize SCALPEL. Electron projection lithography (EPL) was the term used to collectively describe these efforts. In late 2000, ASML and Applied Materials decided to not pursue commercialization of SCALPEL. In 2002, Nikon and IBM remained committed to EPL and are actively developing the technology with many other partners.

In 1989, the first papers were published by researchers at the US Department of Energy national labs in California [180] and NTT in Japan [181] on the use of soft x-rays with wavelength between 4 and 20 nanometers for projection lithography. AT&T Bell Labs also began development of the technology in this time frame [182]. Development was supported by the US government in the early 1990s, and the technology was renamed extreme ultraviolet lithography (EUVL) in 1996. In 1997, Intel founded the Extreme Ultraviolet Lithography Limited Liability Company (EUV LLC). Motorola, AMD and later Micron, Infineon and IBM joined to support development of the technology at Lawrence Berkeley, Lawrence Livermore and Sandia National Labs. Several commercial suppliers, including ASML, Canon and Nikon, are working on developing exposure tools for EUVL. Ion projection lithography (IPL) was developed in Vienna, Austria in the early 1990s. The US government supported an ion beam lithography effort in the mid-1990s, and later Infineon formed an alliance with IMS Nanofabrication GmbH to develop IPL. A prototype tool was completed in 2001.

Due to the difficulty of fabricating masks for advanced optical lithography and emerging lithography options, maskless lithography has been proposed. IBM has long been a leader in deploying e-beam lithography for fabricating semiconductor devices. However, e-beam lithography is a serial patterning process and is quite slow, limiting its use to low volume and custom applications. Many research labs continue to use e-beam lithography for prototyping. To increase throughput, IBM developed shaped-beam systems so that rectangular shapes could be patterned with each exposure event rather than using a single point beam [183]. Even with shaped beam systems, throughput is too low for cost effective high volume manufacturing. Pain et al. [184] describe examples of using shaped e-beam lithography for prototyping advanced ICs in a mix-and-match mode with optical lithography. Cell projection e-beam lithography has also been used for DRAM development where a stencil mask of a small area of a repetitive array is exposed with e-beam and de-magnified to form an image on the wafer [185]. To increase throughput, the use of low energy electrons has been proposed. Resists are more sensitive to low energy electrons due to their small penetration depth. A new direct write e-beam technology based on exposure with electron energy <1 kV is being developed. However, this requires the use of <50-nm thickness resist films.

To exploit the high resolution of e-beam lithography, to increase its throughput, and to avoid the need for masks, IBM started development of microcolumn electron optics to form an array of e-beam writers. Others began developing methods of using multiple e-beams in a single column. In the mid-1990's, the Semiconductor Research Corporation and the Defense Advanced Research Programs Agency co-sponsored a program at several universities to develop maskless lithography. These universities began investigating multiple beams in a single column, arrays of zone plates using ∼1 nm x-rays, arrays of atomic force microscope (AFM) tips, programmable arrays of micro-mirrors illuminated with 13.4-nm radiation to replace the mask in EUVL systems, and arrays of ion beams in a single ion beam column. The group also started to develop methods to handle the large data volume required to realize maskless lithography with high throughput [186].

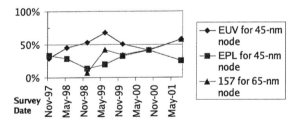

Fig. 51. Plot of the fraction of survey respondents at various International SEMATECH workshops that favored various technology options for the 65nm node on the ITRS roadmap as a function of time.

With many options to be supported with limited industry resources, International SEMATECH assumed the role of facilitating a choice among the many alternatives to be pursued [187]. For four consecutive years, they held workshops where technology champions for direct write e-beam, proxim-

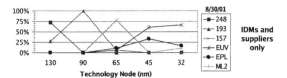

Fig. 52. Plot of the fraction of survey respondents at various International SEMATECH workshops that favored various technology options for the different technology nodes on the ITRS roadmap.

ity x-ray, ion beam projection, electron beam projection and extreme ultraviolet lithographies presented progress in technology development. Participants in the workshops represented IC companies, nationals labs, universities, and suppliers of masks, resist and exposure tools. At each workshop, a survey of the participants was conducted to gauge which technologies were deemed most promising. Participants also developed a list of critical issues for each technology. At each workshop, technology champions were encouraged to address progress against this list of critical issues. Figure 51 shows the historical progression of the fraction of respondents in the survey that favored 157nm, EPL or EUVL for the 65-nm node in the ITRS. Figure 52 shows the fraction of respondents that favored a particular lithography technology for each ITRS node. The workshops helped narrow the options that were supported by the lithography industry so that the most promising options would receive sufficient support for timely development. EUVL and EPL were identified as the most promising emerging lithography options to commercialize for high volume manufacturing of ICs. These two technologies will be described in more detail in the following sections. Other technologies such as maskless lithography and imprint lithography are also being investigated in corporate research labs and at universities and will be discussed in more detail as well.

21. Extreme Ultraviolet Lithography (EUVL)

Extreme UltraViolet Lithography (EUVL) is a projection imaging method [152,188] (Fig. 53) that uses 13.4 to 13.5-nm wavelength radiation–

114

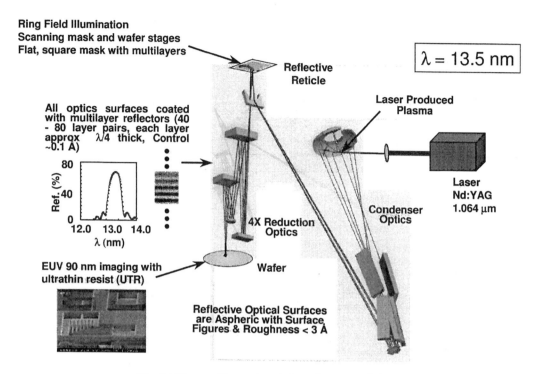

Fig. 53. Diagram illustrating the key elements of EUVL.

which is in the soft x-ray or extreme ultraviolet spectrum–for lithography. The principle of operation of EUVL is similar to that of optical lithography with a condenser and projection optical system and a mask. However, radiation at EUV wavelengths is highly absorbed by all materials; therefore, the reflective optical surfaces, including the mask, are coated with multilayers. The use of reflective optics limits the numerical aperture to relatively small values. EUVL exposure tools have NA of approximately 0.1 to 0.25 in a 4X reduction system. The Engineering test Stand (ETS), which is a full field scanning lithography tool, is described by Tichenor et al [189].

Because EUV radiation is highly absorbed by all materials, the reflective optical surfaces, including the mask, are coated with multilayers. The multilayers form a resonant Bragg reflector that has high reflectivity over a band of wavelengths around the operating wavelength of the exposure tool. The EUVL mask blank consists of a low thermal expansion material substrate with six inch square form factor that is coated with Mo/Si multilayers. The

multilayer-coated mask blank is coated with a patterned absorbing layer stack that defines the integrated circuit (IC) pattern to be imaged on the wafer at 4X demagnification with projection optics. Multilayer coatings for EUVL are distributed Bragg reflectors made of periodic groups of layers designed to maximize reflectivity at specific wavelength [188,190,191]. The reflectivity varies with the angle of incidence and with wavelength. The fractional bandwidth of the reflectivity with respect to wavelength typically ranges from 3.5 to 4% full width at half maximum (FWHM). At EUV wavelengths (5-20 nm), all materials have a real index of refraction near unity with a small imaginary component. Henke et al. [192] have tabulated the complex index of refraction for most elements in terms of:

$$\tilde{n} = 1 - \delta + i\beta \qquad (12)$$

Two or more materials are typically chosen to comprise the multilayer, and one material is chosen to have a large value of δ at the operating wavelength to maximize reflectivity. This material must also have relatively small loss coefficient,s βsgo minimize absorption. Because the normal incidence reflectivity of all materials is small for EUV wavelengths, many layers are arrayed and each is positioned so its reflected component adds in phase with that of the other layers. Therefore, the period of the multilayer stack is chosen to have constructive interference among the light reflected from each layer that has large δ. The other material is chosen to have small δ and small β to maximize index contrast between the layers and to minimize absorption. In the wavelength band around 13.5 nm, Mo and Ru have large δ with relatively small β and Si has small δ and small β. Other materials might be added to the stack to further optimize its reflectivity. The maximum reflectivity occurs at the Bragg or peak wavelength. Spiller, Barbee and Attwood presented the Bragg equation corrected for refraction. The Bragg equation for a multilayer made with layers from two materials and further corrected for absorption is given by:

$$\lambda_B = \frac{2d}{m} \sin(\theta_m) \sqrt{1 - \frac{2\bar{\delta} - \bar{\delta}^2 + \bar{\beta}^2}{\sin^2(\theta_m)}} \qquad (13)$$

where: $\bar{\delta} = \frac{d_1 \delta_1 + d_2 \delta_2}{d_1 + d_2}$ and $d_2 = \Gamma d$ $\bar{\beta}$ is analogously defined like $\bar{\delta}$, m is the diffraction order, d is the multilayer period, and d_1 and d_2 are the thickness values for the two layers comprising each period. Γ is the ratio of the thickness of the layer with higher absorption to the multilayer period. This ratio can be optimized theoretically to maximize peak reflectivity as described by Vinogradov and Underwood. In practice, the ratio is often optimized empirically because the theoretical index of refraction values do not perfectly match that of actual films, and the interface between the layers comprising the stack might not be abrupt.

Despite its similarity to optical lithography, implementing EUVL for high volume manufacturing of integrated circuits has many new technology challenges, including:

- The mask and optics are coated with multilayers to provide high reflectivity around 13.5-nm wavelength.
- The optical path and wafer being exposed must be in a vacuum of ∼1 mTorr pressure with the residual gas species being composed of Ar or He.
- A source of radiation with high long-term power

output must be used. Leading candidates use lasers or electric discharge to create a plasma of Xe gas that emits strongly in the wavelength band from 10.5 to 14 nm. The size and brightness of the source must be matched to the design of a reflective condenser system.

- The plasma in the source must neither degrade the Xe delivery and discharge apparatus nor the nearby collecting optics.
- The optical surfaces must have highly accurate figure since the wavelength is so short. Figure accuracy must be on the order of 0.1 to 0.25 nm. This accuracy requires highly accurate intereferometric metrology to enable fabrication of the surfaces.
- The optical surfaces must have low roughness on the order of 0.1 to 0.25 nm root mean square over the spatial wavelength range between 1-10 microns and 1-10 mm. The exact specifications of this mid-spatial frequency roughness (MSFR) are determined by the optical design. Roughness in this range scatters light out of the secular image; however, unlike stray light in transmission optical projection systems, the scattered EUV radiation from this roughness reaches the image plane causes an undesirable background intensity.
- The periodic multilayer coatings on the optics and mask must have highly accurate and uniform average period on the order of 0.1%. The period must be accurate to match the bandpass of the reflectivity of the optics to each other, to the mask and to the maximum spectral output of the EUV source. The interfaces between layers in the multilayer must be highly sharp to maximize reflectivity. The optical substrate must also be highly smooth over spatial wavelengths <1-10 microns to not scatter EUV radiation out of the pupil of the projection optics, effectively lowering reflectivity. This high spatial frequency roughness (HSFR) must be less than 0.1 to 0.2 nm rms, depending on the method of multilayer coating.
- The multilayer coatings must not oxidize or be coated with films under long-term radiation in a vacuum environment.
- The multilayer coating on the mask must be deposited with low printable defect density to

provide an economical supply of mask blanks [150,195].

- The mask must be protected from defects during exposure and during handling within the exposure tool since a conventional membrane pellicle cannot be used. A membrane pellicle with thickness great enough for mechanical stability would absorb a large fraction of EUV radiation.

- Because the multilayer coatings absorb roughly 40% of the incident radiation, the optics and mask need to be fabricated on low thermal expansion material substrates. These substrates must also be cooled during exposure.

- The resist must have <150-nm thickness in order for the EUV radiation to adequately expose the resist through its depth [196]. Use of this thin resist poses challenges to integration into the CMOS fabrication process [197] .

Despite these significant challenges, EUVL shows great potential for extending the resolution of lithography to 22-nm half pitch and perhaps below. High quality images have been produced with an initial set of full field optics [198], and significant progress has been made on all aspects of the technology as described recently by Gwyn [154]. Although considerable technological challenges remain, it appears that EUV will emerge as the main method of patterning for the 45-nm half pitch device generation and below.

22. Electron projection lithography (EPL)

Electron projection lithography (EPL) uses a mask that scatters electrons. The unscattered electrons are transmitted to expose the resist [155,156] (Fig. 54). Because electron optics have large geometric aberration, the mask is exposed in sections to limit the field size of the electron optics. A shaped beam of 100 KeV electrons is scanned across or illuminates a section of the mask. Electrons unscattered by the reticle pass through an aperture and are electromagnetically imaged onto the wafer. Scattered electrons are blocked by the aperture and prevented from reaching the wafer. The reticle is composed of a patterned metallic scattering material on a thin substrate membrane

with ∼100-nm thickness or of patterned stencil openings in a membrane with ∼1-2 micron thickness. The reticle requires periodic silicon struts for physical support [199]. These struts prevent imaging at their locations; therefore, no reticle features are placed above them. Design patterns bisecting a strut must be stitched together during exposure.

EPL technology has many benefits, which include: high patterning resolution due to the essential lack of diffraction which results from the deBroglie wavelength of ∼4 pm for 100 keV electrons, large depth of focus due to the use of small numerical aperture (∼0.001 to 0.01) optics, 4X reduction masks, use of presently available mature resists, use of standard silicon wafers as reticle substrates, reuse of some E-beam mask writing tool technology, and the reuse of optical 4X reduction scanning technology. Another key advantage is the ability to translate charged particle beams with electric fields to compensate for mask distortions or wafer overlay errors. The main drawbacks of this technology are the requirement for reticle field exposure stitching, the difficulty of manufacturing thin membrane reticles and the limited throughput. EPL is best suited to pattern specific challenging darkfield layers, such as contacts or via levels. A summary list of key challenges includes:

- High beam current is required to rapidly write the pattern and hence have high throughput. Achieving high throughput is difficult due to the fact that electrons are charged particles. The electrons must be focused tightly at the aperture where scattered electrons are stopped. At this point, the electrons repel each other, deflecting from their desired trajectory and blurring the image. The blurring increases as beam current increases [200]. The relationship between resolution and beam current is a nonlinear function of several variables [201]. The exponents of the terms depend on the design of the electron optical column. Decreasing the column length, increasing the field size of the electron optics, and increasing the numerical aperture of the electron optics reduce the distance over which the electrons are close to each other. Increasing the beam voltage also decreases this space charge blurring. Fabricating shorter columns requires more challenging design and high electric and

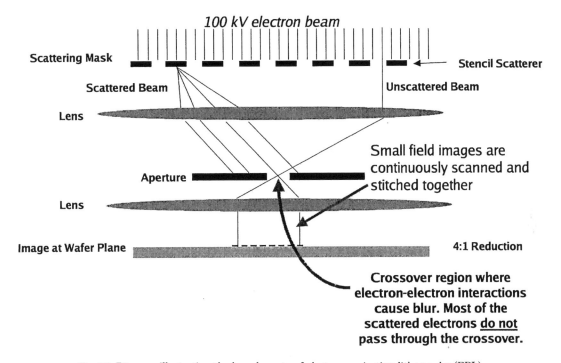

Fig. 54. Diagram illustrating the key elements of electron projection lithography (EPL).

magnetic field values. Increased numerical aperture and field size must be traded for increased geometric aberration, which also decreases resolution. A practical optimum can be determined using simulations [202]. Decreasing the area of the mask pattern which does not scatter electrons reduces the current at the aperture, and increases resolution. Ions can potentially also be added into the column to neutralize the space charge and reduce stochastic beam blur [203]. Therefore, darkfield layers such as contacts or vias can be exposed with higher throughput at a given resolution.

- The space charge not only decreases resolution, it also introduces aberrations [204]. These space charge aberrations may be corrected to some degree by the electron optics [205]; however, in order to be compensated, they must be predicted in advance of the exposure by a computer algorithm. The most important space charge aberration to correct is defocus, which will vary with the pattern density in each subfield of the mask.
- The mask pattern must be divided into subfields

[206].
- Accurate stitching of the exposure sections must be achieved. Stitching errors manifest themselves as CD errors and overlay errors.
- Due to the high current required for high throughput and due to the high transparency of resist to 100 kV electrons, wafer heating must be managed. Compensation strategies must be implemented [207].
- Electrons with high energy elastically scatter in the substrate, scattering to regions of the resist far from the desired imaging location. This proximity effect affects CD control and needs to be compensated by modifying the mask CD values.
- Economically manufacturing defect-free stencil or membrane masks is a significant challenge [208].

Maintaining defect free masks without a protective cover or pellicle during exposure will also be challenging. Both sides of the mask need to be protected.

Electron projection lithography is well suited to layers with low pattern density, and it might play

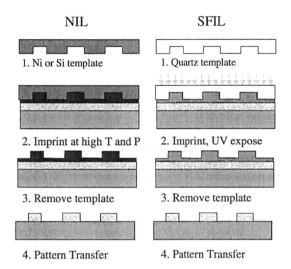

NIL   SFIL

1. Ni or Si template   1. Quartz template

2. Imprint at high T and P   2. Imprint, UV expose

3. Remove template   3. Remove template

4. Pattern Transfer   4. Pattern Transfer

Fig. 55. Diagram of the process for nanoimprint lithography (NIL) and for step and flash imprint lithography (SFIL).

a significant role in patterning contacts and vias for integrated circuits. The first full field exposure tool will be completed in late 2002. With this tool and later tools many of the key challenges can be addressed and the technology can be fully demonstrated.

23. Imprint Lithography

Imprint lithography, also called "step and squish" or step and flash, is a low cost contact printing (1X) method borrowed from the CD-ROM production industry [175-178]. A master mold is used as an imprint mask or template to mold a pattern into a polymeric coating on a wafer. Figure 55 shows three varieties of imprint lithography. In two varieties, temperature and pressure are used to compression mold the resist with the template. In the third variety, called step and flash, the template is filled with a low viscosity polymeric liquid. The liquid is cured in the mold by exposure to ultraviolet light. The template is transparent at the exposure wavelength. Early work has demonstrated 25nm features reproduced over areas up to 30mm by 30mm [176,209,210]. The benefits of this technology are readily apparent. It is low cost, requiring no optics. There is

some reuse of current resist technology and the method has very high potential throughput, especially if entire wafers can be imprinted at once. The drawbacks of this technique are the production of 1X master mold patterns, mold lifetime and stability, defectivity during the release of the mold from the resist [211], and the need for highly accurate mechanical alignment and contact.

24. Maskless lithography

Maskless lithography offers the promise of high throughput lithography without the need to fabricate a complex and expensive master mask. The pattern data is stored in a computer and transferred to the exposure apparatus as needed. To achieve high throughput, many writing elements are arrayed in parallel. The writing elements may be electron beam columns, ion beam columns, photon beam optics or atomic force microscope (AFM) tips. The various options are described above. With the exception of the electron-beam writing columns and arrayed AFM tips, the writing arrays have not been demonstrated for these techniques. The basic technology for all of the writing elements for these methods is still under development.

Besides building the arrays of writing elements, the data handling for high throughput will be a significant challenge. Data transfer rates of 1 to 10 Terabits/second are required to achieve 40 to 60 wafer per hour throughput. Present fiber optic data transfer methods might offer up to 10 Gb/s transfer rates. Clearly, many transfer channels must be implemented in parallel. To help reduce the data transfer burden, Dai and Zakhor [186] have been developing methods to compress data for this transfer. Compression ratios of up to 800 have been obtained, but more compression is required. Furthermore, not all layout data can be compressed to the same degree with the same compression algorithm. In summary, some of the key challenges for implementing maskless lithography, besides developing the basic technology for the writing elements, include:

– CD control: Matching of dose and temporal stability of individual writers affects CD control. Redundancy is probably required in the writing strategy.

- Zero defects: Faulty elements in the writing array create defects. The mean time to failure (MTTF) for any writing element and mean time to re-calibrate the array have a strong impact upon cost.
- Throughput: Data handling for high throughput requires 2 to 10 Terabits/second data rate. Bus technology might limit the data rate. The physical size of the array of writing elements and the complexity of the electronics system increase with increasing throughput.

Despite these challenges, maskless lithography is being actively developed by many groups around the world since fabricating masks for lithography might limit its future extension for cost effective manufacturing. Another reason for developing maskless lithography is for rapid prototyping or for circuit customization. For rapid prototyping, maskless lithography can be used to develop various process modules before a lithography method with higher throughput that uses masks is available. This technique has been applied in research [184]. One issue with this approach is that the resist process and resulting resist profiles might be different with the maskless lithography than with the production lithography using masks. The etch process in particular is sensitive to the etch selectivity of the resist, the resist profile and the layers used underneath the resist. For maskless prototyping to significantly speed development, the resist process, ARC layers and resist profiles must closely match the results obtained with the production lithography. Furthermore, introduction of the production lithography that uses masks requires development of the reticle enhancement technology and resist process for that lithography. Therefore, maskless prototyping might not actually save much development time or cost.

25. Conclusion

Optical lithography has been successful at continually improving the functionality and cost of integrated circuits. This success has extended for 30 years over many circuit and wavelength generations. Optical lithography improvements have played a major part in the development of the current information age. The end is now in sight for optical lithography due to some hard physical limits. We have discussed how optical lithography works, how it can be extended, what it costs, how far it may go, and what the main replacement candidates are. Within five years optical lithography will likely be replaced as the leading edge patterning process, at least in process development. However, the tremendous technical and economic advantages of optical lithography will continue to keep this technology in widespread use for many years to come.

Acknowledgements

The authors would like to gratefully acknowledge the contributions to this work of figures from Dr. Martin McCallum of Nikon Precision Europe, Dr. Steward Robertson of Shipley Corp., and Dr. Ajay Singhal of Motorola as well as generous contributions in editing and review by many members of Motorola's Advanced Product Development and Research Lab lithography team. A significant portion of the text and figures in this work were previously published in the Wiley Encyclopedia of Electrical and Electronic Engineering (ISBN 0471-13946-7). This material is used by the kind permission of John Wiley & Sons, Inc.

References

[1] G.E. Moore, "Microprocessors and integrated electronic technology", *Proc. IEEE*, **64** (6): 837, 1976.

[2] "Lithography", *The International Technology Roadmap for Semiconductors*, San Jose, CA, Semiconductor Industry Association, 2001. http://public.itrs.net/.

[3] J. H. Bruning, "Optical imaging for microfabrication," Journal of Vacuum Science and Technology 17, 1147-55 (1980).

[4] R. K. Watts, "Optical lithography," in VLSI Electronics: Microstructure Science, Vol. 16, edited

by N. G. Einspruch and R. K. Watts, Academic Press, Inc., New York, (1987).

[5] Burn J. Lin, "Computer simulation study of images in contact and near-contact printing," Polymer Engineering and Science 14, 498-508 (1974).

[6] Burn J. Lin, "Optical methods for fine line lithography," in Fine Line Lithography, edited by R. Newman, 108-41, North Holland, New York, (1980).

[7] Janus S. Wilczynski, "Optical lithographic tools: Current status and future potential," Journal of vacuum Science and Technology B 5, 288-92 (1987).

[8] Stanley Myers, "ITRS: How you can participate," Semiconductor Magazine, 16, (October 2002).

[9] Paolo Gargini, "Worldwide technologies and the ITRS in the current economic climate," Emerging Lithographic Technologies VI, SPIE Proceedings volume 4688, edited by Roxann Engelstad, 25-28, 2002.

[10] Harry J. Levinson, Principles of Lithography, SPIE press Vol. PM97, Chapter 7, July 2001.

[11] L.F. Thompson, "Resist Processing", Introduction to Microlithography: Second Edition, Washington, D.C.: ACS Professional Reference Book, 1994. M. Bowden, L. Thompson, C. Willson, editors.

[12] H.J. Levinson, W.H. Arnold, "Optical Lithography", Handbook of Microlithography, Micromachining, and Microfabrication, Volume 1: Microlithography, p11-138, Bellingham, WA: SPIE Optical Engineering Press, 1997. P. Rai-Choudhury, editor.

[13] R.N. Singh, A.E. Rosenbluth, G.L. Chiu, J.S. Wilczynski, "High-numerical-aperture optical designs", IBM Journal of Research and Development, 41(1/2), 39, 1997.

[14] C.G. Willson, "Organic Resist Materials", Introduction to Microlithography: Second Edition, Washington, D.C.: ACS Professional Reference Book, 1994. M. Bowden, L. Thompson, C. Willson, editors.

[15] R.R. Dammel, Diazonapthoquinone-based Resists, Bellingham, WA: SPIE Optical Engineering Press, 1993.

[16] J.M. Shaw, M. Hatzakis, Performance characteristics of diazo-type photoresists under e-beam and optical exposure, IEEE Transactions on Electron Devices, ED-25 (4), 425-430, 1978.

[17] K.J. Orvek, M.L. Dennis, "Deep UV and thermal hardening of novalak resists", Proc. SPIE, 771, 281-8, 1987.

[18] L.J. Lauchlan, D. Nyyssonen, N. Sullivan, "Metrology Methods in Photolithography", Handbook of Microlithography, Micromachining, and Microfabrication, Volume 1: Microlithography, P475-596, Bellingham, WA: SPIE Optical Engineering Press, 1997. P. Rai-Choudhury, editor.

[19] S.M. Sze, VLSI Technology, New York: McGraw-Hill Book Company, 1988.

[20] N. Weste, K. Eshraghian, Principles of CMOS VLSI design, Reading, MA: Addison-Wesley, 1988.

[21] W. Maly, Atlas of IC Technologies, Menlo Park, CA: Benjamin/Cummings Publishing Co., 1987.

[22] C. Mead, L. Conway, Introduction to VLSI systems, Reading, MA: Addison-Wesley, 1980.

[23] K. Jeppson, S. Christensson, N. Hedenstierna, "Formal definitions of edge-based geometric design rules", IEEE Transactions on Computer Aided Design, 12 (1), 59, 1993.

[24] R. Razdan, A. Strojwas, "A statistical design rule developer", IEEE Transactions on Computer Aided Design, 5 (4), 508, 1986.

[25] Katherine Derbyshire, "Design and manufacturing," Semiconductor Magazine, 30-35, August 2002.

[26] T.R. Farrel, R. Nunes, D.J. Samuels, A. Thomas, R.A. Ferguson, A. Molless, A.K. Wong, W. Conley, D.C. Wheeler, S. Credendino, M. Naeem, P. Hoh, "Challenge of 1-Gb DRAM development when using optical lithography", Proc. SPIE, 3051, 333, 1997.

[27] M. McCord, M. Rooks, "Electron Beam Lithography", Handbook of Microlithography, Micromachining, and Microfabrication, Volume 1: Microlithography, p139-250, Bellingham, WA: SPIE Optical Engineering Press, 1997. P. Rai-Choudhury, editor.

[28] Steffen Schulze, "A new definition of 'fracturing'", BACUS News, volume 18 (5), October 2002.

[29] F. Abboud, C. Sauer, W. Wang, M. Vernon, R. Prior, H. Pearce-Percy, D. Cole, M. Mankos, "Advanced electron-beam pattern generation technology for 180nm masks", Proc. SPIE, 3236, 19, 1997.

[30] "C.A.T.S. version 12 release notes," Los Gatos, CA, Transcription Enterprises, 1996.

[31] J.G. Skinner, T.R. Groves, A. Novembre, H. Pfeiffer, R. Singh, "Photomask Fabrication Procedures and Limitations", Handbook of Microlithography, Micromachining, and Microfabrication, Volume 1: Microlithography, p377-474, Bellingham, WA: SPIE Optical Engineering Press, 1997. P. Rai-Choudhury, editor.

[32] P. Buck, "Understanding CD error sources in optical mask processing", Proc. SPIE, 1809, 62, 1992.

[33] M.D. Cerio, "Methods of error source identification and process optimization for photomask fabrication", Proc. SPIE, 2512, 88-98,1995.

[34] R. Dean, C. Sauer, "Further work in optimizing PBS", Proc. SPIE, 2621, 386, 1995.

[35] J. Potzick, "Re-evaluation of the accuracy of NIST photomask linewidth standards", *Proc. SPIE*, **2439**, 232-42, 1995.

[36] E. Abbe, "Contributions to the theory of microscope and microscopic perception," *Archiv. Fur Mikroskopische Anatomie* 9, 413-68, 1873.

[37] Joseph Goodman, *Introduction to Fourier Optics*, McGraw-Hill, New York, Chapters 3 and 4, 1968.

[38] Eugene Hecht, *Optics*, 2nd edition, Addison-Wesley, Reading, MA, Chapter 10, 1987.

[39] Lord Rayleigh, "On the theory of optical instruments, with special reference to the microscope,," *Philosophy Magazine* 42, 167-95, 1896.

[40] K. Yamanaka, H. Iwasaki, H. Nozue, K. Kasama, "NA and σ optimization for high-NA I-line lithography," *Proc. SPIE*, **1927**, 320-31, 1993.

[41] C.A. Mack, *Inside PROLITH: A Comprehensive Guide to Optical Lithography Simulation*, Austin, TX, FINLE Technologies, Inc., 1997.

[42] B. Lin, "The optimum numerical aperture for optical projection microlithography", *Proc. SPIE*, **1463**, 42-53, 1991.

[43] R. E. Tibbets and J. S. Wilczynski, "Design and fabrication of microelectronic lenses," *International Lens Design Conference*, SPIE Proceedings vol. 237, 321-8, 1980.

[44] J. Braat, "Quality of microlithographic projection lenses," Optical Microlithographic Technology for Integrated Circuit Fabrication and Inspection, SPIE Proceedings, vol. 811, 22-30, 1987.

[45] D. M. Williamson, "Compensator selection in the tolerancing of a microlithographic lens," *Recent Trends in Optical System Design II*, SPIE Proceedings, vol. 1049, 178-86, 1989.

[46] E. Glatzel, "New lens for microlithography," *International Lens Design Conference*, SPIE Proceedings, vol. 237, 310-20, 1980.

[47] A. R. Philips and M. J. Buzawa, "High resolution lens system for submicron photolithography," *International Lens Design Conference*, SPIE Proceedings, vol. 237, 329-36, 1980.

[48] H. L. Stover, "Chairman's Pre-Conference overview: Lens specifications and distortions," *Optical Microlithography VI*, SPIE Proceedings, vol. 772, 2-4, 1987.

[49] A. Offner, "New concepts in projection mask aligners," *Optical Engineering*, vol. 14, no. 2, 130-2, 1975.

[50] Fritz Zernike, "Specification and metrology of surface figure and finish for Microlithography optics," *OSA Trends in Optics and Photonics on Extreme Ultraviolet Lithography*, vol. 4, edited by Glenn Kubiak and Don Kania, 94-7, 1996.

[51] H. N. Chapman and D. W. Sweeney, "A rigorous method for compensation selection and alignment of microlithographic optical systems," Proc. SPIE 3331, 102-113 (1998).

[52] P. Yan, J. Langston, "Mask CD Control Requirement at 0.18 ?m Design Rules for 193 nm Lithography," Proc. SPIE, 3051, 164-9, 1997.

[53] R.W. McCleary, P.J. Tompkins, M.D. Dunn, K.F. Walsh, J.F. Conway, R.P. Mueller, "Performance of a KrF excimer laser stepper," Proc. SPIE, 922, 396-9, 1988.

[54] Burn Lin, "The k3 coefficient in nonparaxial ?/NA scaling equations for resolution, depth of focus, and immersion lithography," Journal of Microlithography, Microfabrication and Microsystems 1(1), 7-12, April 2002.

[55] M. Brink, H. Jasper, S. Slonaker, P. Wijnhoven, F. Klaassen, "Step-and-Scan and Step-and-Repeat, A Technology Comparison," Proc. SPIE, 2726, 734-53, 1996.

[56] R. Schenker, F. Piao, W.G. Oldham, "Durability of Experimental Fused Silicas to 193-nm-Induced Compaction," *Proc. SPIE*, **3051**, 44-53, 1997.

[57] M. Rothschild, D.J. Ehrlich, D.C. Shaver, "Effects of excimer laser irradiation on the transmission, index of refraction, and density of ultraviolet grade fused silicas," *Appl. Phys. Lett.*, **55**(13), 1276-78, 1989.

[58] R. Schenker, F. Piao, W.G. Oldham, "Material Limitations to 193-nm Lithographic System Lifetimes," *Proc. SPIE*, **2726**, 698-706, 1996.

[59] M. Rothschild, "Photolithography at Wavelengths Below 200nm," *Proc. SPIE*, **3278**, 222-228, 1998.

[60] S. Hirukawa, K. Matsumoto, K.Takemasa, "New projection optical system for beyond 150nm patterning with KrF and ArF sources", *Proc. SPIE*, **3334**, 414-422, 1998.

[61] R. Schenker, F. Piao, W.G. Oldham, "Material Limitations to 193-nm Lithographic System Lifetimes," *Proc. SPIE*, **2726**, 698-706, 1996.

[62] K. Brown, "SEMATECH and the national technology roadmap: needs and challenges", *Proc. SPIE*, **2440**, 33, 1995.

[63] Richard S. Muller and Theodore I. Kamins, *Device Electronics for Integrated Circuits*, John Wiley and Sons, New York, Chapters 9 and 10, 1986.

[64] Chenming Hu, "Ultra-large-scale integration device scaling and reliability," *Journal of Vacuum Science and Technology B* **12(6)**, 3237-41, Nov/Dec 1994.

[65] B. T. Murphy, "Cost-size optima of monolithic integrated circuits," *Proceedings of the IEEE*, December 1964, 1537-43.

[66] H. Stapper, "Fact and fiction in yield modeling," *Microelectronics Journal*, vol. 20, Elsevier, 129-151 (1989).

[67] B.S. Stine, V. Mehrotra, D.S. Boning, J.E. Chung, D.J. Ciplickas, "A simulation methodology for assessing the impact of spatial/pattern dependent interconnect parameter variation on circuit performance", IEDM Technical Digest, 133, 1997.

[68] D.G. Chesebro, J.W. Adkisson, L.R. Clark, S.N. Eslinger, M.A. Faucher, S.J. Holmes, R.P. Mallette, E.J. Nowak, E.W. Sengle, S.H. Voldman, T.W. Weeks, "Overview of gate linewidth control in the manufacture of CMOS logic chips", IBM Journal of Research and Development, 39(1/2), 189, 1995.

[69] W. Maly, "Modeling of lithography related yield losses for CAD of VLSI circuits", IEEE Transactions on Computer-aided Design, CAD-4(3), 166, 1985.

[70] Li Chen, Linda S. Milor, Charles H. Ouyang, Wojciech Maly and Yeng-Kaung Peng, "Analysis of the impact of proximity correction algorithms on circuit performance," *IEEE Transactions on Semiconductors Manufacturing*, volume 12(3), 313-22, August 1999.

[71] Michael Orshansky, Linda Milor, Pinhong Chen, Kurt Keutzer and Chenming Hu, "Impact of systematic spatial intra-chip gate length variability on performance of high-speed digital circuits," *IEEE Transactions on Computer-Aided Design of Integrated Circuits and Systems* vol.21, no.5, 544-53, May 2002.

[72] Gary Escher, "The importance of mask technical specifications on the lithography error budget," *Photomask Technology and Management*, SPIE vol. 2322, 409-20, 1994.

[73] Govil, P.K.; Tsacoyeanes, J.; Eron, R.; Walters, D., "Contributors to focal plane nonuniformity and their impact on linewidth control in DUV step and scan system," *Proceedings of the SPIE 3334* p.92-103 1998.

[74] Chris J. Progler, "Simulation-enabled decision making in advanced lithographic manufacturing," *Lithography for Semiconductor Manufacturing II, Proceedings of the SPIE 4404*, 68-79, 2001.

[75] M.D. Levenson, D. S. Goodman, S. Lindsey, P. W. Bayer, H.A. Santini, "The phase shifting mask II: imaging simulations and submicrometer resist exposure", *IEEE Transactions on Electron Devices*, **ED-31**, 753, 1984.

[76] Y. Ham, Y. Kim, I. Hur, K. Park, H. Kim, D. Ahn, S. Choi, "Fundamental analysis on fabrication of 256MB DRAM using the phase shift mask technology", *Proc. SPIE*, **2197**, 243, 1994.

[77] G. Wojcik, J. Mould Jr., R. Ferguson, R. Martino, K.K. Low, "Some image modeling issues for I-line, 5X phase shifting masks," Proc. SPIE, 2197, 455-65, 1994.

[78] Y.T. Wang, Y.C. Pati, J.W. Liang, T. Kailath, "Systematic design of phase-shifting masks," Proc. SPIE, 2197, 377-87, 1994.

[79] Lars Liebmann, Jennifer Lund, Fook-Luen Heng and Ioana Graur, "Enabling alternating phase shifted mask designs for a full logic level," *Journal of Microlithography, Microfabrication and Microsystems* 1(1), 31-42, April 2002.

[80] R. Schmidt, C. Spence, Z. Krivokapic, B. Geh, D. Flagello, "Impact of Coma on CD control for Multiphase PSM Design," *Proc. SPIE*, **3334**, 15-24, 1998.

[81] Richard Schenker, "Comparison of single and dual exposure phase shift mask approaches for poly gate patterning," BACUS Symposium on Photomask Technology and Management, SPIE volume 3546, 242-52, September 1998.

[82] H.Y. Liu, Y.T. Wang, L. Kirkland, Y.C. Pati, "Application of alternating phase-shifting masks to 140-nm gate patterning: II. Mask design and manufacturing tolerances," Proc. SPIE, 3334, 2-14, 1998.

[83] B.J. Lin, "The attenuated phase-shifting mask", Solid State Technology, 43-47, January 1992.

[84] Yao-Ching Ku, Erik H. Anderson, Mark L. Schattenburg and Henry I. Smith, "Use of a pi-phase shifting x-ray mask to increase the intensity slope at feature edges," Journal of Vacuum Science and Technology B 6(1), 150-3, 1988.

[85] B. Smith, S. Turgut, "Phase-shift mask issues for 193 nm lithography," *Proc. SPIE*, **2197**, 201-10, 1994.

[86] T. Chijimatsu, T. Higashi, Y. Tabata, N. Ishiwata, S. Asai, I. Hanyu, "Implementation of Attenuated PSMs in DRAM Production," *Proc. SPIE*, **2726**, 461-72, 1996.

[87] D. Van Den Broeke, J. Fung, T. Laidig, S. Hsu, K. Wampler, R. Socha, J. Peterson, "Complex 2D pattern lithography at lambda/4 resolution using chromeless phase lithography (CPL)," *Proc. SPIE*, **4691**, 196-214, 2002

[88] K. Douki, T. Kajita, S. Iwanaga, "A study for the design of I-line photoresist capable of sub-quarter micron lithography: The effects of end group control of novel phenolic resins," *Proc. SPIE*, **3333**, 384-392, 1998.

[89] S.V. Postnikov, M.D. Stewart, H.V. Tran, M. Nierode, D.R. Mederios, T. Cao, J. Byers, S.E. Webber, C.G. Willson, "Study of resolution limits due to intrinsic bias in chemically amplified photoresists", Journal of Vaccum Science and Technology B17(6), 1999

[90] M.D. Stewart, S.V. Postnikov, H.V. Tran, D.R. Medeiros, M.A. Nierode, T.Cao, J. Byers, S.E. Webber, C.G. Willson, "Measurement of acid diffusivity in thin polymer films above and below Tg", ACS Polymeric Material Science and Engineering Vol.81, 1999

[91] M.D. Stewart, M.H. Somervell, H.V. Tran, S.V. Postnikov, C.G. Willson. "Study of acid transport using IR spectroscopy and SEM", *Proc. SPIE, 3999*, 2000, p.665-674

[92] M.D. Stewart, G.M. Schmid, S.V. Postnikov, C.G. Willson, "Mechanistic understanding of line end shortening", *Proc. SPIE, 4345*, 2001, p.10-18

[93] Q. Lin, A. Katnani, T. Brunner, C. DeWan, C. Fairchok, D.L. Tulipe, J. Simons, K. Petrillo, K. Babich, D. Seeger, M. Angelopoulos, R. Sooriyakumaran, G. Walraff, D. Hofer, "Extension of 248 nm Optical Lithography: a Thin Film Imaging Approach", *Proc. SPIE*, **3333**, 278-288, 1998.

[94] S.V. Postnikov, M.H. Somervell, J. Byers, A. Qin, C.L. Henderson, S. Katz, C.G. Willson, "Top Surface Imaging Through Silylation", *Proc. SPIE*, **3333**, 997-1008, 1998.

[95] R.D. Allen, W.E. Conley, R.R. Kunz, "Deep-UV Resist Technology: The Evolution of Matrials and Processes for 250nm Lithography and Beyond," *Handbook of Microlithography, Micromachining, and Microfabrication, Volume 1: Microlithography*, p321-376, Bellingham, WA: SPIE Optical Engineering Press, 1997. P. Rai-Choudhury, editor.

[96] K. Patterson, U. Okoroanyanwu, T. Shimokawa, J.D. Byers, S. Cho, C.G. Willson, "Improving the performance of 193nm photoresists based on alicyclic polymers," *Proc. SPIE*, **3333**, 425-437, 1998.

[97] R.D. Allen et. al., "Design of an etch-resistant cyclic olefin photoresist," Proc. SPIE, **3333**, 463-471, 1998.

[98] T. Steinhausler, D. White, A.J. Blakeney, D.R. Stark, G.K. Rich, K.R. Dean, "Optimization of etch conditions for a silicon-containing methacrylate-based bilayer resist for 193nm lithography," *Proc. SPIE*, **3333**, 122-131, 1998.

[99] R.R. Dammel, S. Fincer, J.E. Oberlander, D.N. Khanna, D.L. Durham, "Lithographic performance of an etch-stable methacrylate resist at 193nm," *Proc. SPIE*, **3333**, 144-151, 1998.

[100] B. Schwartz, H. Robbins, "Chemical Etching of Silicon: IV. Etching Technology," *J. Electrochem. Soc.*, **123**, 1903, 1976.

[101] S. Pang, "Applications of Dry Etching to Microsensors, Field Emitters, and Optical Devices," *Handbook of Microlithography, Micromachining, and Microfabrication, Volume 2: Micromachining and Microfabrication*, p99-152, Bellingham, WA: SPIE Optical Engineering Press, 1997. P. Rai-Choudhury, editor.

[102] Petre C. Logofatu and John R. McNeil, "Sensitivity analysis of fitting for scatterometry," Metrology, Inspection, and Process Control for Microlithography XIII, SPIE volume 3677, 177-83, 1999.

[103] W. Howard, V. Wiaux, M. Ercken, B. Bui, J. Byers, M. Pochkowski,, "Tuning and simulating a 193-nm resist for 2D applications," Proc. SPIE, 4691, 1190-98, 2002.

[104] L. Reimer, Scanning Electron Microscopy, New York, NY: Springer Verlag, 1995.

[105] David Joy, "The future of the CD-SEM: A possible agenda," Microlithography World, www.pennwell.com, 4-6, August 2002.

[106] M.W. Cresswell, J.J. Sniegowski, R.N. Ghoshtagore, R.A. Allen, L.W. Linholm, J.S. Villarrubia, "Electrical Test Structures replicated in silicon-on-insulator material," *Proc. SPIE*, **2725**, 659-76, 1996.

[107] J.L. Sturtevant, J. Allgair, M.W. Barrick, C. Fu, K.G. Green, R.R. Hershey, L.C. Litt, J.G. Maltabes, C. Nelson, B.J. Roman, J. Singelyn, "Full-field CD control for sub-0.20 μm patterning," *Proc. SPIE*, **3051**, 137-45, 1997.

[108] W. L. Stevenson, "A new reticle set for electrical measurement of resolution, proximity, topography, sidewall spacer, and stacked-gate structures," *Integrated Circuit Metrology, Inspection, and Process Control II*, SPIE volume 921, 152-63, 1988.

[109] P. Boher, J.L. Stehle, J.P. Piel, C. Defranoux, L. Hennet, "Precise Measurement of ARC optical indices in the deep-UV range by variable-angle spectroscopic ellipsometry," *Proc. SPIE*, **3050**, 205-14, 1997.

[110] J.N. Hilfiker, R.A. Synowicki, "Employing Spectroscopic Ellipsometry for Lithography Applications," *Semiconductor Fabtech*, 5, October, 1996.

[111] R.A. Synowicki, J.N. Hilfiker, R.R. Dammel, C.L. Henderson, "Refractive Index Measurements of Photoresist and Antireflective Coatings with Variable Angle Spectroscopic Ellipsometry," Proc. SPIE, **3332**, 384-390, 1998.

[112] C.J. Progler, "Optical lens specifications from the user's perspective," *Proc. SPIE*, **3334**, 256-268, 1998.

[113] John D. Armitage, Jr. and Joseph P. Kirk, "Analysis of overlay distortion patterns," *Integrated Circuit Metrology, Inspection and Process Control II*, Proceedings of SPIE volume 921, 1988, 207-22.

[114] M. Perkins, J. Stamp, "Intermix Technology: the key to optimal stepper productivity and cost efficiency", *Proc. SPIE*, **1674**, 559, 1992.

[115] M. A. van den Brink, C. G. M. de Mol and J. M. D. Stoeldraijer, "Matching of multiple wafer steppers for 0.35 um lithography using advanced optimization schemes," *Integrated Circuit Metrology, Inspection and Process Control VII*, Proceedings of SPIE vol. 1926, 1993, 188-207.

[116] C.M. Yuan, A.J. Strojwas, "Modeling of optical alignment and metrology schemes used in integrated circuit manufacturing," *Proc. SPIE*, **1264**, 209, 1990.

[117] K. Lucas, C. Yuan, A. Strojwas, "A rigorous and practical vector model for phase shifting masks in optical lithography", *Proc. SPIE*, **1674**, 253, 1992.

[118] J. Sturtevant, B. Roman, "Antireflection strategies for advanced photolithography," *Microlithography World*, **Autumn**, 13, 1995.

[119] K.D. Lucas, J.A. Vasquez, A. Jain, S.M. Filipiak, T. Vuong, C.F. King, B.J. Roman, "Plasma Anti-reflective Coating Optimization Using Enhanced Reflectivity Modeling," *Proc. SPIE*, **3050**, 194-204, 1997.

[120] K.D. Lucas, M. McCallum, M.E. Kling, J.G. Maltabes, "Manufacturability of sub-wavelength features using reticle and substrate enhancements," *Proc. SPIE*, **3332**, 391-402, 1998.

[121] M. McCallum, K.D. Lucas, "Sub-wavelength contact and trench characterization using lithography simulation," *Future Fab International*, **1** (3), 1997.

[122] A.R. Neureuther, C.A. Mack, "Optical Lithography Modeling," *Handbook of Microlithography, Micromachining, and Microfabrication, Volume 1: Microlithography*, p597-680, Bellingham, WA: SPIE Optical Engineering Press, 1997. P. Rai-Choudhury, editor.

[123] C.L. Henderson, S.N. Pancholi, S.A. Chowdhury, C.G. Willson, R.R. Dammel, "Photoresist Characterization for Lithography Simulation Part 2: Exposure Parameter Measurements," *Proc. SPIE*, **3049**, 816-28, 1997.

[124] V.V. Ivin, D.Y. Larin, K.D. Lucas, T.M. Makhviladze, A.A. Rogov, S.V. Verzunov, "Fast modeling of 3D planar resist images for high NA projection lithography", *Proc. SPIE* ,*3051*, 1997, p.567-577

[125] E.W. Charrier, C.A. Mack, C.J. Progler, "Comparison of simulated and experimental CD-limited yield for submicron i-line process", *Solid State Technology*, **38** (11), 105-112, November 1995.

[126] A. Erdmann, M. Arnz, J. Baselmans, M. Maenhoudt, "Lithographic process simulation for scanners", *Proc. SPIE*, **3334**, 164-175, 1998.

[127] J.P. Stirniman, M.L. Rieger, "Spatial-filter models to describe IC lithographic behavior," *Proc. SPIE*, **3051**, 469-78, 1997.

[128] J. Rey, "Terrain: deposition and etch simulation", *TMATimes*, **VIII** (4), 6, 1996.

[129] S.V. Postnikov, K. Lucas, K. Wimmer, B. Roman, "Re-evaluating simple lambda based design rules for low K1 lithography process control", *Proc. SPIE*, *3998*, 2000, p.901-912

[130] S.V. Postnikov, K. Lucas, K. Wimmer, "Impact of optimized illumination upon simple lambda based design rules for low K1 lithography", *Proc. SPIE*, *4344*, p.797-808

[131] N. Cobb, A. Zakhor, E. Miloslavsky, "Mathematical and CAD framework for proximity correction", *Proc. SPIE*, *2726*, 1996, p.208-222.

[132] Y. Granik, "Correction for etch proximity: new models and applications", *Proc. SPIE*, *4346*, 2001, p.98-112

[133] A. R. Neureuther, "Understanding lithography technology issues through simulation," *Univ. Calif. Berkeley Electronics Research Lab Memo.*, **UCB/ERL 93-40**, 1993

[134] H. Chuang, P. Gilbert, W. Grobman, M. Kling, K. Lucas, A. Reich, B. Roman, E. Travis, P. Tsui, T. Vuong, J. West, "Practical applications of 2-D optical proximity corrections for enhanced performance of $0.25\mu m$ random logic devices," *IEDM Technical Digest*, 483, 1997.

[135] M.L. Rieger, J.P. Stirniman, "Using behavior modeling for proximity correction," *Proc. SPIE*, **2197**, 371-6, 1994.

[136] R.C. Henderson, O.W. Otto, "Correcting for proximity effect widens process latitude," Proc. SPIE, 2197, 361-70, 1994.

[137] V. Axelrad, N. Cobb, M. O'Brien, V. Boksha, T. Do, T. Donnelly, Y. Granik, "Efficient full-chip yield analysis methodology for OPC corrected VLSI designs", Proceedings of the IEEE First International Symposium on Quality Electronic Design , 2000, p. 461-466.

[138] L.W. Liebman, J.A. Bruce, W.Chu, M.Cross, I.C. Graur, J.J. Krueger, W.C. Leipold, S.M. Mansfield, A. E. McGuire, D.L. Sundling, "Optimizing style options for sub-resolution assist features", *Proc. SPIE*, *4346*, 2001, p.141-152

[139] J. Torres, F. Schellenberg, O. Toublan, "Model-assisted double dipole decomposition", Proc. SPIE, 4691, 407-417, 2002.

[140] M. Kling, K. Lucas, A. Reich, B. Roman, H. Chuang, P. Gilbert, W. Grobman, E. Travis, P. Tsui, T. Vuong, J. West, "$0.25\mu m$ logic manufacturing using

proximity correction," Proc. SPIE, 3334, 204-214, 1998.

[141] F. M. Schellenberg, "Adoption cost and hierarchy efficiency for 100 nm and beyond," *BACUS News*, volume 18(3), August 2002.

[142] Alfred Wong, *Resolution Enhancement Techniques*, SPIE Press, Bellingham, WA, (2001).

[143] H. Eisenmann, T. Waas, H. Hartmann, "PROXECCO-proximity effect correction by convolution", Journal of Vacuum Science Technology, B11 (6), 2741, 1993.

[144] Tom Newman, Jan Chabala, B. J. Marleau, Frederick Raymond II, Olivier Toublan, Mark Gesley and Frank Abboud, "Evaluation of OPC mask printing with a raster scan pattern generator," *BACUS News*, volume 18(6), June 2002.

[145] T. Fahey, J. McClay, M. Hansen, B. Tirri, M. Lipson, "SVG 157nm Lithography Technical Review," Proc. SPIE, 4346, 72-80, 2001.

[146] R. Hung, et al., "Resist materials for 157-nm microlithography: an update," Proc. SPIE, 4345, 385-95, 2001.

[147] T.L. Perkinson, L.C. Litt, J.G. Maltabes, S. Murphy, R.R. Hershey, "Who needs I-line," Future Fab International, 1 (3), 179, 1997.

[148] K. Early and W. H. Arnold, "Cost of ownership for X-ray proximity lithography," Electron-Beam, X-Ray, and Ion-Beam Submicrometer Lithographies for Manufacturing IV, Proceedings of SPIE vol. 2194, 22-33, 1994.

[149] Walt J. Trybula, Kurt R. Kimmel and Brian J. Grenon, "Financial impact of technology acceleration on semiconductor masks," *BACUS News*, volume 18(4), April 2002.

[150] Scott D. Hector, Patrick Kearney, Claude Montcalm, James Folta, Chris Walton, William Tong, John Taylor, Pei-Yang Yan and Chuck Gwyn, "Predictive model of the cost of extreme ultraviolet lithography masks," *20th Annual SPIE Symposium on Photomask Technology and Management, SPIE Proceedings* volume 4186, 2000, 733-48.

[151] W. Trybula and D. Dance, "Cost of mask fabrication," Proceedings of SPIE, Vol. 3048, 211-215 (1997). Also, private communication with W. Trybula of International SEMATECH.

[152] C. Gwyn, R. Stulen, D. Sweeney, & D. Attwood, "Extreme ultraviolet lithography," *J. Vac. Sci. Tech. B*, **16**, pp.3142-49, (1998).

[153] Charles W. Gwyn et al., "Extreme Ultraviolet Lithography," unpublished white paper prepared by the EUV LLC, 1999.

[154] Charles W. Gwyn, "EUV lithography update," *SPIE's OE Magazine*, June 2002, 22-24.

[155] L. R. Harriott, *Journal of Vacuum Science and Technology B*, vol. 15, no. 6, Nov/Dec 1997, 2130-5.

[156] Hans C. Pfeiffer, R. S. Dhaliwal, S. D. Golladay, S. K. Doran, M. S. Gordon, T. R. Groves, R. A. Kendall, J. E. Lieberman, P. F. Petric, D. J. Pinckney, R. J. Quickle, C. F. Robinson, J. D. Rockrohr, J. J. Senesi, W. Stickel, E. V. Tressler, A. Tanimoto, T. Yamaguchi, K. Okamoto, K. Suzuki, T. Okino, S. Kawata, K. Morita, S. C. Suzuki, H. Shimizu, S. Kojima, G. Varnell, W. T. Novak, D. P. Stumbo and M. Sogard, "Projection reduction exposure with variable axis immersion lenses: Next generation lithography," *Journal of Vacuum Science and Technology B*, 17(6), 1999, 2840-6.

[157] T. Miura, T. Sato, M. Miyazaki, K. Hada, Y. Sato and M. Tokunaga, "Nikon EPL tool development summary," *Emerging Lithographic Technologies VI, SPIE Proceedings* volume 4688, edited by Roxann Engelstad, 527-34, 2002.

[158] Hans C. Pfeiffer et al., "PREVAIL: latest electron optics results," *Emerging Lithographic Technologies VI, SPIE Proceedings* volume 4688, edited by Roxann Engelstad, 535-46, 2002.

[159] Rainer Kaismeier, Hans Loeschner, Gerhard Stengl, John C. Wolfe and Paul Ruchhoeft, "Ion projection lithography: International development program," *Journal of Vacuum Science & Technology B (Microelectronics and Nanometer Structures)* vol.17, no.6, 3091-7, Nov/Dec 1999.

[160] H. Loeschner et al., "Large-field ion optics for projection and proximity printing and for maskless lithography (ML2)," *Emerging Lithographic Technologies VI, SPIE Proceedings* volume 4688, edited by Roxann Engelstad, 595-606, 2002.

[161] L. P. Muray, J. P. Spallas, C. Stebler, K. Lee, M. Mankos, Y. Hsu, M. Gmur and T. H. P. Chang, "Advances in arrayed microcolumn lithography," *Journal of Vacuum Science & Technology B (Microelectronics and Nanometer Structures)* vol.18, no.6, 3099-3104, Nov/Dec 2000.

[162] Masato Muraki and Susumu Gotoh, "New concept for high-throughput multielectron beam direct write system," Journal of Vacuum Science & Technology B (Microelectronics and Nanometer Structures) vol.18, no.6, 3061-6, Nov/Dec 2000.

[163] Parker, N.W.; Brodie, A.D.; McCoy, J.H., "A high throughput NGL electron beam direct-write lithography system," *Emerging Lithographic Technologies IV*, Proceedings of the SPIE - The International Society for Optical Engineering vol.3997, 713-20, 2000.

[164] http://www.mapperlithography.com

[165] Ulric Ljungblad, Tor Sandstrom, Hans Buhre, Peter Durr and Hubert Lakner, "New architecture for laser pattern generators for 130 nm and beyond," 20th Annual BACUS Symposium on Photomask Technology, edited by Brian Grenon and Giang Dao, Proceeding of SPIE, volume 4186, 16-21, 2001.

[166] Shroff, Y.; Yijian Chen; Oldham, W." Fabrication of parallel-plate nanomirror arrays for extreme ultraviolet maskless lithography," Journal of Vacuum Science & Technology B (Microelectronics and Nanometer Structures) vol.19, no.6, 2412-15, Nov. 2001.

[167] Carter, D.J.D.; Gil, D.; Menon, R.; Mondol, M.K.; Smith, H.I.; Anderson, E.H , "Maskless, parallel patterning with zone-plate array lithography," .Journal of Vacuum Science & Technology B (Microelectronics and Nanometer Structures) vol.17, no.6 p.3449-52 Nov. 1999.

[168] Kathryn Wilder, Hyongsok T. Soh, Abdullah Atalar, and Calvin F. Quate, "Nanometer-scale patterning and individual current-controlled lithography using multiple scanning probes," Review of Scientific Instruments, volume 70, number 6, June 1999, 2822-27.

[169] K. L. Scott, "Maskless Ion Beam Lithography Using Microcolumn Arrays," M. S. Thesis, Univ. of California/Berkeley, 27-Mar-2001.

[170] M. L. Schattenberg, C. R. Canizares, D. Dewey, K. A. Flanagan, M. A. Hamnett, A. M. Levine, K. Lum, R. Manikkalingam and T. H. Markert, "Transmission grating spectroscopy and the advanced x-ray astrophysics facility," Optical Engineering, volume 30, number 10, 1590-99, 1991.

[171] Bruce D. MacLeod, Adam F. Kelsey, Mark A. Leclerc, Daniel P. Resler, Sergey Liberman and James P. Nole, "Fully automated interference lithography," Emerging Lithographic Technologies VI, SPIE Proceedings volume 4688, edited by Roxann Engelstad, 910-21, 2002.

[172] Scott Hector, "Status and future of x-ray lithography," Microelectronic Engineering, 41/42, 1998, 25-30.

[173] Henry I. Smith, Mark L. Schattenburg, Scott D. Hector, Euclid E. Moon, Isabel Y. Yang, Martin Burkhardt, "X-ray nanolithography: Extension to the limits of the lithographic process," Microelectronic Engineering, vol. 32, 1996, 143-58.

[174] Takao Utsumi, "Low energy electron-beam proximity projection lithography: Discovery of a missing link," Journal of Vacuum Science & Technology B (Microelectronics and Nanometer Structures) vol.17, no.6, 2897-2902, Nov/Dec 1999.

[175] Colburn, M.; Bailey, T.; Choi, B.J.; Ekerdt, J.G.; Sreenivasan, S.V.; Willson, C.G., "Development and advantages of step-and-flash lithography," Solid State Technology vol.44, no.7, 67-78 July 2001.

[176] S.Y. Chou, P.R. Krauss, P.J. Renstrom, "Imprint Lithography with 25-Nanometer Resolution," Science, 272, 85-7, 1996.

[177] J. Haisma, M. Verheijen, K. van den Heuvel, J. van den Berg, "Mold-assisted nanolithography: A process for reliable pattern replication," J. Vac. Sci. Technol. B, 14(6), 4124-8, 1996.

[178] S.Y Chou, P.R. Krauss, "Imprint Lithography with Sub-10nm Feature Size and High Throughput," Microelectronics Engineering, 35, 237-40, 1997.

[179] Aaron Hand, "X-ray lithography: Back for round 2," Semiconductor International, 34, August 2002.

[180] Hawryluk, A.M.; Seppala, L.G., "Soft X-ray projection lithography using an X-ray reduction camera," Journal of Vacuum Science & Technology B vol.6, no.6, 2162-6, Nov.-Dec. 1988.

[181] Kinoshita, H.; Kurihara, K.; Ishii, Y.; Torii, Y., "Soft X-ray reduction lithography using multilayer mirrors," Journal of Vacuum Science & Technology B, vol.7, no.6, 1648-51, Nov.-Dec. 1989.

[182] Berreman, D.W.; Bjorkholm, J.E.; Eichner, L.; Freeman, R.R.; Jewell, T.E.; Mansfield, W.M.; MacDowell, A.A.; O'Malley, M.L.; Raab, E.L.; Silfvast, W.T.; Szeto, L.H.; Tennant, D.M.; Waskiewicz, W.K.; White, D.L.; Windt, D.L.; Wood, O.R., II; Bruning, J.H. , "Soft X-ray projection lithography: printing of 0.2-um features using a 20:1 reduction," Optics Letters vol.15, no.10, 529-31, 15 May 1990.

[183] Sturans, M.A.; Hartley, J.G.; Pfeiffer, H.C.; Dhaliwal, R.S.; Groves, T.R.; Pavick, J.W.; Quickie, R.J.; Clement, C.S.; Dick, G.J.; Enichen, W.A.; Gordon, M.S.; Kendall, R.A.; Kostek, C.A.; Pinckney, D.J.; Robinson, C.F.; Rockrohr, J.D.; Safran, J.M.; Sen, "EL5: One tool for advanced x-ray and chrome on glass mask making," Journal of Vacuum Science & Technology B (Microelectronics and Nanometer Structures) vol.16, no.6, 3164-7, Nov.-Dec. 1998.

[184] L. Pain, M. Charpin, Y. Laplanche and D. Henry, "Shaped e-beam lithography integration work for advanced ASIC manufacturing: progress report," Emerging Lithographic Technologies VI, SPIE Proceedings volume 4688, edited by Roxann Engelstad, 607-618, 2002.

[185] N. Saitou and Y. Sakitani, "Cell projection electron beam lithography," Electron-Beam X-Ray, and Ion-Beam Submicrometer Lithographies for Manufacturing IV, Proc. SPIE volume 2194, 11-21, 1991.

[186] Vito Dai and Avideh Zakhor, "Lossless compression techniques for maskless lithography data," *Emerging Lithographic Technologies VI, SPIE Proceedings* volume 4688, edited by Roxann Engelstad, 583-594, 2002.

[187] Giang T. Dao, R. Scott Mackay, Phil K. Seidel, "NGL process and the role of International SEMATECH," *Emerging Lithographic Technologies VI, SPIE Proceedings* volume 4688, edited by Roxann Engelstad, 134-149, 2002.

[188] David Attwood, *Soft X-Rays and Extreme Ultraviolet Radiation: Principles and Applications*, Cambridge University Press, Chapter 4, 1999.

[189] Daniel A. Tichenor, Glenn D. Kubiak, William C. Replogle, Leonard E. Klebanoff, John B. Wronosky, Layton C. Hale, Henry N. Chapman, John S. Taylor, James A. Folta, Claude Montcalm, Russell M. Hudyma, Kenneth A. Goldberg and Patrick Naulleau, "EUV Engineering Test Stand," *Emerging Lithographic Technologies IV*, Proceedings of SPIE, vol. 3997, 48-69 (2000).

[190] Eberhard Spiller, *Soft X-Ray Optics*, SPIE, Bellingham, WA, Chapter 7, 1994.

[191] Troy W. Barbee, Jr., "Multilayers for x-ray optics," *Applications of Thin-Film Multilayered Structures to Figured X-Ray Optics*, SPIE vol. 563, 2-28, 1985.

[192] B. Henke, E. Gullikson, and J. Davis, "X-Ray interactions: photoabsorption, scattering, transmission, and reflection at E=50-30,000 eV, Z=1-92," *Atomic Data and Nuclear Data Tables*, 54, 181-342, 1993. See also http://www-cxro.lbl.gov/optical_contants/.

[193] V. Vinogradov and B. Ya. Zeldovich, "X-ray and far uv multilayer mirrors: Principles and possibilities," *Applied Optics*, vol 16, no. 1, 89-93, 1977.

[194] J. H. Underwood and T. W. Barbee, Jr., "Layered synthetic microstructures as Bragg diffractors for X rays and extreme ultraviolet: theory and predicted performance," *Applied Optics*, vol. 20, no. 17, 3027-34, 1981.

[195] Scott D. Hector, "EUVL Masks: Requirements and Potential Solutions," *Emerging Lithographic Technologies VI, SPIE Proceedings* volume 4688, edited by Roxann Engelstad, 134-149, 2002.

[196] Jonathan Cobb, S. Dakshina-Murthy, Colita Parker, Eric Luckowski, Arturo Martinez, Richard D. Peters, Wei Wu, and Scott D. Hector, "Integration of UTR processes into MPU IC manufacturing flows," *Advances in Resist Technology and Processing XIX*, Theodore H. Fedynyshyn, Editor, *Proceedings of SPIE* volume 4690, 277-286, 2002.

[197] J. Cobb, S. Dakshina-Murthy, C. Parker, E. Luckowski, A. Martinez, R. Peters, W. Wu. And S. Hector, "Integration of UTR processes into MPU IC manufacturing flows," *Proceedings of the SPIE* vol. 4690, 277, 2002.

[198] P. P. Naulleau et al., "Static microfield printing at the Advanced Light Source with the ETS Set-2 optic," *Emerging Lithographic Technologies VI, SPIE Proceedings* volume 4688, edited by Roxann Engelstad, 64-71, 2002.

[199] J.A. Liddle, C.A. Volkert, "Mechanical stability of thin-membrane masks", *J. Vac. Sci. Technol.*, **B12**, 3528, 1994.

[200] Liddle, J.A.; Blakey, M.I.; Bolan, K.; Farrow, R.C.; Gallatin, G.M.; Kasica, R.; Katsap, V.; Knurek, C.S.; Li, J.; Mkrtchyan, M.; Novembre, A.E.; Ocola, L.; Orphanos, P.A.; Peabody, M.L.; Stanton, S.T.; Teffeau, K.; Waskiewicz, W.K.; Munro, E, "Space-charge effects in projection electron-beam lithography: Results from the SCALPEL proof-of-lithography system," Journal of Vacuum Science & Technology B (Microelectronics and Nanometer Structures) vol.19, no.2, 476-81 March 2001.

[201] Liqun Han, R. Fabian Pease, W. Dan Meisburger, Gil. I. Winograd and Kimitoshi Takahashi, "Scaled measurements of global space-charge induced image blur in electron beam projection system," Journal of Vacuum Science & Technology B (Microelectronics and Nanometer Structures) vol.18, no.6, 2999-3003, Nov/Dec 2000.

[202] H. Liu, X. Zhu, E. Munro and J. A. Rouse, "Tolerancing of electron beam lithography columns," Microelectronic Engineering, 41/42, 163-6, 1998.

[203] Kimitoshi Takahashi, Liqun Han, R. Fabian Pease and W. Dan Meisburger, "Stochastic Coulomb interaction effect in ion-neutralized electron-bema projection optics," Journal of Vacuum Science & Technology B (Microelectronics and Nanometer Structures) vol.19, no.2, 2572-80, Nov/Dec 2001.

[204] Gregg M. Gallatin, "Analytic evaluation of the intensity point spread function," Journal of Vacuum Science & Technology B (Microelectronics and Nanometer Structures) vol.18, no.6, 3023-8, Nov/Dec 2000.

[205] M. S. Gordon, W. A. Einchen, S. D. Golladay, H. C. Pfeiffer, C. F. Robinson and W. Stickel, "PREVAIL: Dynamic correction of aberrations," Journal of Vacuum Science & Technology B (Microelectronics and Nanometer Structures) vol.18, no.6, 3079-83, Nov/Dec 2000.

[206] Hiroshi Yamashita, Kunio Takeuchi and Hidecki Masaoka, "Mask split algorithm for stencil mask in electron projection lithography," Journal of Vacuum Science & Technology B (Microelectronics and Nanometer Structures) vol.19, no.2, 2478-82, Nov/Dec 2001.

[207] N. fares, S. Stanton, J. Liddle and G. Gallatin, "Analytical-based solution for SCALPEL wafer heating," Journal of Vacuum Science & Technology B (Microelectronics and Nanometer Structures) vol.18, no.6, 3115-21, Nov/Dec 2000.

[208] Scott Hector and Pawitter Mangat, "Predictive model of the cost of continuous membrane masks for electron projection lithography", *BACUS News*, volume 17, no. 3, March 2001, 1-12.

[209] D. J. Resnick et al., "High-resolution templates for step and flash imprint lithography," Emerging Lithographic Technologies VI, SPIE Proceedings volume 4688, edited by Roxann Engelstad, 205-13, 2002.

[210] Bailey, T.C.; Resnick, D.J.; Mancini, D.; Nordquist, K.J.; Dauksher, W.J.; Ainley, E.; Talin, A.; Gehoski, K.; Baker, J.H.; Choi, B.; Johnson, S.; Colburn, M.; Meissl, M.; Sreenivasan, S.; Ekerdt, J.G.; Willson, C.G., "Template fabrication schemes for step and flash imprint lithography," Microelectronic Engineering vol.61-62, 461-7, July 2002.

[211] T. Bailey, B. Smith, B. J. Choi, . Colburn, M. Meissl, S. V. Sreenivasan, J. G. Ekerdt, and C. G. Willson, "Step and flash imprint lithography: Defect analysis," Journal of Vacuum Science & Technology B (Microelectronics and Nanometer Structures) vol.19, no.2, 2806-10, Nov/Dec 2001.

Nano and Giga Challenges in Microelectronics
Greer at al (Editors)
© 2003 Elsevier Science B.V. All rights reserved

Experimental Investigations of the Stability of Candidate Materials for High-K Gate Dielectrics in Silicon-Based MOSFETs

Susanne Stemmer [a] and Darrell G. Schlom [b]

[a] *Materials Department, University of California, Santa Barbara, CA 93106-5050, U.S.A.*

[b] *Department of Materials Science and Engineering, Penn State University, University Park, PA 16802-6602, U.S.A.*

Abstract

Due to the continuous decrease in feature size, silicon devices are approaching a number of fundamental limits. In particular, the gate oxide in modern integrated circuits has reached atomic dimensions. When the thickness of the traditionally used gate dielectric material, SiO_2, falls below ~ 1 nm, it has excessive leakage currents due to direct tunneling. Use of alternative gate oxides with higher dielectric constant (K) could permit similar transistor performance with drastically reduced leakage currents due to the greater physical thickness of the gate dielectric. Although they have been predicted to be thermodynamically stable in contact with silicon, these alternative oxides show a number of stability problems, in particular when exposed to high temperature anneals, which are typically around 1000°C. This chapter provides a description of the thermodynamic basics of high-K / silicon interface stability and of phase separation in silicate alloys. In addition, analytical experimental methods with high spatial resolution are essential in understanding the physical mechanisms of interfacial reactions and stability in these ultrathin layers. Such methods are discussed and experimental investigations of the stability of high-K / silicon interfaces are presented.

Key words: Gate oxide, alternative gate dielectrics, thin films, ZrO_2, HfO_2, silicate, CMOS, phase diagrams, spinodal decomposition, transmission electron microscopy

1. Introduction

Due to the continuous decrease in feature size (Moore's Law), silicon devices are approaching a number of fundamental limits. In particular, the gate oxide in modern integrated circuits has reached atomic dimensions. When the thickness of the traditionally used gate dielectric material, SiO_2, falls below ~ 1 nm (the exact thickness limit depends on the application), it has excessive leakage currents due to direct tunneling. Use of a different gate oxide with higher dielectric constant could permit similar transistor performance with drastically reduced leakage currents due to the greater physical thickness of the gate dielectric. In addition to a permittivity significantly greater than that of SiO_2 ($K_{SiO_2} = 3.9$), alternative gate dielectrics need to (1) have a large band gap and large band offsets with silicon for low leakage, (2) be thermodynamically stable in contact with silicon at typical device fabrication temperatures, (3) have a low density of electrically-active defects at the dielectric / silicon interface, (4) have a film morphology that permits low leakage currents, (5) be compatible with the gate electrode and, (6) show sufficient reliability [1-3].

ZrO$_2$ and HfO$_2$, and their alloys with SiO$_2$ (silicates) and Al$_2$O$_3$ (aluminates) are potential candidates to replace SiO$_2$ as the gate dielectric. Although they have been predicted to be thermodynamically stable in contact with silicon [4], these oxides show a number of stability problems, in particular when exposed to high temperature rapid thermal anneals (typically around 1000 °C). Such high temperatures are required for dopant activation in a conventional, self-aligned transistor fabrication process that uses a polycrystalline silicon ("polysilicon") gate electrode (Fig. 1). Materials stability problems include, for example, interfacial reactions and phase separation. There is significant disparity in the literature with respect to high-K stability. For example, zirconium silicide formation during polysilicon gate electrode deposition is reported for ZrO$_2$ under some experimental conditions [5-7], whereas other deposition conditions render a stable interface [8,9].

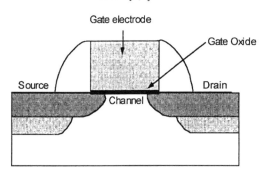

Fig. 1. Schematic cross-section of a MOSFET. The gate electrode is currently made of heavily doped polycrystalline silicon. Due to high-K- poly-silicon integration issues and gate depletion, metal electrodes might be used in future. Metal electrodes are also attractive because the threshold voltage could potentially be adjusted without more threshold implants.

This chapter has two goals. The first is to provide a correct description of the thermodynamic basics of high-K/ silicon interface stability and of phase separation in silicates. We will focus our discussions on ZrO$_2$ and zirconium silicates, because more thermodynamic data are available for these dielectrics. Similar calculations can be carried out for many of the other candidate oxides, such as

HfO$_2$ and Al$_2$O$_3$, as more reliable thermodynamic data become available. Secondly, we will show that analytical experimental methods with high spatial resolution are essential in understanding the physical mechanisms of interfacial reactions and stability in these ultrathin layers.

2. Thermodynamics

An obvious way to avoid reactions between a high-K dielectric and silicon is to select a dielectric that is thermodynamically stable in contact with silicon. The thermodynamic stability of all binary oxides (MO_x) in contact with silicon was investigated by *Hubbard* and *Schlom* [4,10]. Their approach is summarized in Sec. 2.1. As shown in that section, ZrO$_2$, HfO$_2$ and Al$_2$O$_3$ fall in the category of oxides that are stable in contact with silicon up to 1500 K.

In real high-K dielectric gate stacks, however, thermodynamic stability based solely on reactions between condensed phases is often an insufficient criterion for predicting the absence of interfacial reaction layers. Typical gate dielectrics are ultrathin layers and it is possible for gaseous species (e.g., SiO and oxygen) to diffuse through the layer at high temperature. For example, even for gate oxide materials that are thermodynamically stable in contact with silicon there is a driving force for SiO$_2$ to form at the interface between silicon and the gate oxide if excess oxygen is transported through the gate oxide during processing. Interfacial reactions involving gaseous species will be discussed in Sec. 2.2. Furthermore, under conditions under which the gate dielectric may be reduced, reaction is seen at the interface. The thermodynamic basics of reactions under reducing conditions will be discussed in Sec. 2.3. With respect to the thermal stability of the silicates, phase separation has been observed under typical device annealing conditions, and will be discussed in Sec. 2.4 in the context of metastable phase diagrams.

2.1. The thermodynamic stability of the interfaces between binary oxides and silicon

The thermodynamic stability of all binary oxides (MO_x) in contact with silicon was investigated recently [4,10]. All binary oxides that are solid at 1000 K and are not radioactive were considered. The key concept was that any reaction between silicon and a binary oxide that lowered the Gibbs free energy of the system resulted in the elimination of that binary oxide from further consideration. The reaction did not need to be the most favorable, as **any** reaction with $\Delta G < 0$ implies that the interface between silicon and that binary oxide is thermodynamically unstable. Even though there are many possible reactions, several key reactions were identified by considering the metal-oxygen-silicon phase diagrams. Initially, phase diagrams with only one binary oxide and no ternary phases were considered. For the binary oxide to be stable in contact with silicon a tie line must exist between the binary oxide and silicon. For example, Fig. 2 shows the ternary phase diagram for Zr-Si-O in the temperature range 700 °C – 950 °C, as determined in the literature [11,12]. Tie lines exist between ZrO_2 and silicon, and ZrO_2 and $ZrSiO_4$, indicating that at these temperatures ZrO_2 is stable in contact with both substances.

To determine if a tie line exists, the Gibbs free energies of reactions were calculated. If the system passed these reactions, that is $\Delta G > 0$ for each reaction, then additional reactions were checked if the system contained more than one binary oxide or contained ternary oxides, to the extent that thermodynamic data existed. It is important to realize the presence of simplifying assumptions in this thermodynamic analysis method. Only volume free energies were included; interfacial energies were neglected. As the gate dielectric layer becomes thinner, the interfacial free energies become more important, and conceivably they could alter the sign of ΔG for reactions in which its magnitude is close to zero. Notwithstanding these assumptions, this method serves to cull unsuitable dielectrics from consideration and identify the best candidates for detailed study.

The results of these comprehensive studies on

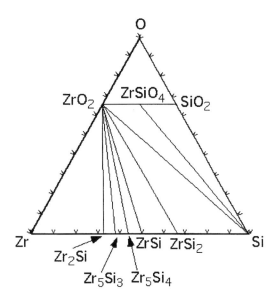

Fig. 2. Ternary phase diagram for Zr-Si-O in the temperature range 700 – 950°C with tie lines, re-drawn after [11]. Composition ranges due to solubility are not shown.

the thermodynamic stability of binary oxides in contact with silicon are graphically summarized on the periodic table in Fig. 3. Elements M having no binary oxide (MO_x) that is stable or potentially stable in contact with silicon are crossed out by diagonal lines, and the reason for their elimination is given. An element M is only crossed out when **all** of its binary oxides have been eliminated. A summary table of all those binary oxides that may be thermodynamically stable in contact with silicon and their values (the free energy change of the key reactions at a temperature of 1000 K) appears in Ref. [10].

ZrO_2 is an example of a binary oxide for which there are sufficient thermodynamic data to complete calculations and conclude its thermodynamic stability in contact with silicon, as shown below:

$$Si + ZrO_2 \longrightarrow Zr + SiO_2$$
$$\Delta G^\circ_{1000K} = +177.684 \frac{kJ}{mol} \qquad (1)$$

$$3Si + ZrO_2 \longrightarrow ZrSi_2 + SiO_2$$
$$\Delta G^\circ_{1000K} = +24.742 \frac{kJ}{mol} \qquad (2)$$

132

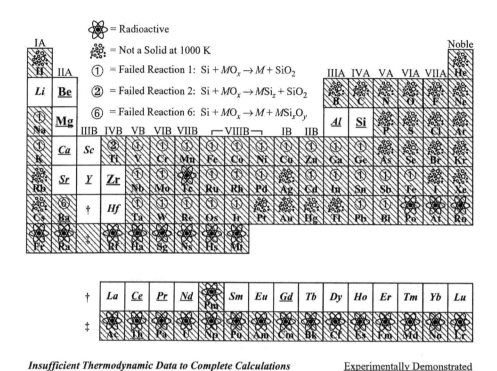

Fig. 3. Summary of which elements M have an oxide (MO_x) that *may* be thermodynamically stable in contact with silicon at 1000 K, based on Ref. [4,10]. Elements M having no thermodynamically stable or potentially thermodynamically stable oxide (MO_x) are crossed out by diagonal lines and the reason for their elimination is given. The elements M for which sufficient thermodynamic data exist to complete all necessary calculations to establish the thermodynamic stability of MO_x in contact with silicon appear in bold. The elements M having an oxide (MO_x) that has been experimentally demonstrated to be stable in direct contact with silicon are underlined. Performing the thermodynamic analysis over the full range of temperatures for which relevant thermodynamic data are available (as much as 300-1600 K) does not alter the conclusions shown.

$$\frac{3}{2}\text{Si} + \text{ZrO}_2 \longrightarrow \frac{1}{2}\text{ZrSi}_2 + \frac{1}{2}\text{ZrSiO}_4$$

$$\Delta G^{\circ}_{1000K} = +9.129\,\frac{\text{kJ}}{\text{mol}} \qquad (3)$$

$$\frac{1}{2}\text{Si} + \text{ZrO}_2 \longrightarrow \frac{1}{2}\text{Zr} + \frac{1}{2}\text{ZrSiO}_4$$

$$\Delta G^{\circ}_{1000K} = +85.600\,\frac{\text{kJ}}{\text{mol}} \qquad (4)$$

For each of these reactions ΔG°_{1000K} is the free energy change of the system when each reaction between reactants and products, all taken to be in their standard state (the meaning of the ° superscript), proceeds in the direction indicated at a temperature of 1000 K [13]. As we are able to complete all necessary calculations to establish the thermodynamic stability of ZrO_2 in contact with silicon, zirconium appears in bold on the periodic table in Fig. 3. Additional details, including all of the thermodynamic reactions considered, are given elsewhere [4,10].

Use of the recently reported value of $\Delta H_{f,298K}$ for ZrSi$_2$ [14], which differs by 21 kJ/mol (13%) from the value in Ref. [13], could, however, affect our conclusion that ZrO$_2$ is thermodynamically stable in contact with silicon. Unfortunately only $\Delta H_{f,298K}$ was reported in this recent study of ZrSi$_2$ and not its Gibbs free energy. But if we use the heat capacity and entropy values of ZrSi$_2$ from Ref. [13] in combination with the revised value $\Delta H_{f,298K}$ from Ref. [14], the value of ΔG°_{1000K} for reaction (2) becomes $\Delta G^{\circ}_{1000K} = -1.616$ kJ/mol. Although negative (which would imply that ZrO$_2$ is not stable in direct contact with silicon [6]), the magnitude of ΔG°_{1000K} is within the approximate uncertainty of the thermodynamic data [4], making the result of this calculation inconclusive. Note that for a thin ZrO$_2$ film on silicon, ZrSi$_2$ is the silicide in equilibrium with silicon and unless kinetic constraints exist for its formation, the other silicides (Fig. 2) do not need to be considered further for determining the stability of ZrO$_2$ on silicon.

Reactions involving gaseous species were not considered in the thermodynamic analysis in this section, as the idea is to test the stability of the interface between two solids (silicon and the dielectric), and there is no free volume in which a gas can exist. However, since the gate dielectrics are thin and possibly permeable to gaseous species, we will discuss such reactions in Sec. 2.2.

2.2. The thermodynamic stability of the silicon / high-K interface involving diffusing gas phase species

2.2.1. Interfacial SiO$_2$ formation

Oxides with low permeability to oxygen are desired for gate oxides for MOSFETs because even for gate oxide materials that are thermodynamically stable in contact with silicon there is a driving force for SiO$_2$ to form at the interface between silicon and the gate oxide if excess oxygen is transported through the gate oxide during sample processing. The driving force for this is illustrated by the reaction below:

$$Si + \text{Stable Gate Oxide} + O_2 \longrightarrow$$
$$SiO_2 + \text{Stable Gate Oxide}$$
$$\Delta G^{\circ}_{(5)} = -730.256\frac{kJ}{mol}, \tag{5}$$

where $\Delta G^{\circ}_{(5)} = -730.256\frac{kJ}{mol}$ is the free energy change of the system when the reaction proceeds in the direction indicated at a temperature of 1000 K [13]. Such reactions are common [15-18], even in the growth of thermodynamically stable oxides on silicon, because growth is typically performed with excess oxygen in order to fully oxidize the gate dielectric. The amount of SiO$_2$ formed depends on how much oxygen is transported through the gate oxide and will be highest for materials with high oxygen permeability and for high growth or annealing temperatures.

For example, the oxygen diffusivity for monoclinic, nanocrystalline zirconia follows an Arrhenius behavior $D_v = D_0 \exp\left(-Q_v/k_BT\right)$ with an activation energy $Q_v = (2.29 \pm 0.1)$eV and a pre-exponential factor $D_0 = (2.5 \pm 1.5) \times 10^{-7}\text{m}^2/\text{s}$ [19]. At 900 °C and 20 sec annealing time, the diffusion length $\sqrt{D_v t}$ is 27 nm, much greater than the typical gate dielectric thickness. Furthermore, grain boundary diffusion coefficients in nanocrystalline zirconia are four orders of magnitude faster than volume diffusion coefficients [19]. It should be noted that Eqn. 5 will proceed in the direction indicated during annealing steps if the annealing atmosphere contains a sufficient partial pressure of oxygen (e.g., from oxygen impurities in nitrogen gas), which typically is the case. For example, annealing at 1 atm in nitrogen gas containing 1 ppm residual oxygen corresponds to annealing in an oxygen partial pressure of $\sim 10^{-3}$ torr, which is sufficient for SiO$_2$ growth [20,21].

Furthermore, Busch et al. observed that the interfacial SiO$_2$ growth rates underneath ZrO$_2$ films exceeded the thermal growth rate of SiO$_2$ on bare Si in O$_2$ [16]. The authors suggested that the ZrO$_2$ catalyzes the decomposition of the O$_2$ (or H$_2$O) and transports atomic oxygen to the film, which are better oxidizers than O$_2$ molecules.

2.2.2. Interfacial SiO desorption

At lower oxygen partial pressures and sufficiently high temperatures, SiO desorption from bare or SiO_2 covered silicon surfaces is a well-known phenomenon, and is described by the following reaction equations

$$2Si + O_2 \rightarrow 2SiO(g) \qquad (6a)$$

$$Si + SiO_2 \rightarrow 2SiO(g) \qquad (6b)$$

The oxygen partial pressure and temperature ranges under which SiO_2 growth (Eqn. 5) and the competing SiO etching (Eqn. 6a) reaction occur on a bare silicon wafer have been determined experimentally [20-22]. Therefore, the reaction described by Eqn. 5 might ideally not be a concern for a silicon-compatible gate oxide material during the high-temperature annealing step, if a reducing environment is used during annealing, and if SiO can desorb through the high-K film. A reducing environment may, however, cause a reduction or decomposition reaction of the gate oxide material, as will be discussed in Sec. 2.3. Furthermore, SiO_2 decomposition is usually spatially inhomogeneous, and might result in interfaces of poor quality [23,24].

SiO desorption and outdiffusion through a thin high-K film has been reported by several authors [25,26]. However, kinetic limitations may exist, and the rate of desorption at a given temperature and partial pressures depends on the gate dielectric thickness and the specific gate material [27].

It should also be noted that at low oxygen partial pressures, the interfacial SiO_2 may also thin by decomposition and subsequent filling of oxygen vacancies in the high-K dielectric, i.e., without SiO desorbing at the surface.

2.2.3. Interfacial reactions involving gaseous species

If gaseous species (e.g., SiO, O_2) can diffuse through a thin dielectric layer at high temperatures, they should be included in the thermodynamic stability analysis. As an example of such an analysis, we determine the maximum pressure tolerated by a system consisting of the components ZrO_2, $ZrSi_2$, $SiO(g)$, $O_2(g)$, and silicon, in which

ZrO_2 is to be reduced to $ZrSi_2$. For simplicity, we assume no SiO_2 at the interface [for example it has already been desorbed from the interface by reaction (6b)]. In addition to Eqn. 6a, the following reactions **may** occur:

$$ZrO_2 + 2Si \rightarrow ZrSi_2 + O_2(g) \qquad (7)$$

$$ZrO_2 + 4Si \rightarrow ZrSi_2 + 2SiO(g) \qquad (8)$$

At 1200 K, $\Delta G^0_{(7)}$ for reaction (7) is large and positive ($\Delta G^0_{(7)} = 720.6$ kJ/mol) [13], yielding an equilibrium partial pressure p_{O_2} of 4.3×10^{-32} atm or 3.3×10^{-29} torr, as calculated from the equilibrium constant K_{eq} ($\Delta G^o_{(7)} = -RT \ln K_{eq} \approx -RT \ln p_{O_2}$). Therefore, silicide formation by reaction mechanism (7) will **not** occur for any viable oxygen partial pressures in deposition processes. Accessing this processing window directly with oxygen partial pressure is not feasible. It is common to use H_2/H_2O mixtures or CO/CO_2 mixtures to attain such low equivalent partial pressures of oxygen. It should be noted, however, that polysilicon gate electrode deposition using a silane or related process, causes very low oxygen partial pressures due to the hydrogen gas involved.

At 1200 K, $\Delta G^0_{(8)}$ for reaction (8) is also positive ($\Delta G^0_{(8)} = 313.2$ kJ/mol) [13], yielding an equilibrium SiO partial pressure p_{SiO} of 1.5×10^{-7} atm or 1.2×10^{-4} torr, as calculated from the equilibrium constant ($\Delta G^o_{(8)} = -RT \ln K_{eq} \approx -RT \ln p^2_{SiO}$) [13]. Therefore, for SiO partial pressures of less than 1.2×10^{-4} torr, reaction (8) **may** occur.

At 1200 K, $\Delta G^0_{(6a)}$ for reaction (6a) is negative ($\Delta G^0_{(6a)} = -407.4 kJ/mol$) [13]. With $\Delta G^0_{(6a)} = -RT \ln K_{eq} \approx -RT \ln p^2_{SiO}/p_{O_2}$, this yields

$$\frac{p^2_{SiO}}{p_{O_2}} = 5.4 \times 10^{17} \text{atm}, \qquad (9)$$

consistent with the partial pressures obtained from reactions (7) and (8), i.e., $(1.5 \times 10^{-7} \text{atm})^2 / 4.3 \times 10^{-32} \text{atm} = 5.4 \times 10^{17} \text{atm}$.

Therefore, for ZrO_2, Si, and $ZrSi_2$ to be in equilibrium at 1200 K with a gaseous atmosphere of SiO and O_2, the atmosphere must be of composition $p_{SiO} = 1.5 \times 10^{-7}$ atm and $p_{O_2} = 4.3 \times 10^{-32}$ atm. If $p_{O_2} + p_{SiO}$ is less than $\sim 1.5 \times 10^{-7}$ atm, the silicide reaction is favorable.

We next consider the more realistic scenario when the oxygen partial pressure is greater than 4.3×10^{-32} atm and assume that enough excess silicon is available for it not to be immediately consumed by reaction (6a). In this case, reaction (6a) will be shifted to the right, i.e. silicon will be etched and the SiO partial pressure increases, shifting both reaction (7) and (8) to the left, i.e. no silicide will be formed. Therefore, silicide formation according to reaction (8) is only favorable if SiO formed by reactions (6a) and (8) is continuously **removed** from the interface and p_{SiO} is locally reduced below 1.5×10^{-7} atm.

Some authors have suggested continuous recycling of SiO via continuous SiO$_2$ formation and etching by reactions of ZrO$_2$ with SiO to explain the formation of silicide [28,29]. Given the typical conditions of either flowing nitrogen or pumping, however, a SiO "recycling" reaction seems unlikely.

In summary, these reactions can result in silicide formation under kinetic boundary conditions that should be verified by additional experimental studies. They also cannot explain why silicide reactions are observed in capped stacks, i.e., after deposition of the polysilicon gate electrode. In the next section we consider an alternative mechanism of silicide formation.

2.3. The thermodynamic stability of the silicon/high-K interfaces involving high-K nonstoichiometry

2.3.1. Oxygen deficiency

High-K oxides are stable down to very low partial pressures, as shown in the Ellingham diagram in Fig. 4 [10,13]. Materials such as ZrO$_2$ can become oxygen deficient when exposed to low oxygen partial pressures. For example, heating zirconia single crystals in ultra high vacuum (UHV) causes darkening of the crystals presumably due to formation of an oxygen deficient ZrO$_{2-x}$ surface layer and associated reduction of Zr^{4+} to Zr^{3+} [30]. Furthermore, nanocrystalline thin films can have three orders of magnitude higher oxygen vacancy concentrations than microcrystalline materials due to a lower oxygen vacancy formation enthalpy [31]. High-K films deposited by physical va-

por deposition techniques, such as metal oxidation, reactive evaporation and sputtering, maybe oxygen deficient [32], particularly if deposition conditions are used that are intended to avoid extensive interfacial SiO$_2$ formation [33]. Polysilicon gate electrode deposition by silane (SiH$_4$) causes extremely reducing atmospheres, i.e. very low oxygen partial pressures that can reduce high-K oxides such as ZrO$_2$. Silicide formation during this process has been reported [5-7]. Therefore, oxygen nonstoichiometry in the dielectric is an important consideration for high-K stability.

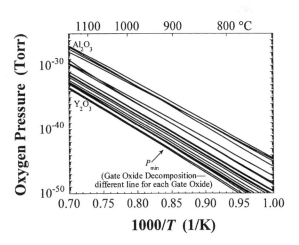

Fig. 4. Thermodynamic stability lines showing the minimum partial pressure of oxygen as a function of reciprocal temperature. Each binary oxide is thermodynamically stable at an oxygen partial pressure above its stability line and is thermodynamically unstable at oxygen pressures below its stability line. The order of the binary oxides between Y$_2$O$_3$ and Al$_2$O$_3$ at the left edge of the figure is Y$_2$O$_3$, Sc$_2$O$_3$, Er$_2$O$_3$, CaO, Ho$_2$O$_3$, Lu$_2$O$_3$, Tm$_2$O$_3$, Tb$_2$O$_3$, Dy$_2$O$_3$, ThO$_2$, Gd$_2$O$_3$, Sm$_2$O$_3$, Yb$_2$O$_3$, BeO, Nd$_2$O$_3$, Pr$_2$O$_3$, EuO, Ce$_2$O$_3$, La$_2$O$_3$, MgO, SrO, HfO$_2$, Li$_2$O, BaO, UO$_2$, ZrO$_2$, and Al$_2$O$_3$. Based on thermodynamic data from Ref. [13].

Recently, an alternative mechanism for the silicide formation, involving oxygen nonstoichiometry of the ZrO$_2$ film, has been proposed [8]. An oxygen deficient ZrO$_{2-x}$ film has a higher free energy than a stoichiometric film. The difference between the free energy of the stoichiometric film and the non-stoichiometric film provides the driving force for the silicide reaction. The free energy of the oxygen deficient film can be lowered by oxidation, if oxygen is available, or by outdiffusion of excess zirconium, which can then react with silicon. Employing the reaction equation used by *Perkins et al.* [8] this process can be expressed as follows:

$$ZrO_2 + 4e' + 2V_O^{\bullet\bullet} + 2Si \rightarrow ZrSi_2 + 2O_O^x \quad (10)$$

The physical reactions behind Eqn. (10) are the decomposition of part of the nonstoichiometric ZrO_2, the reaction of zirconium with silicon to form $ZrSi_2$, and the filling of two oxygen vacancies in the other part of the nonstoichiometric ZrO_2 by the oxygen of the decomposition reaction [8].

At 1473 K, using the free energies for ZrO_2, $ZrSi_2$ [13] and for the reduction reaction for tetragonal zirconia, as given by *Olander et al.* [34], the Gibbs free energy for this reaction, $\Delta G^o_{(10)}$, is -392 kJ/mol. Due to lack of similar thermodynamic data for any of the other ZrO_2 polymorphs, we will restrict our discussion to temperatures where tetragonal ZrO_2 is stable.

Using $\Delta G^o_{(10)}$, we can now determine at the oxygen nonstoichiometry for which the silicide reaction (10) would become thermodynamically favorable. With

$$\begin{aligned} \Delta G^o_{(10)} &= -RT \ln K_{eq} \\ &= -RT \ln \left(\frac{1}{16 \cdot [V_O^{\bullet\bullet}]^6} \right) \end{aligned} \quad (11)$$

the equilibrium concentration of oxygen vacancies is determined to be 0.3%. Therefore, for any ZrO_{2-x} film with oxygen nonstoichiometry x greater than 0.003, the silicide reaction may proceed. A possible indication of oxygen deficiency in thin films may be that the tetragonal phase, rather than the low-temperature equilibrium monoclinic phase, is often observed [35,36]. In zirconia (and likely hafnia) oxygen vacancies stabilize the higher symmetry phases at low temperatures [37]. As the tetragonal structure may also be stabilized by small crystallite sizes, experimental methods that can accurately determine oxygen deficiency

in thin films are required to distinguish between possible mechanisms.

2.3.2. *Oxygen excess and/or hydroxide absorption*

Exposing high-K dielectrics to an atmosphere after deposition can lead to significant hydroxide absorption. In particular, rare earth and related oxides strongly absorb and react with water [38,39]. Hydroxide species may also be present in low temperature chemical vapor deposition processes and can be incorporated in the high-K films during deposition. In the case of rare-earth binary oxides, such as Y_2O_3, Gd_2O_3 and La_2O_3, predicted to be stable in contact with silicon, excess oxygen present in the films or in the annealing atmosphere can lead to extensive interfacial silicate formation [40-43]. Furthermore, in these oxides interfacial silicate also forms easily by consuming interfacial SiO_2 layers [44,45]. For example, annealing of La_2O_3 in oxygen containing atmospheres results in competing reactions of interfacial SiO_2 formation by oxygen diffusion through the high-K film and silicon oxidation, and reaction of this SiO_2 to lanthanum silicate [15,17]. Stoichiometric films show no interfacial silicate layers [41] as expected from the thermodynamics analysis described in Sec. 2.1, and epitaxial growth of rare oxide films on silicon is possible [46-50].

Interfacial silicate formation in rare earth oxides with absorbed OH species can be understood as proceeding by the following reaction processes. Water molecules are strongly absorbed on the surface and react with the bulk oxide layer to form the hydroxide, i.e., $Y(OH)_3$ [38]. In the extreme case of a completely hydrated film the following reaction may take place during annealing:

$$2Y(OH)_3 + \frac{3}{2}Si \rightarrow Y_2O_3 + \frac{3}{2}H_2 \quad (12)$$

Recent thermodynamic data shows that the binary oxides, Y_2O_3 and SiO_2, are unstable with respect to the reaction to form yttrium-silicates [51]. For example, the reaction enthalpy of Y_2SiO_5 from the oxides is large and negative (-53 ± 5 kJ/mol) and the reaction is also favored by entropy of mixing. Therefore, the reaction products of Eqn. 12 will react form the silicate:

$$Y_2O_3 + \frac{3}{2}SiO_2 \rightarrow Y_2SiO_5 + \frac{1}{2}SiO_2 \qquad (13)$$

This reaction should result in a 1.4 nm thick SiO$_2$ film for a 6 nm Y$_2$O$_3$ film, however the silicate reaction may proceed further:

$$\frac{1}{2}Y_2SiO_5 + \frac{1}{2}SiO_2 \rightarrow \frac{1}{2}Y_2Si_2O_7 \qquad (14)$$

Therefore, interfacial SiO$_2$ in uncapped or *ex-situ* capped CVD samples that contain hydroxide or are exposed to oxygen containing atmospheres probably forms once the silicate reaction is completed and enough excess oxygen is available, or if silicate formation is kinetically limited, for example by the rate of silicon diffusion. The rate of silicon diffusion from the substrate may be reduced by nitriding the silicon-surface prior to deposition, resulting in reduced interfacial silicate formation [40].

2.4. Phase separation of hafnium and zirconium silicate alloys

The desired microstructure of the high-K gate dielectric, i.e., amorphous, polycrystalline or single-crystalline, is still a subject of scientific debate. In the near future, amorphous gate dielectrics might be preferred [2]. Low temperature deposition processes usually yield amorphous films, but device fabrication temperatures of up to 1000 °C are above the crystallization temperatures of the binary oxides. Alloying the binary oxides with SiO$_2$ will increase the crystallization temperature of the dielectric [2,52,53], albeit at the cost of lowering the dielectric constant. Silicate films may also be more resistant to oxygen diffusion.

Many investigators have reported that an amorphous silicate annealed for short times at elevated temperatures, transforms to a phase separated microstructure that consists of a crystalline HfO$_2$ or ZrO$_2$-rich phase embedded in an amorphous silica-rich matrix [52,54,55]. For some compositions, the onset of phase separation is reported to precede the onset of crystallization of the zirconia or hafnia rich phase [55]. Very dilute alloys are reported not to phase separate [53]. The equilibrium crystalline silicates ZrSiO$_4$ or HfSiO$_4$, predicted by the equilibrium phase diagram (Fig. 5), were never ob-

served under the annealing conditions used to process dielectric films [52,54].

Kingon and *Maria* were the first to explain how the observed microstructures are due to the presence of a liquid miscibility gap in the ZrO$_2$-SiO$_2$ system (Fig. 5) [15,56]. This region of liquid immiscibility can extend as a metastable miscibility gap to lower temperatures and cause an amorphous film to lower its free energy by separating into two phases with compositions defined by the extension of the metastable liquidus lines. In addition, kinetic suppression of the crystalline silicate (HfSiO$_4$ or ZrSiO$_4$, respectively) has also been suggested [15]. The pathways of microstructure evolution that determine crystallization and chemical inhomogeneity can be understood by applying concepts of metastable phase equilibria as is commonly used in rapid solidification [57]. Next we will discuss this approach.

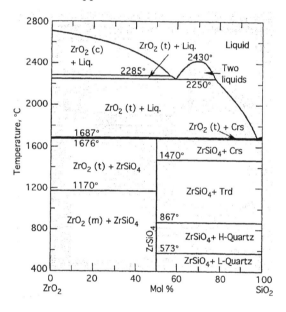

Fig. 5. The pseudo-binary ZrO$_2$-SiO$_2$ phase diagram, redrawn after [58].

The ZrO$_2$-SiO$_2$ phase diagram has been characterized experimentally and computationally [58,59], while the HfO$_2$-SiO$_2$ phase diagram is only partially known [60]. In particular, the high temperature region of the only published HfO$_2$-SiO$_2$ phase diagram cannot be correct, because it does not show the cubic to the tetragonal phase transition of HfO_2, which is correctly represented in the ZrO_2-SiO_2 phase diagram shown in Fig. 5. However, the reliable features in the HfO_2-SiO_2 phase diagram coincide with features of the ZrO_2-SiO_2 phase diagram and hafnium and zirconium, consistent with the very similar chemical behavior of Hf and Zr. We assume that the differences between the two systems are minor and will use the better-developed pseudo-binary ZrO_2-SiO_2 phase diagram for our discussions. Both diagrams contain one compound, $ZrSiO_4$ or $HfSiO_4$, respectively, that either first transforms to solid ZrO_2 and SiO_2 on heating before any SiO_2-rich melt is formed ($ZrSiO_4$ [58]) or melts incongruently (as suggested for $HfSiO_4$, but there is insufficient data available to assess this region). No noticeable solid solution is found in either system. No liquid immiscibility has been reported for the HfO_2-SiO_2 system, but insufficient experimental data is available for this region of the phase diagram. According to an empirical rule for silicates, an immiscibility gap above the liquidus is present if the ratio z/r is greater than about 1.6, where z is the formal ionic charge of the metal cation and r its ionic radius in Å [61]. Both ZrO_2 and HfO_2 meet this condition.

In a metastable phase equilibrium one or more phases are absent [62]. In the HfO_2-SiO_2 or ZrO_2-SiO_2 system, this is the crystalline silicate. Metastable phase diagrams and phase hierarchies are used widely in rapidly solidified alloys [63,64] and can be applied to amorphous materials synthesized from precursors (see Ref. [65] and references cited therein). The amorphous film represents a state of high free energy. The film can lower its free energy by crystallizing into a metastable phase assembly, in which structurally complex phases that require extensive atomic rearrangement are kinetically suppressed in favor of structures with simpler atomic arrangement and/or tolerant of disorder, such as the fluorite derived structures. Metastable phase boundaries can be inferred graphically from the equilibrium phase diagram, as shown in Fig. 6 (a) for a schematic phase diagram that omits the liquid miscibility gap. The kinetically limited $ZrSiO_4$ (or $HfSiO_4$) phase is excluded and a metastable eutectic reaction $L \rightarrow SiO_2 + ZrO_2$ (or $L \rightarrow SiO_2 + HfO_2$) occurs,

resulting in a two-phase assembly above the glass transition temperature, consisting of crystalline ZrO_2 (or HfO_2) and amorphous SiO_2. It should be noted that the formation of a metastable eutectic does not depend on whether the compound is incongruently or congruently melting; the condition is simply that the compound is kinetically excluded in the crystallization process.

As shown in Fig. 6 (b), the situation is more complicated for the ZrO_2-SiO_2 system (and most likely for HfO_2-SiO_2) due to the presence of the liquid miscibility gap and the thermodynamically required spinodal [57]. Within the metastable extensions of the spinodal, the metastable extensions of the liquidus lines shown in Fig. 6(a) become unstable [62,63]. Upon rapid thermal annealing ("upquenching"), amorphous films with compositions that lie within the metastable extension of the spinodal, are **unstable** and must unmix, as any composition fluctuation immediately decreases the free energy. The microstructural evolution of these films will involve an amorphous, phase-separated microstructure below the crystallization temperature (T_g). The initial zirconia (or hafnia) -rich phases will shed more silicon over time, until the two phases have reached the compositions given by the metastable extensions of the miscibility gap. At a given annealing temperature, if during demixing the zirconia (or hafnia) -rich phase reaches a composition for which the glass transition temperature for this phase is exceeded, crystalline ZrO_2 (or HfO_2) will form. Outside the spinodal, but within the miscibility gap, films will probably also phase separate before they crystallize [66].

Outside the metastable extension of the miscibility gap, the film will lower its free energy by nucleation and growth of crystallites of ZrO_2 (or HfO_2) if the annealing temperature exceeds T_g of the homogeneous, non-separated phase, as shown in Figs. 6(a) and (b). These compositions will lie on the ZrO_2 (or HfO_2)-rich side of the phase diagram, because the miscibility gap is on the silica-rich side. The glass transition temperatures on this side of the phase diagram are probably lower than the required temperatures for device processing ($\sim$ 1000 °C). With respect to the crystallizing polymorph of the HfO_2 (or ZrO_2) phase, predictions can also be made from the phase diagram. It has

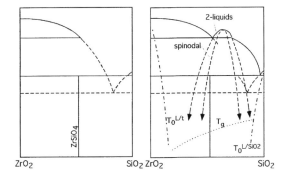

Fig. 6. (a) Metastable eutectic between ZrO_2 and SiO_2 when the $ZrSiO_4$ phase is excluded (assuming a phase diagram with no liquid immiscibility). Some of the stable phase fields were excluded for clarity. (b) metastable phase diagram with the extensions of the liquid miscibility gap and spinodal (dashed), the glass transition temperature (dotted line), and T_0 curves (dash-dotted line). The metastable extensions of the liquidus lines become unstable in the spinodal and are not shown. Figure to be published in [57].

been shown previously that in systems with a two-phase field between the liquid and the tetragonal phase, the crystallization of the tetragonal phase is associated with a larger decrease in free energy compared to the cubic phase [67]. The tetragonal phase can then transform to monoclinic. This explains why the cubic polymorph is never observed [52,57].

Another feature of metastable phase diagrams are T_0 curves [62], which are the upper bounds of metastable single phase fields, and indicate the locus in the temperature-composition diagram where the liquid and the solid phase have the same free energy at a given composition. Systems such as HfO_2-SiO_2 and ZrO_2-SiO_2, with almost no solid solubility, typically show T_0 curves that rapidly plunge and do not intersect [an example how they might look like is included in Fig 6(b)]. Between the T_0 curves glass formation is possible, but crystallization to a single phase is prohibited. In summary, the application of metastable phase diagrams predicts that after rapid thermal annealing, phase separated microstructures should be observed across the phase diagram, even outside the miscibility gap, with the exception of compositions under the T_0 curves.

To predict, as a function of composition and an-

nealing temperature, whether single-phase amorphous films will phase separate by spinodal decomposition, the location of the metastable extensions of the liquid miscibility gap and the spinodal are required. Several authors have recently calculated the metastable extensions of this miscibility gap [57,68], with very similar results.

As an example of such a prediction, Fig. 7 shows two calculations for the ZrO_2-SiO_2 system, calculated using commercially available thermodynamic modeling software (MTDATA) [57]. Details regarding the oxide database and models are described elsewhere [59,69,70]. The metastable extension of the liquid miscibility gap for the ZrO_2-SiO_2 was calculated by including only the liquid phase in the calculation of the equilibrium phases. One set of curves shown in Fig. 7 is based on data published previously [59], the other set of curves uses data of a more recent oxide database [69,71]. Both sets of curves are shown in Fig. 7 to indicate the degree of uncertainty in the calculations. The calculations show that on the zirconia-rich side, compositions with greater than ~ 80 mol% ZrO_2 lie outside the metastable liquid miscibility gap (and therefore outside the spinodal) at temperatures above 1000 K. On the silica rich side only compositions with greater than ~ 90 mol% SiO_2 lie outside the spinodal at typical annealing temperatures of 1000 °C [57].

3. Experimental Observations of High-K Stability

3.1. Experimental Methods

One of the most challenging aspects of investigating high-K stability is the need for physical characterization techniques that are able to determine the composition and structure of the high-K oxide and interfacial layers. As high-K dielectrics are ultrathin layers, ideally, a spatial resolution approaching the Angstrom level, with sensitivity better than one monolayer, is required. To date, no one technique can meet these requirements and complementary techniques are usually employed. Many physical characterization techniques that are

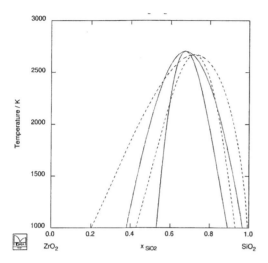

Fig. 7. Calculated metastable extensions of the liquid miscibility gap and spinodals to lower temperatures in ZrO_2-SiO_2. One s t of calculated curves is based on an earlier database [59] (hairline), and the other uses data of a more recent oxide database (dashed lines) [69,71].

used to study high-K oxides are traditional surface techniques, such as x-ray photoelectron spectroscopy (XPS). Several excellent review articles of new developments in physical characterization techniques for high-K dielectric films have been published in literature [72,73].

Among the most powerful emerging physical characterization techniques for the study of high-K thin films are medium energy ion scattering (MEIS) and scanning transmission electron microscopy (STEM) based analytical techniques, in particular high-angle annular dark-field imaging (HAADF or Z-contrast) and electron energy-loss spectroscopy (EELS).

Medium-energy ion scattering (MEIS) is a high-resolution version of Rutherford backscattering spectroscopy (RBS) [74,75]. MEIS uses ion energies lower than RBS, and is thus useful for examining films with thicknesses between 1 to 15 nm. The near surface depth resolution of MEIS can be a few Angstrom. For thicker films the resolution is limited by energy straggling (the stochastic broadening of the ion energy distribution with increasing depth). For thin films (< 100 Å) subnanometer depth resolution can be achieved. MEIS can be used to detect oxygen with a sensitivity of 0.1

monolayer [74], is nondestructive and is suited for isotope studies. In recent years, MEIS has successfully solved a number of important issues in growth and stability of SiO_2, oxynitrides and high-K thin films [76,77]. One requirement for MEIS is that the "stopping power" (the rate of energy loss of the incident ion) is known. The stopping power depends on sample composition and density, so at least some a-priori understanding of the sample and ion scattering is required.

STEM, as conventional high-resolution transmission electron microscopy (HRTEM), requires a thin TEM sample. However, the principles of image formation in STEM and HRTEM are fundamentally different. In STEM, the atomic-resolution HAADF image is formed scanning a small electron beam (≤ 2Å in diameter) and collecting the signal on an annular dark field detector, whereas an HRTEM image is acquired by parallel illumination. The HAADF image is also used to position the probe for atomic resolution EELS [78-80]. To date, STEM/EELS can routinely be performed using state-of-the-art field emission transmission microscopes originally designed for conventional TEM [81,82]. Using the combination of STEM/EELS in such a TEM/STEM, it is possible to measure the chemical composition across a gate stack with a spatial resolution of less than 2Å.

HRTEM images are coherent, phase-contrast images that are subject to contrast reversals, whereas HAADF images are essentially incoherent and intuitively interpretable. For thin samples, the image contrast in HAADF is roughly proportional to Z^2 (atomic number), thus providing the additional benefit of a chemically sensitive image. For example, thin unintentional low-K SiO_2 interfacial layers in amorphous high-K stacks might be difficult to detect in conventional HRTEM, but are easily visible in HAADF images due to the lower atomic number of silicon compared with zirconium and hafnium. Recently, single Sb dopant atoms have been imaged in silicon by HAADF [83]. HAADF imaging fundamentals have been reviewed by *Nellist et al.* [84]. In addition to HAADF imaging, exit wave reconstruction, a recent development in conventional HRTEM [85-87], yields more directly interpretable conventional HRTEM images.

An instructive example demonstrating the importance of the development of directly interpretable atomic resolution imaging techniques are amorphous oxide/silicon interfaces, where interfacial roughness can complicate measurements of oxide thickness by conventional HRTEM. *Baumann et al.* have simulated the effect of interfacial roughness on HRTEM images to illustrate the complications in image contrast interpretation [88]. Lattice fringes in HRTEM appear at interfaces and can make rough interfaces appear smooth and TEM objective lens defocus and sample thickness dictate the apparent abruptness. Recently, *Diebold et al.* [89] developed a reproducible method for gate dielectric thickness determination by advanced TEM techniques, including HAADF imaging.

As each element has a unique EELS binding energy, EELS absorption edges can be used to identify the elements present in gate stacks and even in very thin interfacial layers. Furthermore, the near-edge fine-structure of EELS edges reflects the unoccupied density of states at the exited atom [90], thus additional information of phases, bonding and electronic structure can be obtained in parallel. An example of the capabilities of atomic resolution EELS applied to gate dielectric interfaces is *Muller et al.*'s study of the SiO_2/silicon interface that revealed a 1.6 Å wide sub-stoichiometric oxide at the interface and shifts in the oxygen K-edge that were interpreted as states within the band gap [91].

In the following we present exemplary experimental studies of high-K stability, using HRTEM, STEM, and EELS.

3.2. Interfacial SiO_2 and silicide formation (HfO_2 and ZrO_2)

A number of studies have established the relationship between oxygen partial pressure and interfacial silicide or SiO_2 formation, respectively. For example, ZrO_2 films react with silicon to form a silicide when annealed in vacuum [28,92,93], whereas interfacial SiO_2 forms under oxidizing conditions [15,16,94]. There is some discussion in literature as to whether the interfacial layer formed between ZrO_2 or HfO_2 and silicon under oxidizing conditions is SiO_2 or a silicate. Both types of interfacial layers have been reported [28,92,95-98]. It should be noted that interfaces tend to be rough, and intermixing can be difficult to distinguish from roughness. An interfacial SiO_2 is consistent with a mechanism that involves oxygen diffusion through the high-K layer and oxidation of silicon. Amorphous layers may subsequently interdiffuse, up to concentrations reaching the boundaries of the metastable extensions of the spinodal (shown in Fig. 7 for ZrO_2-SiO_2). Once the layers are crystallized, little interdiffusion is expected (at least into the grain interiors), due to the limited solid solubility in equilibrium.

To experimentally investigate the processing window between interfacial SiO_2 and silicide formation, respectively, *Maria et al.* annealed ZrO_2 thin films at 1000 °C in nitrogen under different oxygen partial pressures (p_{O_2}) in the ranges of approximately 10^{-4} torr to 10^{-7} torr, adjusted by the pressure of nitrogen gas that contained oxygen [15]. Figure 8 shows conventional high-resolution TEM (HRTEM) images of the as-deposited and annealed ZrO_2/Si stacks [18]. After annealing at a p_{O_2} of about 10^{-4} torr, the entire oxide stack has grown to a thickness of about 4.1-4.4 nm [Fig. 8(b)] due to growth of the interfacial SiO_2-like layer to a thickness of about 2.2 nm. The ZrO_2 film annealed at a p_{O_2} in the range of 10^{-5} torr showed only a thin ($\sim$1 nm) interfacial SiO_2 layer [Fig 8(c)].

All samples annealed at 1000 °C below $\sim$$10^{-4}$ torr lie in the region in which reaction (6) dominates for bare silicon [20]. Therefore, interfacial oxide growth of the ZrO_2 sample annealed $\sim$$10^{-4}$ torr may be due to a somewhat higher oxygen partial pressure during the experiment, probably because the oxygen level in the nitrogen gas was greater than the estimated 1 ppm, or residual oxygen was present on the furnace walls. Furthermore, the kinetics of the relative rates of SiO(g) escape through the overlying ZrO_2 layer versus the rate of SiO_2 formation could also explain this result. If it is indeed such a kinetic issue, then at a fixed temperature the minimum oxygen pressure at which SiO_2 forms beneath a ZrO_2 layer on Si would be expected to depend on the thickness (and microstructure) of the overlying ZrO_2 film.

For the sample annealed at $p_{O_2} \sim$$10^{-5}$ torr, neither extensive growth nor etching was observed.

142

Fig. 8. Cross-section HRTEM images of ZrO_2 grown be reactive evaporation on silicon substrates: (a) the as-deposited stack, (b) after annealing at 1000 C and 10^{-4} torr, (c) after annealing at 1000 C and 10^{-5} torr. Micrographs (a) and (b) were recorded from areas of the samples that were covered by the patterned Pt electrode. Figure first published in [18]. Reproduced with permission.

The film annealed at the lowest p_{O_2}, about 10^{-7} torr, shows a markedly different microstructure (Fig. 9). $ZrSi_2$ precipitates were observed along the interface and protrude into the silicon surface.

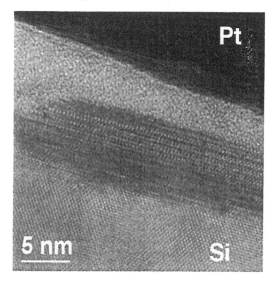

Fig. 9. Cross-section HRTEM images of ZrO_2 grown be reactive evaporation on silicon substrates after annealing at 1000 C and 10^{-7} torr. The image was recorded from an area of the samples that was covered by the patterned Pt electrode. Figure first published in [18]. Reproduced with permission.

Consistent with the trends observed in these annealing experiments, other authors show that silicide reaction between ZrO_2 films and polysilicon

electrodes can be avoided by using less reducing conditions during the gate electrode deposition [8,9]. Silicide formation is consistent with a reaction mechanism that includes oxygen deficiency of the ZrO_2 films. As the oxygen partial pressures are reduced, the metal oxide films may become more oxygen deficient, shifting the driving force for reaction (10) to the right. In uncapped annealing experiments, such as described above, SiO formation, discussed in Sec. 2.2, may be an additional driving force.

With respect to the differences between HfO_2 and ZrO_2, in particular, that HfO_2 films are reported to be more resistant to silicide formation during polysilicon gate deposition than ZrO_2 [99,100], the experimental constants of the reduction reaction [i.e., the analogy of reaction (10)] need to be determined for HfO_2. HfO_2 and ZrO_2 behave chemically similar, but hafnium forms a stronger bond with oxygen than zirconium [101]. Electronic structure calculations show that the oxygen vacancy formation energy is greater in HfO_2 than in ZrO_2 [102,103]. Thus, HfO_2 films may be less oxygen deficient under the same experimental conditions.

3.3. Phase Separation of HfO_2–SiO_2 and ZrO_2–SiO_2 Mixtures

There is overwhelming experimental evidence of phase separation in these films, in particular at intermediate concentrations [52,54,55,57,104]. Phase separation at very low concentrations (around 10 mol%) of HfO_2 or ZrO_2, respectively, as is expected from calculations [57,68], has not yet been observed. At the silica-rich side of the phase diagram, the volume fraction of a phase-separated hafnia- or zirconia-rich phase will be small and experimentally very difficult to detect.

Figure 10 shows plan-view HRTEM micrographs of films with 80 mol% HfO_2, 60 mol% HfO_2 and 40 mol% HfO_2 after annealing at 1000 °C for 10 s in flowing nitrogen [57]. All three films show a two-phase microstructure consisting of crystalline and amorphous regions, as predicted in Sec. 2.4. The volume fraction of crystalline phase increased with increasing HfO_2 concentration. Crystallite sizes ranged between 5- 10 nm (80 mol% HfO_2) and $\sim$ 5 nm (40 mol% HfO_2). The onset of crystallization was composition dependent, as observed by other authors [2,52,104]. The film with 80 mol% HfO_2 began to crystallize below 800 °C, the film with 60 mol% HfO_2 crystallized between 800 °C and 900 °C, and the film with 40 mol % HfO_2 crystallized between 900 °C and 1000 °C. The insets in Fig. 10 are electron diffraction patterns of the film with 80 mol% HfO_2 and 60 mol% HfO_2, respectively. The ring patterns indicate the polycrystalline film microstructure. The diffraction spots in the film with 80 mol% HfO_2 can be indexed as monoclinic hafnia, in particular weak spots corresponding to spacings of $\sim$ 5.1 Å and 3.7 Å, not found in any of the other polymorphs, are present in Fig. 10 (a). The diffraction ring spacings in the film with 60 mol% HfO_2 can be assigned to tetragonal or monoclinic hafnia. Diffraction spots corresponding to lattice spacings greater than $\sim$ 2.9 Å were not observed in this sample. Therefore, hafnia crystals in this sample are likely tetragonal. The sample with 40 mol% HfO_2 did not contain enough crystalline phase to obtain ring patterns in electron diffraction. As mentioned above, tetragonal hafnia is predicted to be the first crystallizing phase, which

can then transform to monoclinic. Transformation to the monoclinic phase might be prohibited by the small crystallite sizes of the samples with 60 and 40 mol% HfO_2. It is well-known, for example from dispersed zirconia ceramics, that tetragonal zirconia particles constrained by a surrounding matrix must be of a critical size for the transformation to the monoclinic polymorph to occur. Films annealed at lower temperatures, below the onset of crystallization, showed regions of dark and bright contrast in the amorphous phase, also visible in the amorphous regions in Figs. 10 (b) and (c). This contrast can be interpreted as phase separation in the amorphous material or as clustering associated with the onset of the nucleation of crystallites.

Despite the likely very different pathways, i.e. nucleation and growth vs. spinodal decomposition, the resulting microstructures are very similar. This is likely due to the high undercooling, which prohibits extensive coarsening of the microstructure of the films outside the spinodal, although the film with 80% HfO_2 show slightly larger grain sizes. Furthermore, kinetic studies using small angle x-ray scattering are required to investigate whether selective amplification of one wavelength in the phase separated microstructure and diffuse interfaces, as predicted by the theory of spinodal decomposition, are present. With respect to the application of these films as gate dielectrics, the precise morphology of the phase-separated microstructure will determine the effect on the dielectric properties. For example, a parallel-connected assembly of hafnia and silica-rich phases will be less detrimental for the dielectric constant than a hafnia-rich phase embedded in a low-K silica-rich phase.

3.4. Silicate Formation in Rare Earth Oxide Thin Films

Several experimental studies demonstrated that excess oxygen present in rare earth and related oxide films (such as Y_2O_3, Gd_2O_3 and La_2O_3) as OH or a related species, or excess oxygen in the annealing atmosphere, can lead to extensive silicate formation at the silicon interface [40-43]. Interfacial rare earth silicates can also form by consuming

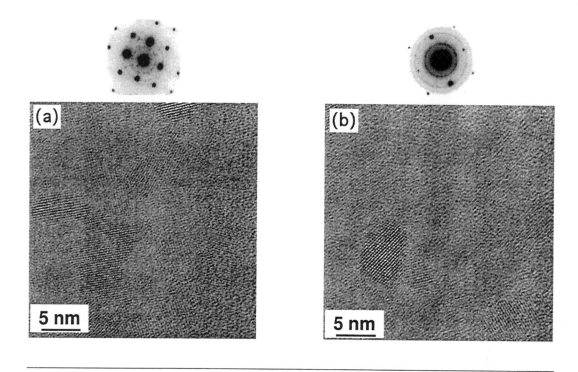

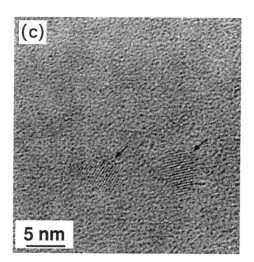

Fig. 10. Plan-view high-resolution micrograph of (a) the film with 80 mol% HfO_2, (b) of the film with 60 moland (c) film with 40 mol% HfO_2 after rapid thermal annealing at 1000°C for 10 s. Crystallites in (c) are indicated by arrows. The closely spaced lattice fringes in the upper part of (c) are due to the underlying silicon substrate not removed by ion-milling in thicker parts of the TEM sample. The insets in (a) and (b) are electron diffraction patterns. The sharp, bright diffraction spots are due to the underlying silicon, the diffraction rings, marked along the arrow, are due to the film. The lattice spacings in (a) are 5, 3.7, 2.9, 2.5, 1.8, and 1.5 Å. The lattice spacings in (b) are $\sim$ 2.9, 2.5, 1.8 and 1.5 Å. Figure to bepublished in [57].

interfacial SiO_2 layers [44,45].

Figure 11 shows a STEM/EELS study of interfacial reactions between a Y_2O_3 film deposited either directly on clean silicon, or after a nitridation step, respectively, by oxygen plasma-assisted chemical vapor deposition (CVD) [40]. Figures 11 (a) and (b) show HRTEM and HAADF images, respectively, of a $\sim$14 nm thick film on bare silicon. The film is mostly amorphous. A less than 0.5 nm wide dark layer at the film/substrate interface can be observed in Fig. 11(b) and is due to an extremely thin interfacial SiO_2 layer that is not visible in HRTEM. The EELS line profile of the silicon L-edge across the dielectric [Fig 11(c)] shows that the concentration of silicon in the film decreases away from the silicon substrate surface and then increases again near the interface with the polysilicon capping layer, indicating significant reaction of the Y_2O_3 film with both silicon interfaces to form an amorphous yttrium silicate during the anneal. Figure 11 (d) shows an HRTEM micrograph of the film grown on nitrided silicon under identical deposition conditions as the film on bare silicon. The entire thickness of the dielectric is about 11 nm, which is $\sim$3 nm less than the film on bare silicon. The interior of this film is crystallized, suggesting little silicon in Y_2O_3, whereas the interfacial regions with silicon and polysilicon are amorphous, and about 1.8 nm and 2.6 nm wide, respectively. The concentration profile of silicon across the dielectric is also shown in Fig. 11 (c). In the film interior (crystallized region) the silicon concentration is below the detection limit of the method, consistent with the observation that the film crystallizes if the silicon concentration is low. Furthermore, the silicon concentration away from the silicon substrate interface decreases more quickly than at the polysilicon interface. Taking the position where the silicon counts are above 10% of the bulk silicon value as the boundary of the reaction layer, the interfacial reaction layer is about 2 nm wide at the silicon interface, and about 2.5 nm wide at the polysilicon interface, consistent with the widths of the crystallized parts of the film.

These results show that *in situ* capping layers can successfully prevent extensive SiO_2 formation at the interface with silicon. The extremely thin (< 0.5 nm) SiO_2 layer observed here might have

formed after substrate cleaning, or be due to a reaction with small amounts of excess oxygen in the films or the deposition process. During annealing, and possibly already during deposition, Y_2O_3 reacts with both the silicon substrate and the polysilicon electrode to form a silicate. The silicon profiles and crystallization behavior show that this reaction can be impeded at the substrate interface by a pre-deposition nitridation step that likely slows the silicon diffusion from the substrate. The reduced overall film thickness of the dielectric on nitrided silicon compared to the film on bare silicon is a further indication of the impeded reaction of the pre-nitrided samples, as less silicate is formed for this film. The silicate formation reaction requires excess oxygen. If excess oxygen introduced during the CVD deposition process is responsible for the silicate formation, it is not sufficient to cause extensive SiO_2 formation. Interfacial SiO_2 in uncapped or *ex situ* capped CVD samples probably forms once the silicate reaction is completed and enough excess oxygen is available, or if silicate formation is kinetically limited, for example by the rate of silicon diffusion from the substrate.

Kinetic barriers to the formation of a crystalline silicate also exist in the Y_2O_3-SiO_2 system. The reaction of a crystalline Y_2O_3 film with the silicon substrate results in the formation of an amorphous silicate interlayer. Such amorphization reactions are typical for a high kinetic of the crystalline phase [105].

3.5. *Exploratory investigations of new materials for gate dielectrics*

Due to limitations in the stability and performance of binary metal oxides and their silicates, the search for alternative materials is the subject of ongoing research. One approach involves alloying silicates with other elements that potentially improve stability against phase separation and crystallization [106]. Other researchers are investigating ternary high-K oxides [10]. To evaluate the stability of these new materials, the high-resolution analytical capabilities afforded by MEIS, HRTEM and STEM are essential, as little thermodynamic data exists for multicomponent oxides.

146

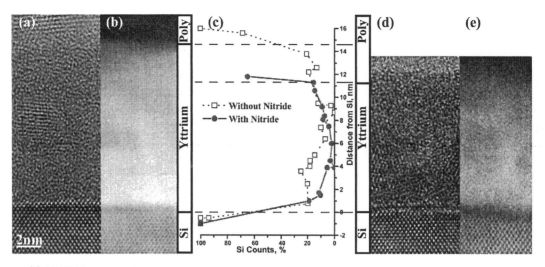

Fig. 11. (a) HRTEM micrograph of the film grown on bare silicon. (b) Annular dark-field image (filtered) of the film on bare silicon. (c) EELS line profiles of across (b) and (e) for the silicon L edge as a function of distance from the silicon substrate interface. Open squares are from the film on bare silicon, circles from the film on nitrided silicon. Note that the films have different thickness (the position of the interfaces and layers are indicated as a guide). (d) HRTEM of the film on nitrided silicon. (e) Annular dark-field image of the film on nitrided silicon. Figure published in [40]. Reproduced with permission.

A new material that is of interest is $LaAlO_3$ because of its high dielectric constant ($K_{11} \approx K_{33}$ = 24.1±0.2 at 145 GHz), its optical bandgap of 5.6 eV, and its favorable heterojunction bandgap lineup with silicon [107-109]. To investigate the stability of the $LaAlO_3$/Si interface, silicon films were grown by MBE under UHV conditions and temperatures at or above 800 °C on $LaAlO_3$ single crystals. The goal of these experiments is to investigate the stability of direct, single crystalline interfaces, not easily obtained by growing oxides on silicon. Figure 12 shows a high-resolution image of a Si/$LaAlO_3$ interface annealed at 1026 °C in nitrogen gas. The film is fully epitaxial (although it shows twinning), and the interface remained stable [110]. The study shows that the search for alternative gate dielectric materials is far from complete.

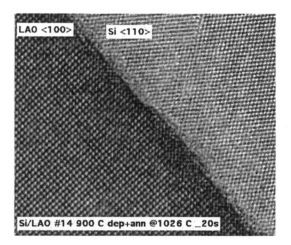

Fig. 12. Epitaxial silicon film grown by MBE at T_{sub} 850 °C on $LaAlO_3$. Even after standard Motorola implant activation anneal for MOSFETs at 1026 °C the interface remained stable [110].

4. Conclusions

High-K oxide stability and silicon compatibility are just two of the many requirements that a successful alternative gate dielectric material for use in silicon MOSFETs must possess, but nevertheless prove to be challenging tasks for both materials sci-

entists and the semiconductor industry. Binary oxides that are stable or potentially stable in contact with silicon can show thermal and interface stability problems due to high-K nonstoichiometry and high oxygen diffusivities. Novel high-resolution analytical capabilities, such as STEM and MEIS, are essential for characterizing the stability of new gate dielectrics. These techniques further our understanding of the physical mechanisms that determine high-K structure and stability, and may indicate ways in which the thermal stability can be improved. Of particular future interest are experimental techniques to measure and elucidate the role of point defects, such as oxygen vacancies and impurities, in these ultrathin films.

Acknowledgements

The results presented in this chapter were obtained in collaboration with Zhiqiang Chen, K. Eisenbeiser, Brendan Foran, John Gisby, Kevin Hubbard, Angus Kingon, James Lettieri, Hao Li, Pat Lysaght, Jon-Paul Maria, Greg Parsons, Venugopalan Vaithyanathan, Yi Wei, and Yan Yang. We gratefully acknowledge discussions with Rick Garfunkel, Carlos Levi, Dave Muller, Jim Speck, Ravi Droopad, and Robby Beyers, and the financial support of the Semiconductor Research Corporation (SRC) and SEMATECH through the SRC/SEMATECH FEP Center.

References

[1] G. D. Wilk, R. M. Wallace, and J. M. Anthony, J. Appl. Phys. **89**, 5243-5275 (2001).

[2] A. I. Kingon, J.-P. Maria, and S. K. Streiffer, Nature **406**, 1032-1038 (2000).

[3] MRS Bulletin **27**, 186-229 (2002).

[4] K. J. Hubbard and D. G. Schlom, J. Mater. Res. **11**, 2757-2776 (1996).

[5] M. A. Gribelyuk, A. Callegari, E. P. Gusev, M. Copel, and D. A. Buchanan, J. Appl. Phys. **92**, 1232-1237 (2002).

[6] M. Gutowski, J. E. Jaffe, C.-L. Liu, M. Stoker, R. I. Hedge, R. S. Rai, and P. J. Tobin, Appl. Phys. Lett. **80**, 1897-1899 (2002).

[7] T. Z. Ma, S. A. Campbell, R. Smith, N. Hoilien, B. Y. He, W. L. Gladfelter, C. Hobbs, D. Buchanan, C. Taylor, M. Gribelyuk, M. Tiner, M. Coppel, and J. J. Lee, IEEE Trans. Electron Dev. **48**, 2348-2356 (2001).

[8] C. M. Perkins, B. B. Triplett, P. C. McIntyre, K. C. Saraswat, and E. Shero, Appl. Phys. Lett. **81**, 1417-1419 (2002).

[9] A. Callegari, E. Gousev, T. Zabel, D. Lacey, M. Gribelyuk, and P. Jamison, Appl. Phys. Lett. **81**, 4157-4158 (2002).

[10] D. G. Schlom, C. A. Billmann, J. H. Haeni, J. Lettieri, P. H. Tan, R. R. M. Held, S. Völk, and K. J. Hubbard, Appl. Phys. A (to be published) (2003).

[11] S. Q. Wang and J. W. Mayer, J. Appl. Phys. **64**, 4711-4716 (1988).

[12] R. Beyers, R. Sinclair, and M. E. Thomas, J. Vac. Soc. Sci. Technol. B **2**, 781-784 (1984).

[13] I. Barin, *Thermochemical Data of Pure Substances*, Vol. I and II, 3rd ed. (VCH, Weinheim, 1995).

[14] S. V. Meschel and O. J. Kleppa, J. Alloys Compounds **274**, 193-200 (1998).

[15] J. P. Maria, D. Wicaksana, A. I. Kingon, B. Busch, H. Schulte, E. Garfunkel, and T. Gustafsson, J. Appl. Phys. **90**, 3476-3482 (2001).

[16] B. W. Busch, W. H. Schulte, E. Garfunkel, T. Gustafsson, W. Qi, R. Nieh, and J. Lee, Phys. Rev. B **62**, R13290-R13293 (2000).

[17] S. Stemmer, J. P. Maria, and A. I. Kingon, Appl. Phys. Lett. **79**, 102-104 (2001).

[18] S. Stemmer, Z. Chen, R. Keding, J.-P. Maria, D. Wicaksana, and A. I. Kingon, J. Appl. Phys. **92**, 82-86 (2002).

[19] U. Brossmann, R. Würschum, U. Södervall, and H.-E. Schneider, J. Appl. Phys. **85**, 7646-7654 (1999).

[20] J. J. Lander and J. Morrison, J. Appl. Phys. **33**, 2089-2092 (1962).

[21] F. W. Smith and G. Ghidini, J. Electrochem. Soc. **129**, 1300-1306 (1982).

[22] D. Starodub, E. P. Gusev, E. Garfunkel, and T. Gustafsson, Surf. Rev. Lett. **6**, 45-52 (1999).

[23] R. Tromp, G. W. Rubloff, P. Balk, F. K. LeGoues, and E. J. v. Loenen, Phys. Rev. Lett. **55**, 2332-2335 (1985).

[24] Y.-K. Sun, D. J. Bonser, and T. Engel, Phys. Rev. B **43**, 14309-14312 (1991).

[25] M. Copel, Appl. Phys. Lett. **82**, 1580-1582 (2003).

[26] S. Sayan, E. Garfunkel, T. Nishimura, W. H. Schulte, T. Gustafsson, and G. D. Wilk, to be published (2003).

[27] E. Garfunkel, personal communication.

[28] T. S. Jeon, J. M. White, and D. L. Kwong, Appl. Phys. Lett. **78**, 368-370 (2001).

[29] K. Muraoka, Appl. Phys. Lett. **80**, 4516-4518 (2002).

148

[30] V. A. Loebs, T. W. Haas, and J. S. Solomon, J. Vac. Soc. Technol. A **1**, 596-599 (1983).

[31] I. Kosaki, V. Petrovsky, and H. U. Anderson, J. Am. Ceram. Soc. **85**, 2646-2650 (2002).

[32] S. Ramanathan, D. A. Muller, G. D. Wilk, C. M. Park, and P. C. McIntyre, Appl. Phys. Lett. **79**, 3311-3313 (2001).

[33] J. Lettieri, J. H. Haeni, and D. G. Schlom, J. Vac. Sci. Technol. A **20**, 1332-1340 (2002).

[34] W.-E. Wang and D. R. Olander, J. Am. Ceram. Soc. **76**, 1242-1248 (1993).

[35] S. Ramanathan, P. C. McIntyre, J. Luning, P. Pianetta, and D. A. Muller, Phil. Mag. Lett. **82**, 519-528 (2002).

[36] C. Zhao, G. Roebben, M. Heyns, and O. van der Biest, in *Euro Ceramics Vii, Pt 1-3*; Vol. *206-2* (2002), p. 1285-1288.

[37] S. Fabris, A. T. Paxton, and M. W. Finnis, Acta Mater. **50**, 5171-5178 (2002).

[38] Y. Kuroda, H. Hamano, T. Mori, Y. Yoshikawa, and M. Nagao, Langmuir **16**, 6937-6947 (2000).

[39] D. Niu, R. W. Ashcraft, and G. N. Parsons, Appl. Phys. Lett. **80**, 3575-3577 (2002).

[40] S. Stemmer, D. O. Klenov, Z. Q. Chen, D. Niu, R. W. Ashcraft, and G. N. Parsons, Appl. Phys. Lett. **81**, 712-714 (2002).

[41] B. W. Busch, J. Kwo, M. Hong, J. P. Mannaerts, B. J. Sapjeta, W. H. Schulte, E. Garfunkel, and T. Gustafsson, Applied Physics Letters **79**, 2447-2449 (2001).

[42] D. Niu, R. W. Ashcraft, Z. Chen, S. Stemmer, and G. N. Parsons, Appl. Phys. Lett. **81**, 676-678 (2002).

[43] G. A. Botton, J. A. Gupta, D. Landheer, J. P. McCaffrey, G. I. Sproule, and M. J. Graham, J. Appl. Phys. **91**, 2921-2928 (2002).

[44] M. Gurvitch, L. Manchanda, and J. M. Gipson, Appl. Phys. Lett. **51**, 919-921 (1987).

[45] S.-K. Kang, D.-H. Ko, E.-H. Kim, M. H. Cjo, and C. N. Whang, Thin Solid Films **353**, 8-11 (1999).

[46] S. Guha, N. A. Bojarczuk, and V. Narayanan, Appl. Phys. Lett. **80**, 766-768 (2002).

[47] G. Apostolopoulos, G. Vellianitis, A. Dimoulas, M. Alexe, R. Scholz, M. Fanciulli, D. T. Dekadjevi, and C. Wiemer, Appl. Phys. Lett. **81**, 3549-3551 (2002).

[48] J. Kwo, M. Hong, A. R. Kortan, K. L. Queeney, Y. J. Chabal, R. L. Opila, D. A. Muller, S. N. G. Chu, B. J. Sapjeta, T. S. Lay, J. P. Mannaerts, T. Boone, H. W. Krautter, J. J. Krajewski, A. M. Sergnt, and J. M. Rosamilia, J. Appl. Phys. **89**, 3920-3927 (2001).

[49] J. Kwo, M. Hong, A. R. Kortan, K. T. Queeney, Y. J. Chabal, J. P. Mannaerts, T. Boone, J. J. Krajewski, A. M. Sergent, and J. M. Rosamilia, Appl. Phys. Lett. **77**, 130-132 (2000).

[50] J. P. Liu, P. Zaumseil, E. Bugiel, and H. J. Osten, Appl. Phys. Lett. **79**, 671-673 (2001).

[51] O. Fabrichnaya, H. J. Seifert, R. Weiland, T. Ludwig, F. Aldinger, and A. Navrotsky, Z. Metallkd. **92**, 1083-1097 (2001).

[52] D. A. Neumayer and E. Cartier, J. Appl. Phys. **90**, 1801-1808 (2001).

[53] G. D. Wilk, R. M. Wallace, and J. M. Anthony, J. Appl. Phys. **87**, 484-492 (2000).

[54] G. Rayner, R. Therrien, and G. Lucovsky, Mat. Res. Soc. Symp. **611**, C1.3.1-C1.3.9 (2000).

[55] G. B. Rayner, D. Kang, Y. Zhang, and G. Lucovsky, J. Vac. Sci. Technol. B **20**, 1748-1759 (2002).

[56] A. I. Kingon, J.-P. Maria, D. Wicaksana, C. Hoffmann, and S. Stemmer, in *Extended Abstracts of International Workshop on Gate Insulator IWGI* (Japan Publishing Trading Co., Tokyo, 2001), p. 36-41.

[57] S. Stemmer, Z. Chen, C. Levi, P. S. Lysaght, B. Foran, J. A. Gisby, and J. R. Taylor, Jap. J. Appl. Phys. Part 1, No. 6A (2003).

[58] W. C. Butterman and W. R. Foster, Amer. Mineralogist **52**, 880-885 (1967).

[59] R. G. J. Ball, M. A. Mignanelli, T. I. Barry, and J. A. Gisby, J. Nuclear Mat. **201**, 128-249 (1993).

[60] V. N. Parfenenkov, R. G. Grebenshchikov, and N. A. Toropov, Dokl. Akad. Nauk SSSR **185**, 840-842 (1969).

[61] H. A. J. Oonk, T. J. v. Loo, and A. A. Vergouwen, Phys. Chem. Glasses **17**, 10-12 (1976).

[62] J. C. Baker and J. W. Cahn, in *Solidification* (ASM, Metals Park, 1971).

[63] J. H. Prepezko and W. J. Boettinger, Mat. Res. Soc. Symp. Proc. **19**, 223-240 (1983).

[64] W. J. Boettinger, in *Rapidly Solidified Amorphous and Crystalline Alloys*, edited by B. H. Kear, B. C. Giessen, and M. Cohen (Elsevier, 1982).

[65] C. G. Levi, Acta mater. **46**, 787-800 (1998).

[66] J. W. Cahn, J. Amer. Ceram. Soc. **52**, 118-121 (1969).

[67] M. L. Balmer, F. F. Lange, and C. G. Levi, J. Am. Ceram. Soc. **77**, 2069-2075 (1994).

[68] H. Kim and P. C. McIntyre, J. Appl. Phys. **92**, 5094-5102 (2002).

[69] R. H. Davies, A. T. Dinsdale, J. A. Gisby, J. A. J. Robinson, and S. M. Martin, Calphad **26**, 229-271 (2002).

[70] T. I. Barry, A. T. Dinsdale, and J. A. Gisby, JOM **45**, 32-38 (1993).

[71] J. A. Gisby, A. T. Dinsdale, I. Barton-Jones, A. Gibbon, and P. A. Taskinen, in *Sulfide Smelting 2002: Proceedings from the 2002 TMS Annual Meeting in Seattle 2002*, edited by R. L. Stephens and H. Y. Sohn (TMS, Warrendale, 2002), p. 533-545.

[72] B. W. Busch, O. Pluchery, Y. J. Chabal, D. A. Muller, R. L. Opila, J. R. Kwo, and E. Garfunkel, MRS Bulletin **March,** 206-211 (2002).

[73] R. M. Tromp (ed.) *Emerging Analytical Techniques* (IBM Journal of Research and Development **44**, 2000).

[74] M. Copel, IBM J. Res. Develop. **44**, 571-582 (2000).

[75] J. F. van der Veen, Surf. Sci. Rep **5**, 199-287 (1985).

[76] E. P. Gusev, H. C. Lu, T. Gustafsson, and E. Garfunkel, Phys. Rev. B **52**, 1759-1775 (1995).

[77] T. Gustafsson, H. C. Lu, B. W. Busch, W. H. Schulte, and E. Garfunkel, Nucl.Instr. Meth. Phys. Res. B **183**, 146-153 (2001).

[78] P. E. Batson, Nature **366**, 727-729 (1993).

[79] N. D. Browning, M. F. Chisholm, and S. J. Pennycook, Nature **366**, 143-146 (1993).

[80] D. A. Muller, Y. Tzou, R. Raj, and J. Silcox, Nature **366**, 725-727 (1993).

[81] E. M. James, N. D. Browning, A. W. Nicholls, M. Kawasaki, Y. Xin, and S. Stemmer, J. Electron Microsc. **47**, 561-574 (1998).

[82] E. M. James and N. D. Browning, Ultramicroscopy **78**, 125-139 (1999).

[83] P. M. Voyles, D. A. Muller, J. L. Grazul, P. H. Citrin, and H. J. L. Gossmann, Nature **416**, 826-829 (2002).

[84] P. D. Nellist and S. J. Pennycook, in *Advances in Imaging and Electron Physics; Vol. 113* (Academic Press, San Diego 2000), p. 147-203.

[85] A. Thust, W. M. J. Coene, M. O. deBeeck, and D. VanDyck, Ultramicroscopy **64**, 211-230 (1996).

[86] M. A. O'Keefe, C. J. D. Hetherington, Y. C. Wang, E. C. Nelson, J. H. Turner, C. Kisielowski, J. O. Malm, R. Mueller, J. Ringnalda, M. Pan, and A. Thust, Ultramicroscopy **89**, 215-241 (2001).

[87] C. L. Jia and A. Thust, Phys. Rev. Lett. **82**, 5052-5055 (1999).

[88] F. H. Baumann, C.-P. Chang, J. L. Grazul, A. Kamgar, C. T. Liu, and D. A. Muller, Mat. Res. Soc. Symp. **611**, C4.1.2-C4.1.12 (2000).

[89] A. C. Diebold, B. Foran, C. Kisielowski, D. Muller, S. Pennycook, E. Principe, and S. Stemmer, Micros. Microanal. (to be published).

[90] R. F. Egerton, *Electron Energy-Loss Spectroscopy in the Electron Microscope*, second ed. (Plenum Press, New York, 1996).

[91] D. A. Muller, T. Sorsch, S. Moccio, F. H. Baumann, K. Evans-Lutterodt, and G. Timp, Nature **399**, 758-761 (1999).

[92] M. Copel, M. Gribelyuk, and E. Gusev, Appl. Phys. Lett. **76**, 436-438 (2000).

[93] H. Watanabe, Appl. Phys. Lett. **78**, 3803-3805 (2001).

[94] Y.-M. Sun, J. Lozano, H. Ho, H. J. Park, S. Veldman, and J. M. White, Appl. Surf. Sci. **161**, 115-122 (2000).

[95] D. A. Muller and G. D. Wilk, Appl. Phys. Lett. **79**, 4195-4197 (2001).

[96] T. Yamaguchi, H. Satake, N. Fukushima, and A. Toriumi, Appl. Phys. Lett. **80**, 1987-1989 (2002).

[97] C. M. Perkins, B. B. Triplett, P. C. McIntyre, K. C. Saraswat, S. Haukka, and M. Tuominen, Appl. Phys. Lett. **78**, 2357-2359 (2001).

[98] O. Renault, D. Samour, J. F. Damlencourt, D. Blin, F. Martin, S. Marthon, N. T. Barrett, and P. Besson, Appl. Phys. Lett. **81**, 3627-3629 (2002).

[99] S. J. Lee, H. F. Luan, W. P. Bai, C. H. Lee, T. S. Jeon, Y. Senzaki, D. Roberts, and D. L. Kwong, Tech. Dig. Int. Electron Devices Meet. (2000).

[100] L. Kang, K. Onishi, Y. Jeon, B. H. Lee, C. Kang, W.-J. Qi, R. Nieh, S. Gopalan, R. Choi, and J. C. Lee, Tech. Dig. Int. Electron Devices Meet. (2000).

[101] N. I. Medvedeva, V. P. Zhukov, M. Y. Khodos, and V. A. Gubanov, phys. stat. sol. (b) **160**, 517-527 (1990).

[102] A. S. Foster, V. B. Sulimov, F. L. Gejo, A. L. Shluger, and R. M. Nieminen, Phys. Rev. B **6422**, art. no.-224108 (2001).

[103] A. S. Foster, F. L. Gejo, A. L. Shluger, and R. M. Nieminen, Physical Review B **65**, art. no.-174117 (2002).

[104] J. P. Maria, D. Wickaksana, J. Parrette, and A. I. Kingon, J. Mater. Res. **17**, 1571-1579 (2002).

[105] K.-N. Tu, J. W. Mayer, and L. C. Feldman, *Electronic Thin Film Science for Electrical Engineers and Materials Scientists* (Macmillan, New York, 1992).

[106] M. R. Visokay, J. J. Chambers, A. L. P. Rotondaro, A. Shanware, and L. Colombo, Appl. Phys. Lett. **80**, 3183-3185 (2002).

[107] S. G. Lim, S. Kriventsov, T. N. Jackson, J. H. Haeni, D. G. Schlom, A. M. Balbashov, R. Uecker, P. Reiche, J. L. Freeouf, and G. Lucovsky, J. Appl. Phys. **91**, 4500-4505 (2002).

[108] J. Robertson, MRS Bulletin **27**, 217-221 (2002).

150

[109] P. W. Peacock and J. Robertson, J. Appl. Phys. **92,** 4712-4721 (2002).

[110] Y. Yang, Z. Q. Chen, L. F. Edge, V. Vaithyanathan, J. Lettieri, D. G. Schlom, S. A. Chambers, H. Li, Y. Wei, K. Eisenbeiser, and S. Stemmer, to be published.

Nano and Giga Challenges in Microelectronics
Greer at al (Editors)
© 2003 Elsevier Science B.V. All rights reserved

Models of defects in wide-gap oxides: Perspective and challenges

A. L. Shluger, A. S. Foster[†], J. L. Gavartin and P. V. Sushko

Department of Physics and Astronomy, University College London, Gower Street, London WC1E 6BT, UK
[†]Laboratory of Physics, Helsinki University of Technology, P.O. Box 1100, FIN-02015 HUT, Finland

Abstract

In this chapter we will review some of the main issues and recent calculations pertaining mainly to point defects and defect processes in oxides. Several years of exploratory research and hundreds of publications have not lead so far to a final consensus about the materials to be used in DRAM and gate applications in future devices. The prevailing line of thought at the moment seems to favour "something ZrO_2 or HfO_2-based". Although the situation remains fluid, these materials as well as silicates and aluminates emerged as strong contenders and provide good examples of typical defect related issues. Therefore we will use them, were appropriate, to illustrate the present state of the art of calculations and main conclusions relevant to the performance of oxide materials. To some extent we will be following empirical approaches, which rely on knowledge of other materials in similar states to those to be used. Therefore we will draw some analogies with other oxides, such as silica and MgO. The properties we are going to consider will relate mainly to non-stoichiometry and disorder of prospective materials. We will discuss the mechanisms of defect diffusion and will review the relation between defects and transparent conductive behaviour of traditional and newly emerging materials, such as ITO and $12CaO \cdot 7Al_2O_3$.

Key words: defects, oxides, electronic structure theory, electron trapping, transparent conductors

1. Introduction

Defects play a crucial role in the performance and reliability of micro-electronic devices, optical fibres, transparent conductors, catalysts, sensors and many other systems. Achieving a better understanding of defect structures and processes has been a driving force behind development of many powerful theoretical concepts and methods. For example, extensive studies of colour centres in the sixties driven by radiation physics and chemistry helped to shape point defect, polaron and exciton theories and to develop a wide range of scintillators, laser materials and dosimeters. Doping of semiconductors, which proved to be vital for micro-electronics industry, has long been an inspiration for developing computational solid state methods. Problems related to the mechanisms of silicon oxidation, reliability of micro-electronic devices, Bragg grating formation in optical fibres, luminescence of porous silicon and others promoted extensive studies of amorphous materials. More recent examples concern the development of new high-k dielectrics for microelectronics applications, transparent conductive oxides, materials exhibiting magneto-resistant properties, high-T_c superconductors and various applications of quantum dots. Again, most of the work will be done in experimental labs where various combinations of materials are already tried and their properties tested. At the same time, predictive theoretical modelling and simulation of growth and properties of complex and often disordered thin oxide films became one of the challenges of computational materials science. Using new sophisticated methods, theory and computational experiments play an important role in screening new materials and studying and predicting their properties.

A significant component of this research, close to the interests of the authors of this paper, is related to defects in oxides. They are responsible for leakage current, breakdown and degradation of micro-electronic devices, Bragg grating formation, conductivity of electro-active insulators, chemical sensing and many other processes. Defects are likely to play even more important roles as devices continue to scale and reach dimensions where single defects can affect their performance.

Therefore it seems timely to give an overview of established models of some of the typical defects in oxides and to highlight some recent results of theoretical modelling as well as existing problems of computational methods. We will focus mainly on oxides relevant for applications in microelectronics and as transparent conductors. This choice is mainly determined by the general topic of the book and our assertion that these materials are likely to remain in focus of active research for the next few years. However, we hope that this review will provide a useful source of models and understanding of theoretical modelling in a wider area of wide gap oxides.

Successful development of new materials for gate oxides and transparent conducting oxides (TCOs) poses difficult problems for theoretical modelling. Some of the typical problems include predicting stable structures and electronic properties of Si/dielectric systems. This implies calculations of band off-sets; possible interface defects, such as dangling bonds, Si - Si bonds, hydrogen defects etc.; studies of the structure and properties of grain boundaries (they can, for example, accumulate vacancies), and electronic states due to disorder if the dielectric is amorphous. Another important issue concerns predicting the structure and energy levels of possible defects induced by

growth and post-deposition annealing in different ambient conditions, for example, due to incorporation of boron, chlorine, oxygen, nitrogen, hydrogen and other species. Determining the most stable charge states of interface and bulk defect species, and their possible role in static charging of as grown and annealed oxides has long been one of the main challenges of theoretical modelling. An even more difficult problem concerns transient electron and hole traps responsible for leakage current and possible ways of eliminating them. This includes predicting defect diffusion and recharging in applied electric fields. Defect-induced conductivity is the key issue for TCOs, such as indium tin oxide (ITO). Theory should also help in identifying defects by predicting their vibrational, EPR, optical and other properties.

Fig. 1 gives an overview of some of the processes relevant to gate oxide films: creation of defects - incorporation from the environment, creation during growth; charge transfer - defects can change their charge state due to electron and hole trapping; reactions - defects react with each other to form different defects; defect diffusion. Theoretical models discussed in this paper may help to understand the mechanisms of these processes as well as the structure and electronic properties of defects as a function of oxide composition.

The initial stage of studying defects in a new material concerns building structural models of defect species. For example, hydrogen, oxygen and nitrogen species can incorporate into the lattice during growth or annealing of oxide films on Si in atomic and molecular forms in different charge states. Experiments on zirconia [1,2,3] have suggested that oxygen incorporates from the surface and diffuses inside the oxide in atomic form. Both atomic and molecular interstitial oxygen species can exist in silica [4]. Since such experiments are still absent for hafnia, it is interesting to use theory to study what are the most stable forms of interstitial oxygen species in this material.

Another class of problems is related to the energy levels of vacancies and incorporated species. As-grown films are almost always non-stoichiometric. In relatively narrow gap oxides, such as In_2O_3, SnO_2, $Sr(Ba)TiO_3$ oxygen vacancies are the source of electron doping, which makes them n-type con-

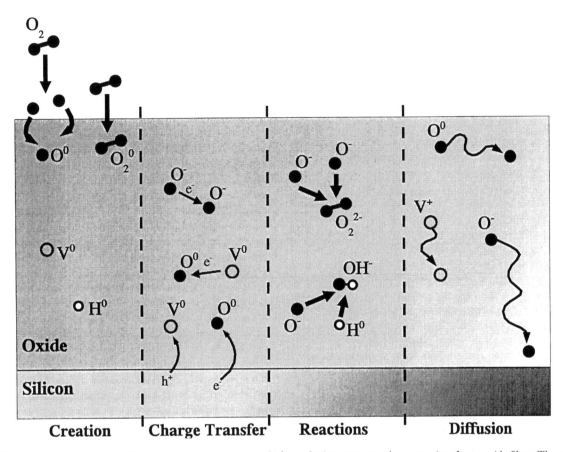

Figure 1. Diagram showing the different defect processes which can be important to the properties of gate oxide films. These include defect creation during growth and post-deposition anneal; electron transfer processes between defects and from/to silicon; reactions between defects leading to formation of more complex species and annihilation of existing defects; defect diffusion.

ductors. On the other hand, electron and hole trapping by interstitial oxygen and oxygen vacancies from Si in gate oxides may lead to oxide charging and affect leakage current. For example, recent studies of plasma deposited zirconia thin films on Si [5] have demonstrated significant electron and hole trapping, and this is supported by theoretical predictions of the trapping properties of vacancies and interstitials in zirconia [6,7]. Charged defects can create strong electric fields and affect band offsets. They can contribute to dielectric loss and their diffusion in an electric field is equivalent to electrolysis (see, for example, ref. [8]).

Building a complete picture of defect processes also includes defect reactions, requiring the parameters of electron transfer between defects, calculations of defect diffusion and recombination. More complex modelling concerns the dynamics of defect processes, mechanisms of energy dissipation and ultimately absolute rates of these processes.

In many cases, reliable and predictive theoretical screening of complex oxide materials takes much longer than the decision regarding their irrelevance for particular applications is reached experimentally. Therefore computer experiments to investigate characteristics related to basic electronic properties, processing and operational performance of these materials are carried out mainly on prototype systems. In the case of high-k dielectrics, these include ZrO_2, HfO_2 and $HfSiO_4$.

The best-studied prototype TCOs include SnO_2 and In_2O_3. SiO_2 has long served and will still continue to serve for some time as a gate dielectric, and the material of optical fibres and optical components in lithography. Therefore many of our examples will concern basic defects in this material. Finally, there are many relevant studies of electron and hole defects in the bulk and at surfaces of MgO, some ferroelectrics and in other oxide materials. They allow one to outline some general defect models and trends from their properties. For this study we will focus mainly on oxygen-, hydrogen- and nitrogen-related defects in these materials due to their ubiquitous nature and general relevance.

Below we will give an overview of common theoretical methods and will discuss their application to calculating properties of perfect materials, such as the electronic structure and dielectric function. These are vital both for performance of oxides in perspective devices and for understanding of the defect properties. Then we will turn to reviewing of the defects and defect processes outlined above. Many of these defects and related processes have been previously studied in bulk, crystalline oxides. Therefore we will discuss the transferability of defect models developed in model oxides to other more complex materials and will consider the effect of disorder in amorphous materials on defect properties.

This paper can inevitably provide only a very specific slice of an extensive and vibrant area of research. Recent excellent reviews related to the topics outlined here may help to broaden the picture. In particular, defects in cubic oxides have been considered in [9,10,11]. Extensive reviews of defects in SiO_2 in relation to Bragg grating formation in optical fibres and in gate dielectrics can be found in ref. [4,12]. Advances in the theory of oxide-semiconductor interfaces have been discussed in ref. [13] and more recently in an extensive collection of papers [14]. Defects in wide band gap II-VI crystals have been reviewed in ref. [15]. Models of radiation-induced point defects in MgO and several ferroelectric perovskites have been considered in refs. [11,16,17]. Challenges for atomic scale modelling of electronic materials and high-k dielectrics have been discussed in refs. [18,19,20]. A significant part of research related to defects in oxides

is carried out in relation to diverse applications of oxide surfaces in catalysis and in solid electrolyte fuel cells as reviewed in ref. [21].

2. Modelling techniques

We will start from briefly discussing some techniques available for modelling of defects in insulators. They exist in the form of computer codes implementing a model of a system with a defect and *methods* for calculations of defect structure and properties. Here, we aim to establish terminology and notations used in further discussions. In particular, studying the system's electronic structure and properties involves an hierarchy of models. Solving the Schrodinger equation (stationary or time-dependent) for a system of electrons and atomic cores requires establishing a *model* for calculation of the perfect and defective crystal. This includes three major components: a total energy expression (Hamiltonian), boundary conditions imposed on a wave-function, and a wave-function representation (basis set). The defect *model* which comes out of calculations includes its main fingerprints, such as the geometric and electronic structure, spectroscopic properties, mechanism of motion (diffusion), reactions with other defects and other parameters which can be studied experimentally and used for ultimate defect identification. Below we consider basic approaches used in defect studies.

2.1. *Periodic and cluster approaches*

Two models are often used in computer simulations of point defects in solids: a *periodic* model, and a finite *cluster* model. They differ by the boundary conditions imposed on a system wave-function. Initially, the periodic model was developed for calculation of electronic structure and properties of ideal crystals, whereas the cluster model was introduced from molecular calculations. Consequently, the terms "periodic model" and "cluster model" for the same defect appeared. It has long been realized that neither of them can be used alone if an accurate picture of a defect is to be established. Therefore, both models were gradually developed towards a more accurate description of defect structure and processes.

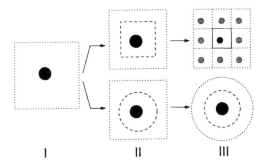

Figure 2. Periodic and embedded cluster approaches for calculation of point defects in crystals. I) a crystal with a point defect; II) the immediate environment of the defect in the periodic approach - supercell (top) and in the cluster approach - quantum cluster (bottom); III) top: periodically translated supercell forming a new ideal crystal, bottom - the quantum cluster with the defect (inner circle) embedded in the host lattice (outer circle).

Periodic model

The periodic model is based on a definition of a unit cell (or more generally - supercell), which can be periodically translated to build an infinite solid as indicated in Fig. 2. There are no limitations, in principle, on the size and shape of the unit cell providing that the host lattice is defined unambiguously. Note that the same approach is often used to build models of amorphous solids too. The periodicity imposed on the atomic structure also applies to the electronic structure: the charge density is the same in every supercell and is matched at the boundary between two neighboring supercells. In the case of the ideal crystals this approach is technically exact.

To calculate the properties of an individual point defect using the periodic model, a supercell is defined which comprises the defect and its immediate environment (see Fig. 2II). The supercell is then periodically repeated as shown in Fig. 2II. The defect in each supercell interacts with an infinite number of similar defects (called images) in all other supercells. Therefore, the application of this

approach to individual defects is adequate only if an artificial defect-image interaction is small.

One of the most attractive features of the periodic model is that it allows one to consider a perfect system and that with a defect on the same footing. Most of the computer codes implementing the periodic model use the basis set of plane-waves or their variations. This provides a very flexible, potentially complete basis set, a quality almost inaccessible in cluster calculations.

An obvious feature of the periodic model, which can be both to its advantage and drawback, is that "defects" are periodically translated and interact between themselves. It can be an advantage if one is interested, for example, in periodic adsorption of molecules at the surface and would like to vary their concentration. It is a drawback if one is interested in the properties of an individual defect in different charge states. In particular, if the defect is charged with respect to the host lattice, the electrostatic interaction of such defects is divergent. Some special techniques developed to deal with this problem are discussed later. Yet another downside of the periodic model is that the defect-induced distortion of the host lattice can only be taken into account within the supercell. This may lead to inaccurate results if these distortions propagate further than cell boundaries, but can be checked by increasing the supercell size. Several recent developments in computational techniques using the periodic model have led to an improvement of their treatment of defects:

– The total energy in the periodic model includes interaction of the defect with its periodic images. This constitutes the problem for modelling charged defects, since a spurious electrostatic interaction between the periodic images diverges. In order to circumvent this problem, an algorithm involving uniform neutralizing background potential is usually applied, which allows for the calculations of energetics of charged systems.

– The problem of spurious electrostatic interaction persists even for neutral defects whose charge distribution is characterized by sizable multipole moments (dipole, quadrupole et ce.). For such systems slowly decaying multipole-multipole interactions between repeated defects affect total energies, and ultimately result in a slow convergence of a defect energetics with the size of the supercell. To improve this convergence and, hence, to reduce the size of the minimum supercell required for the calculations, Makov and Payne [22] proposed the algorithm for correcting total energies in the periodic calculations for the spurious defect-defect electrostatic interaction. Initially formulated for the cubic supercells, this technique was later generalised for arbitrary supercells by Kantorovich [23].

Finally, calculations of electronic excited states in the periodic model present a significant challenge. On one hand, excited states of defects are often more delocalized than their ground state, and hence require much bigger supercells for reliable predictions. On the other hand, computational techniques for calculating excited states in the periodic model are developed to a lesser extent than those available for molecules and cluster models. Having said that, we note that rapid advances in many-body perturbation theory, such as GW-approximation, and in time-dependent density functional theory (TDDFT), result in the development of new potent techniques for the excited state calculations within periodic model. Recent progress in this field is comprehensively discussed in a review by Onida et al. [26].

To summarise, the periodic model is ideal if one is interested in the ground state properties of neutral defects, which weakly perturb the surrounding lattice.

Cluster model

An alternative approach is to consider a point defect and its immediate environment as a large hypothetical molecule usually called a "cluster" (see Fig. 2(II)) . This model is, generally speaking, only applicable if the following two conditions are satisfied: i) the atomic and electronic structure of the host lattice is accurately reproduced by the cluster, and ii) the defect-induced perturbations are essentially localized within the cluster boundaries [27].

To define a cluster means to somehow "cut out" a fragment from an infinitely large host lattice and to consider it separately. The "cutting" could be naturally made by a minimal charge density surface. This is relatively straightforward in atomic, molecular and ionic crystals where structural elements, such as ions or molecules can be well defined and their electron densities separated. As a result, an integer number of molecules or ions respectively can be attributed to a cluster. In the case of polar or covalent crystals the cutting is made through atoms so that the cluster should be terminated by "pseudo-atoms" specially designed to terminate broken bonds (see review [28] for numerous examples).

The range of applications of the cluster model is often complementary to that of the periodic approach. A cluster model is a natural choice for a system containing an isolated well localized defect. Other examples include properties of nanopowders, low-coordinated surface sites or different sites in an amorphous structure, i.e. situations where supercells are difficult to construct or too expensive to treat. One of the main advantages of cluster models is the large number of quantum-chemical methods for electronic structure calculations available from molecular calculations, especially for excited states.

On the downside, cluster model calculations are usually more computationally demanding than periodic ones. In spite of enormous progress in computing power, clusters of only a few tens of atoms can be calculated routinely. This is because the computational expense growths as n^4 where n is the number of atomic basis set functions typically used in these calculations.

Using small molecular clusters for treating larger systems brings a number of problems. i) Polarizability of the "surface" ions, especially O^{2-}, is larger than those in the "bulk" of the cluster [29]. A large ratio of "surface" to "bulk" atoms in a small cluster may result in a overestimated electronic polarization due to the "surface" atoms and, in principle, in an incorrect electronic structure altogether. The latter can manifest itself in the formation of artificial "ghost" states in the band gap. ii) The electrostatic potential due to the host lattice, which is particularly important in ionic compounds, is not adequately treated in small clusters. As a result, positions of energy levels can be wrong. iii) Defect-induced lattice relaxation can be considered only for the atoms nearest to the defect, while atoms at the cluster surface are usually kept fixed. This implies that the lattice response to the presence of a defect as well as the effect of that response on the defect itself can be significantly underestimated.

We finish this section with two examples of cluster models. In the case of highly ionic cubic oxides with non-polar surfaces, such as MgO, choosing a compact symmetric cluster with the cation/anion ratio close to stoichiometric often leads to reliable answers. Popular clusters of this kind are cubic $Mg_{13}O_{14}$ and cuboid $Mg_{32}O_{32}$. In ionic-covalent systems, such as SiO_2, cluster formation always requires breaking chemical bonds. A popular fix in this case is to passivate broken bonds by real hydrogen atoms or by "pseudo-atoms" with parameters specially fitted to fulfil this task. A cluster $Si_2O_7H_6$ including two tetrahedra connected via a common oxygen atom with all other oxygens passivated by hydrogen atoms is a frequent choice. Note a characteristic difference: in the case of MgO both clusters exist as stable entities, whereas in the case of SiO_2 cluster $Si_2O_7H_6$ is an artificial, but convenient construction (see ref. [28] for detailed discussion).

Thus a molecular cluster model tends to completely disregard the effects of the rest of the atoms outside the cluster and applies some semi-empirical fixes to the atoms at the cluster surface to diminish their adverse effect. This major problem is addressed and partially solved in the "embedded cluster" model.

Embedded cluster model

The molecular cluster model can be improved by taking into account the interaction of the quantum cluster with the rest of the host lattice, the perturbation of the lattice by the defect, and the effect of the lattice response on the defect itself. This is achieved by constructing an external potential in which the cluster is then embedded. Such a potential is called an embedding potential and the model, an embedded cluster model. Accurate representation of the embedding potential is a major challenge, with two groups of approaches existing. The first group includes the methods which attempt to "correct" the molecular cluster approach mainly by including the Coulomb interaction with the rest of the system. This can be done at three levels of sophistication:

- The electrostatic potential due to the part of the crystal lattice outside the cluster is included in the calculations by: i) placing point charges of appropriate charge at the lattice sites (see, for example, [30,31]; ii) fitting the Madelung potential by several point charges located at specially selected points [32]; iii) calculating the matrix elements due to the Madelung potential directly and adding these matrix elements to the potential energy matrix [33].

- To further improve the description of the cluster boundary, an interface region between the cluster and the point charges is introduced. Interface atoms can be approximated by effective core pseudo-potentials [34,35] at the cation sites around the cluster or, more accurately by effective potentials calculated self-consistently specifically for the material under study [31]. The latter approach allows one to reduce the cluster size and the computational load in some cases.

- The defect-induced lattice distortion is accounted for by embedding the cluster in an infinite polarizable lattice of classical ions represented using a shell model [36] or a polarizable ion model [37,38]. Unlike the two previous cases, these ions are not kept fixed, but are allowed to adjust their positions to minimize the total energy of the system. This allows one to relax explicitly a large region (several hundred atoms) of the host lattice near the defect. More importantly, embedding into a polarizable lattice can be made in such a way that also accounts for the effect of the lattice response on the cluster electronic structure. This allows one to achieve self-consistency in the calculation of the defect-induced lattice perturbation and the effect of this perturbation on the defect structure and properties. This effect is similar in many ways to a polaron digging its potential well in a polarizable media.

Some of the examples of development of this approach and applications to defects in oxides are discussed in refs. [39,40,41,42,43]. They clearly demonstrate that providing the system is ionic, it can be well treated in the embedded cluster model discussed above. However, if building up a quantum cluster requires breaking covalent bonds between atoms, the Coulomb embedding potential is not sufficient and a more sophisticated treatment is needed. An extension of this model to SiO_2 has recently been described in ref. [44].

A more consistent, but also computationally demanding approach is to use the results of periodic calculations of the ideal crystal to couple the defect region with the infinite lattice self-consistently. Different realizations of this approach are described in [45,46,47,48,49] and others. However, the most successful applications of this approach are again to ionic systems (see, for example, ref. [50] for discussion of applications to more covalent SiO_2).

To summarize, the molecular cluster approach is very easy to implement and can be fairly accurate if one is interested in local defect properties. The embedded cluster model can successfully compete with the periodic model in ionic systems where the embedded cluster model gives quite accurate results. One of the main advantages of the cluster approach is that it is inherently local by nature and can be easily applied to bulk crystals, amorphous materials, surfaces [43,51] and complex interfaces [52].

2.2. Electronic structure calculations: basis sets and methods

The second component of any computational tool is a method used for calculating the total energy and properties of the system. Both periodic and cluster models can be used in combination with different methods. Below we will briefly survey some of the quantum mechanical methods used later in discussing particular systems. Details of these techniques could be found, for example, in [53,54]. Since we are mainly concerned with the electronic properties of materials, we shall not discuss the atomistic simulation techniques based on classical inter-atomic potentials. These methods play a very important role in modelling and simulation of defects in solids, and are discussed in detail in, for example, [55].

Basis sets

The basis set controls the accuracy of calculations for a given method and its boundary conditions. Two types of basis sets are used in most of the calculations: a plane-wave (PW) basis set and a basis of atomic orbitals (AO).

Plane-Wave basis set

The PW basis can only be used in conjunction with periodic boundary conditions. It is very effective in practical calculations because: i) it is an orthogonal basis, and ii) a product of two planewaves is also a plane-wave and their derivatives are products in k-space. Due to these PW properties and very effective computational methods exploiting the periodicity of the system, the number of atoms that can be included in a supercell is generally a few times larger than the number of atoms in cluster calculations for comparable accuracy. The PW basis is independent of atomic positions and spans the whole supercell uniformly. For the given size of the supercell, it can be systematically saturated by increasing a single parameter - the cutoff energy, E_c. The value of E_c is determined by the highest density curvature. This implies that the description of the core electrons requires inaccessibly large values of E_c. To overcome this problem the PW basis is used together with the pseudopotential method (see, for example,[56,57]), or with the projector augmented-wave method (PAW) [58,59] representing atomic cores. On the other hand, regions of small and weakly changing charge density are often described with excessive accuracy. This substantially increases the computational effort in calculations including free space, such as cage systems or surfaces.

Extended character of plane waves comes as disadvantage when essentially local interaction has to be described. In particular, PW basis set is not well suited for the efficient calculations of the exact exchange interactions. As a result, PW basis in DFT is often used in conjunction with the local or semilocal (i.e. GGA) density approximations. The socalled self-interaction error arising from these approximations may lead to qualitatively incorrect predictions in systems where interaction between electrons leads to a formation of localized states [60,61,62,63].

Atomic basis

Some of the problems related to the PW basis can be partially lifted if a localized basis sets is used instead. The basis set of localized *atomic* orbitals (AO) is often used in molecular, cluster and periodic calculations because it is not coupled to particular boundary conditions. However, the AO basis functions are, in general, not orthogonal. Therefore, the basis set optimization procedure is not well defined and the construction of a complete basis set presents a problem of its own. Application of AO basis sets results in many cases in a linear dependent yet incomplete basis and in the Basis Set Superposition Error (BSSE) [64].

The AO basis set is generally localized on atoms and hence depends on atomic positions. This allows for better wave-function representation close to atomic cores where it is needed. Thus an AO basis set can be used for calculation of the core electrons and their electronic transitions. In this sense the AO basis is physically justified. Due to a small overlap of atomic orbitals of distant atoms, the system in question can be partitioned in such a way that the computations of different parts can be carried out independently. This opens a possibility for effective parallelization of computer codes [24]. Also various so-called order-N algorithms were proposed. These methods heavily rely on the localized character of atomic basis-sets in that all non-trivial interactions between the distant atoms can be neglected. This allow one to significantly reduce the computional load and to make it linearly proportional to the number of particles, N, in the system (hence order-N). Systems of several thousand atoms can be considered using these methods [24,25].

An additional advantage stems from the fact that localized functions are better suited to describe the behavior of electron density in free space (e.g. above the surface or inside open voids in crystal structure). There are, however, problems if electrons which are not associated with atomic cores are considered (an F centre is an obvious example, see Sec. 4). The most common examples of AO basis sets include Slater- and Gaussian-type functions [65,66], splines and numerical functions [24,25,67]. Perhaps the most computationally advanced approach is to represent the AO functions as linear combinations of Gaussian-type functions due the their mathematical simplicity. However, Gaussian functions provide poor asymptotic behaviour for the electron density near atomic cores and at large distances from them. Therefore a relatively large number of functions is needed to represent an atomic orbital. Recent approaches where a mixed plane-wave and Gaussian-type basis functions are used [68] are expected to be beneficial for both periodic and cluster calculations.

Methods

The number of theoretical methods for electronic structure calculations is too large to be even briefly described in this section. Instead we concentrate on only three families of methods which are widely used in defect calculations. These families are the Hartree-Fock (HF) and post-HF methods, the Density Functional approach and methods combining advantages of both HF and DFT - so called hybrid DFT methods. Most of the theoretical results described in the following sections were obtained using these methods.

Hartree-Fock and post HF methods

The method suggested by Hartree for electronic structure calculations of N electrons in the field of all nuclei accounts for the electrostatic interaction of all particles composing the system. Each electron of the system moves in a mean field due to the nuclei and other electrons. A way to account for the Pauli exclusion principle for the electrons was later suggested by Fock. In the most simple realization, the wave-function is written as a single determinant build from orthonormal one-electron functions called molecular orbitals (MO). The MO and their energies are found by solving the system of HF equations. In the ground state, the lowest $N/2$ MOs are occupied by electrons and all other are free. In crystal calculations, the difference of one-electron energies of the lowest unoccupied and the highest occupied MOs is called a single-particle band gap. In HF calculations it is usually overestimated by about 100%. A significant advantage of the HF approach is that it combines reasonable accuracy with a relatively modest demand for computing resources and generally fast convergence. It includes the non-local exchange interaction between electrons, but completely neglects electron-electron correlation.

To account for electron correlation post-HF methods have been developed. Most of the post-HF methods are based on different realizations of the Configuration Interaction (CI) approach, where the wave function is written as a linear combination of determinants built from the HF wave-function by multiple substitutions of the occupied MOs by unoccupied ones. The Full CI, for example, takes into account all possible substitutions, whereas a popular CISD method considers only all Single and Double substitutions so that no more that two occupied orbitals are replaced by unoccupied ones in every determinant. These methods are applicable to calculations of both ground and excited states of the system and can provide a very accurate and, in principle, even exact solution for a given basis set. However, they are very computationally demanding and therefore their use is limited to relatively small systems.

Density functional approach

The Density Functional Approach is based on two theorems by Hohenberg and Kohn which state that: i) the ground state energy of a non-degenerate electronic state is a unique functional of its density, $E[\rho]$ (density functional); ii) the energy can be obtained by variation of the universal density functional with respect to the charge density. This implies that calculation of the wave-function of the many-electron system is not required in order to find the total energy, only the charge density is required. However, the exact density functional is not known and some approximate functionals are used instead.

The density functional $E[\rho]$ can be divided into terms which describe the kinetic energy of electrons, their interaction with the external potential, their electrostatic interaction with each other and, finally, the term which accounts for the exchange and electron correlation, E_{xc}. The formally exact method for the self-consistent calculation of the ground state energy E by varying the charge density, was proposed by Kohn and Sham [85,54]. Although the Kohn-Sham method is formally exact, all its practical emplementations require an approximation to be made for the exchange-correlation energy term, E_{xc}, whose exact form is unknown. A widely used Local Density Approximation (LDA) assumes that the exchange-correlation energy E_{xc} for each element $\rho(r)dr$ of the system is the same as for the uniform electron gas of the same density. The exchange energy and potential in LDA can be written in a closed local form, while the correlation term is normally parametrized to more refined atomic and molecular calculations [54].

Although the LDA is clearly a very strong approximation, it proved to be very successful in predicting the geometric structure and ground state electronic properties of many materials. Its obvious deficiency is that it neglects the effect of fast-changing electron density on the exchange-correlation energy, E_{xc}. This results in systematically underestimated equilibrium inter-atomic distances predicted by this method. To account for this effect semilocal functionals were developed which also include the dependence on the gradient of density. This approach is known as Generalised Gradient Approximation or GGA.

Some of the disadvantages of LDA and GGA relevant to further discussion include systematically underestimated single particle band gaps and an overall tendency to delocalize the electron density. This may become particularly problematic in calculations of defects in insulators because too narrow band gaps make it difficult or impossible to describe shallow defect states. Similarly, the tendency to delocalize the electron density can, and often does, lead to qualitatively wrong conclusions regarding defect electronic structure (see, for example, discussion in refs. [60,61,63]). Addressing this issue remains a strong focus of methods research in materials science, and several possibilities exist. GW perturbation theory [69,26] offers highly accurate energy gaps in semiconductors, both quantum Monte Carlo (QMC) [70] and time-dependent DFT [71,?] give improved gaps but at a cost of greatly increased computational resources.

Some of the problems discussed above can also be eliminated if one combines the advantages of DFT and HF in the so-called hybrid density functional approach. Perhaps the most widely used hybrid functional is due to Becke [72] and is called B3LYP. It has demonstrated great improvements in many systems including MgO, alumina and TiO_2 [73]. In this functional, the correlation term in the E_{xc} is the one suggested by Lee, Yang and Parr [74] (hence LYP), and the exchange term is the sum of three contributions including the Hartree-Fock exchange weighed with some parameters (this is what the number 3 in B3LYP stands for). The parameters were fitted to reproduce the ionization potentials and electron affinities of atoms, and dissociation energies of a large number of molecules. An accumulated experience demonstrates that the B3LYP parametrization is good at reproducing both atomic and electronic properties of many ionic materials, and is widely used in both periodic and cluster calculations. However, the fact that B3LYP includes to some extent the HF exchange makes it a non-local functional and it is therefore implemented only in computer codes which use localized basis sets.

Basis vs. Method

To complete this section we will make a few comments on the relation between the methods, basis sets, and models used in calculations.

The HF approach can be used in both periodic and cluster models. In commercially available packages it is usually implemented using Gaussian-type basis functions. However, as mentioned above, recent developments open up the possibility for plane-wave calculations using the HF approach.

The DFT can also be used in both periodic and cluster calculations. In periodic calculations both plane-wave or Gaussian-type basis sets can be used. Cluster calculations usually employ Gaussian basis sets.

2.3. *Calculation of defect properties*

Formation energies

Before discussing some examples, it is important to introduce the concept of formation energy, as it will prove important throughout this discussion. The vacancy formation energies (or equivalently, the oxygen atom/molecule incorporation energies in the next section) $E_{for}(D)$ were calculated as the energy difference between the fully relaxed defected neutral supercell (QM cluster), E_D^0, and the perfect neutral supercell (QM cluster), E_0^0, and an isolated oxygen atom/molecule, E_O, according to:

$$E_{for}(D) = E_D^0 - (E_0^0 \pm E_O). \qquad (1)$$

Here E_O is the total energy of the individual oxygen atom or molecule. It is subtracted for a vacancy and added for an interstitial. The value of E_O depends on *process* being modeled in the calculation [6,75,76,77]. For example, during annealing of an oxide film in oxygen atmosphere, oxygen molecules can dissociate into atoms at the surface or due to irradiation. However, in most cases calculations are concerned only with final possible state of an atom or a molecule in the bulk of a material, and tend to disregard the process of defect formation. If atoms are available directly from an atomic source then an atomic reference for energy E_O is used. If a molecule can dissociate into atoms in the process of incorporation, then half of the molecular energy is taken as a reference to account for the dissociation costs. Thus the formation energy gives just a measure of difficulty of forming a neutral defect in the oxide - the higher the energy, the more difficult it is to create.

Electrical levels

Yet another issue, specific for microelectronics applications, is that the oxide of interest can be in contact with an electrode(s) - semiconducting (e.g. silicon) or metallic - which provides a source of extra electrons. Also an oxide film can be placed in the electric field between two electrodes, which may alter the defect charge states. To define which charge state of an intrinsic defect (e.g. vacancy) or

162

incorporated defect specie is favored, we assume the common case of an oxide film grown on silicon, where the source or sink of electrons is the bottom of silicon conduction band at the Si/SiO$_2$ interface. Hence the chemical potential of the electron can be chosen at the bottom of Si conduction band or at the silicon midgap energy [78]. Electrons are assumed to be able to tunnel elastically or inelastically [79] from/to these states to/from defect states in the oxide and create charged species. Since the interface is not explicitly included in most calculations, one needs to use experimental information [80,81] (e.g. 4.6 eV for the valence band offset at the Si/SiO$_2$ interface, 1.1 eV for the band gap of Si) to estimate the energy of an electron at the bottom of the conduction band of Si with respect to the theoretical zero energy level (see Fig. 3). The calculated energies are readily adjusted if doping, temperature, or bias favour another source.

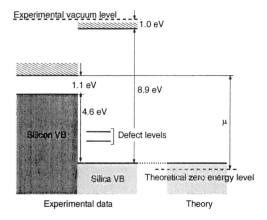

Figure 3. Schematic band offset diagram showing the relative energies of the top of the valence bands and the bottom of the conduction bands of the bulk materials in contact (for the Si/SiO$_2$ interface [80,81]). Note that the zero energy level in theoretical calculations of 3D infinite bulk materials is poorly defined and depends on particular implementation of a computational method. It can be located even below the top of the calculated valence band.

We note that in many DFT packages for periodic calculations the positions of the electron bands stay the same independent on the defect charge state. On the algorithmic level this is achieved by adding a constant shift to the one-electron energies which compensates the potential produced by

the charged defect. As a result the calculation of the electrical levels in periodic DFT is relatively straightforward. This is not the case in cluster calculations where all one-electron levels are shifted respectively up or down if a negative or positive defect is introduced. Therefore the absolute positions of bands are not well defined.

To define the dependence of the defect charge state on the position of the system Fermi energy, it is useful to distinguish two charge-state levels: the thermodynamic level, E_{th}, and the switching level, E_{sw}. The thermodynamic charge-state level corresponds to the system Fermi energy for which a defect changes its charge in thermal equilibrium. E_{th} is defined as a difference of total energies of systems of N and $N+1$ electrons, each in its fully relaxed ground state: $E_{th} = E(N+1) - E(N)$. Thus, if E_F is the energy which is required to remove an electron from some electron reservoir (e.g. Si at the Si/SiO$_2$ interface), E_{th} - E_F is the energy required to add an electron from the reservoir to the lowest unoccupied charge state level.

Taking into account that the electrical measurements and the electronic processes are fast compared with the time needed for the system to come to equilibrium, another level is introduced which is consistent with the Frank-Condon principle. E_{sw} is defined as the energy required to trap an electron or a hole in the geometry of the system as it was prior to the trapping. Thus E_{sw} are obtained as the differences of the total energies of the system with N and $N+1$ electrons, but in the local atomic geometries which correspond to one of these systems. For example, $E_{sw}(0/-) = E(-) - E(0)$ denotes the energy required to add an electron to the neutral defect calculated using the neutral defect geometry.

The analysis of electron tunnelling between semiconductor bands and insulator traps at a semiconductor-insulator interface in these terms is presented in ref. [82]. For insulator traps, which exhibit large lattice relaxation, this model predicts resonant tunnelling followed by atomic relaxation at a defect. It was first used to explain the hysteresis observed by Zvanut et $al.$ in band-to-trap tunnelling at the Si/SiO$_2$ interface [83] and is applicable to other similar systems.

Once the charge-state levels are calculated, the dependence of the defect charge state on the electrode Fermi energy can be introduced via a correction term qE_F, where E_F is varied in the energy range from the of the maximum of the last valence band (VBM) of the oxide to the bottom of the first conduction band (CBM) (providing the band gap is correctly reproduced). The results are presented as plots of (E_{th} vs. E_F) or (E_{sw} vs. E_F) calculated for different values of E_F. The crossing points of the corresponding lines determine the regions where one charge state of a defect is more favorable than the other. Although the Fermi energy remains unknown in the experiment, such diagrams allow one to identify some critical cases and help in the analysis of the experimental data. The same approach is used to determine the dependence of the defect formation energies on the position of the Fermi energy.

3. Results for perfect (non-defective) crystals

To get some feeling of the possibilities and accuracy of various models and techniques, it is instructive to consider non-defective systems. Here we consider the accuracy of different methods for studying both the physical and electronic structures of bulk systems, as well as dielectric properties.

3.1. Periodic DFT calculations: general trends

In the last two decades, DFT [84,85], implementing LDA and GGA functionals, has become almost a reference for determining the atomic structure of a wide variety of bulk materials [86,87,88,89,90,91]. The overall accuracy in calculating electron exchange and correlation contributions to the total energies means that predicted structures are generally within 5% or better of experimental results. This surprising accuracy (even to its original authors [92]) is due to a somewhat fortuitous cancellation in the total exchange-correlation functional of the individual errors in calculating the correlation and exchange energy.

For oxides, which are especially relevant to microelectronics applications, periodic DFT provides very reliable predictions of atomic structure and thorough theoretical studies have demonstrated excellent structural agreement with experiment in, for example, TiO_2 [93], MgO [94], alumina [95] and ceria [96]. The confidence in the reliability of calculated structures, also means that DFT can be used as a tool to predict the stability of different structural phases and to study the process of phase transition itself. We will see a demonstration of this for hafnia and zirconia later, but at this point it is relevant to discuss some of the more extensive works in the literature. For example, DFT has played an important role in understanding phase transitions in alumina [97], where Al cation migrations are responsible for the transformation from γ- to θ-alumina. In TiO_2, DFT calculations [93] were able to reproduce the experimentally observed order of four complex phase transitions: anatase-columbite; rutile-columbite; columbite-baddeleyite; and baddeleyite-cotunnite.

However, it should be noted that DFT is not a 'panacea', and for many systems it does not provide a correct description: the approximations in the calculation of the correlation means that DFT using LDA/GGA functionals fails in the description of the electronic structure of strongly correlated materials, such as high-Tc superconductors and magnetic materials like NiO. More crucially, it also means that the band gaps of materials are always underestimated severely - sometimes even by 50%. We will discuss some specific examples later, but it is important to understand that this effect is not systematic, and the error can vary significantly with different materials. Further discussion of this issue can be found in Sec. 2.

Silica

Silica (SiO_2) exists in many different phases at ambient pressure, and can exist in several additional metastable phases via rapid cooling from a high temperature phase - silica glass being the most obvious example. At ambient pressure the sequence of silica transitions with temperature are as follows: α-quartz to β-quartz at 847 K, then to HP-tridymite at 1140 K and finally to β-cristobalite at 1743 K. All of these structures are formed from networks of corner-sharing SiO_4 tetrahedra with differing amounts of disorder. Silica, in various phases, has been studied extensively in the literature using a variety of techniques. Some example studies are: periodic Hartree-Fock [88,98]; periodic DFT [98,99,100,101]; molecular cluster [102]; and many other studies referenced throughout this chapter. Hence, in this section we will just briefly mention DFT calculations of α-quartz as a comparison to the next section.

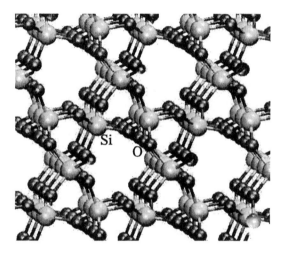

Figure 4. Atomic structure of the α-quartz phase of silica.

As can be seen in Fig. 4, the α-quartz structure of silica is very open, with large interstitial space. This property will prove to be important in the discussion of the incorporation of defects, and contrasts strongly with somewhat more close-packed structures of zirconia and hafnia shown in Fig. 5. The calculations of silica were performed using PW DFT as described in the previous sections, with a pseudopotential for silicon of the form $[Ne]3s^23p^2$. Table 1 shows that the calculations reproduce the experimental lattice properties to within about 2%, although the actual Si-O bond lengths are reproduced to less than 1%. However, the predicted single particle band gap of 6.1 eV, as in all standard DFT calculations, is a large underestimation of the experimental value of 8.9 eV.

Property	Calc.	Exp. [103]
a (Å)	5.026	4.902
b (Å)	5.021	4.902
c (Å)	5.509	5.400
γ (°)	120	120

Table 1
Comparison of theoretical and experimental (taken at 13 K) lattice parameters for α-quartz silica.

Zirconia and hafnia

The homological electronic outer shell configuration of the cations in zirconia (ZrO_2) and hafnia (HfO_2) suggests that they have many similar properties. In fact, both types of atoms are often found together in natural minerals [104]. Both exist in three polymorphs at atmospheric pressure: at low temperatures the monoclinic C_{2h}^5 phase (space group $P2_1/c$), above 2000 K (1400 K for zirconia) the tetragonal D_{4h}^{15} ($P4_2/nmc$) phase, and above 2870 K (2600 K for zirconia) the cubic fluorite O_h^5 ($Fm3m$) phase.

The cubic phase, simplest of the three, consists of oxygen ions forming the simple cubic lattice, with Hf(Zr) occupying alternate volume interstitial positions as shown in Fig. 5a. The tetragonal phase is closely related to the fluorite, and can be obtained by up and down displacement of oxygen ions from their centrosymmetric positions (Fig. 5b) by some value δz. This displacement is associated with the zone boundary X_2^- phonon vibration in the fluorite phase. Both fluorite and tetragonal phases are characterized by similar local coordination - cations are 8-fold coordinated and anions are tetrahedrally coordinated. This is in contrast with the monoclinic phase, where the cation coordination is reduced to seven, while anions occupy two non-equivalent lattice positions - four- and three-fold coordinated (Fig. 5c). It is evident that the transformation between the phases is primarily associated with the change in oxygen internal coordinates. The small change in the cation sublattice is entirely due to a unit cell deformation - the z-axis is slightly elongated in the tetragonal lattice, and elongated and tilted in the monoclinic lattice.

Numerous experimental studies of zirconia and hafnia made available extensive data on their structural [105,106,107,108], mechanical [109,104] and dielectric properties [110,111,112,113], as well as on their phase diagrams and phase transformations. In particular, much attention was given to the martensitic character of the cubic-tetragonal and tetragonal-monoclinic transformations and room temperature stabilization of the cubic and tetragonal zirconia (see recent review by Kelly and Rose [109] and references therein). This interest is primarily driven by wide engineering application of these mechanically tougher phases.

Early electronic structure calculations also focused on the stabilization mechanisms of cubic zirconia [114,115,116,117]. However, the recently emerged potential application of zirconia and especially hafnia as gate dielectrics in CMOS devices, resulted in revived interest and the appearance of higly accurate calculations of zirconia [6,118,119,120,121,122] and hafnia [7,123,124,125] polymorphs. These calculations demonstrate general agreement on the physical and electronic properties of these materials. In particular, a remarkable stability of the monoclinic phase is stipulated by a reduced cation coordination, resulting in a stronger Zr-O bond [126]. This also explains the stabilization of tetragonal phase by three-valent impurities, which produce anion vacancies and effectiveley reduce zirconium coordination. At the same time the chemically pure fluorite phase is understood to be mechanically unstable with respect to the X_2^- phonon. Therefore, the tetragonal to cubic phase transformation can not be described as simply a second order displacive transition [120,127]. As we shall see, the dielectric properties of the high-k oxides are directly related to phase stability.

In this section, we will highlight results of our own calculations of bulk Zirconia and Hafnia [6,7], followed by the discussion of the dielectric properties.

The calculations were performed using the PW DFT implemented in VASP code [128,129]. The spin-polarized DFT at the GGA level with the density functional of Perdew and Wang [130] was used together with the ultrasoft Vanderbilt pseudopotentials [56,57]. The pseudopotential for the zirconium atom was generated in the electron configuration [Kr]$4d^35s^1$, hafnium atom in

[Xe.$4f^{14}$]$5d^36s^1$ and that for the oxygen atom in [$1s^2$]$2s^22p^4$, where the core electron configurations are shown in square brackets. For each system, total energy convergence was tested within a k-point range between 1 and 80 (60) k-points and a plane wave cutoff energy range between 200 and 700 eV. Convergency to within 10 meV was achieved with 20 k-points and a cutoff energy of 400 eV. The bulk unit cell lattice vectors and atomic coordinates were then relaxed at a series of fixed volumes. The obtained energies were fitted with a Murnaghan equation of state [131] to give the equilibrium volume and the minimum energy.

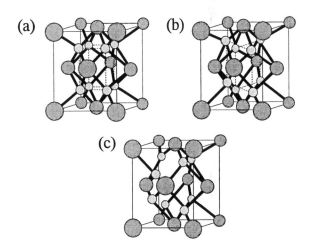

Figure 5. Diagram showing the three phases of zirconia/hafnia: (a) cubic (flourite), (b) tetragonal and (c) monoclinic (the tilt of the z-axis is not shown for clarity). Large dark grey atoms are metal ions and small light grey are oxygens.

The first stage of any extensive calculations using pseudopotentials is to check their validity for the types of system you are interested in. For zirconia, previous studies had demonstrated that the Zr and O pseudopotentials were valid, however no similar studies were performed for hafnia. Hence, in order to validate both the Hf pseudopotential (and also to find the chemical potential of Hf) calculations were performed on bulk hcp hafnium metal. For the pure metal, the total energy was found to converge to within 10 meV for a plane wave cutoff energy of 250 eV and 252 k-points in the irreducible part of the Brillouin zone (BZ). Table 2 demonstrates that the experimental stuctural parameters of hafnium metal are well reproduced.

Property	Calc.	Exp. [132,133]
a (Å)	3.1797	3.1946
c (Å)	5.0239	5.0510
c/a	1.58	1.58
V_0 (Å^3)	21.99	22.32
B_0 (GPa)	110	110

Table 2
Comparison of calculated and experimental properties of bulk hcp hafnium.

Having established confidence in the pseudopotentials for elemental systems, the next step is to approach the three different phases of zirconia and hafnia. Successfully reproducing the experimental structure and energy hierarchy of the different phases would provide a fundamental base for more complex calculations of defects etc. The static equilibrium cell parameters for zirconia and hafnia are given in table 3, along with experimental values. Once again it is clear that the calculations provide excellent accuracy in predicting the physical structure of these materials in all phases. Equally important is that the calculations also predict the correct ordering of the phases, so that for both zirconia and hafnia the monoclinic phase is most stable (i.e. has lowest energy), with tetragonal and cubic increasing in energy. The energy difference between the phases of zirconia also agree fairly well with experiment. Table 3 also demonstrates general similarity of zirconia and hafnia. We note however, that monoclinic hafnia has a smaller unit cell volume than zirconia. This will become important in later discussions about defects in these materials. Note however, that the experimental data for cubic and tetragonal phases are taken at high temperature, e.g. 2073 K for tetragonal, and are also somewhat dependent on temperature, and therefore comparisons with the 0 K theoretical data are limited.

Despite the accuracy achieved in the structure of bulk zirconia and hafnia, due to the problems discussed previously, a similar level cannot be achieved for the electronic properties of the materials. This is easily demonstrated by considering

Property	Zirconia		Hafnia	
	Calc. [6]	Exp. [105,106]	Calc. [7]	Exp. [104,108,134]
Cubic				
Volume (Å^3)	32.97	32.97	32.49	32.77
a (Å)	5.090	5.090	5.07	5.08
Tetragonal				
Volume (Å^3)	34.55	34.07	33.12	35.075
a (Å)	5.131	5.050	5.06	5.15
c/a	1.023	1.026	1.024	1.027
δz	0.049	0.057	0.051	-
Monoclinic				
Volume (Å^3)	36.05	35.22	34.81	34.62
a (Å)	5.192	5.150	5.1322	5.1187
b/a	1.014	1.012	1.011	1.010
c/a	1.032	1.032	1.034	1.035
β (^{o})	99.81	99.23	99.78	99.18
Zr(Hf)$_x$	0.277	0.275	0.277	0.276
Zr(Hf)$_y$	0.044	0.040	0.044	0.040
Zr(Hf)$_z$	0.209	0.208	0.209	0.207
O1$_x$	0.072	0.070	0.070	0.071
O1$_y$	0.338	0.332	0.333	0.332
O1$_z$	0.341	0.345	0.345	0.344
O2$_x$	0.447	0.450	0.448	0.446
O2$_y$	0.758	0.757	0.758	0.755
O2$_z$	0.479	0.479	0.478	0.480
Energy Differences between Phases (/MO$_2$)				
E^{t-c} (eV)	-0.07	-0.06	-0.08	-
E^{m-c} (eV)	-0.17	-0.12	-0.24	-

Table 3
Comparison of calculated and experimental bulk unit cell parameters for the cubic, tetragonal and monoclinic phases of zirconia and hafnia. δz is the shift in fractional coordinates of oxygen atoms in the tetragonal cell with respect to their ideal cubic positions, β is the angle between lattice vectors a and c in the monoclinic cell and x, y, z are the fractional coordinates of the nonequivalent sites in the m structure. Note that the experimental values are taken from several sources.

the band gaps predicted by theory. The band-gap can be calculated in two different ways [6]: (i) as the energy difference between the highest occupied and lowest unoccupied one-electron states, or (ii) it can be estimated as a difference of the total energies of the system with N, N+1 and N-1 electrons [135]:

$$E_g(theor) = E(per, -1) + E(per, +1) \\ -2 \cdot E(per, 0) \tag{2}$$

Here $E(per, 0)$ is the total energy of the perfect neutral supercell, and $E(per, -1)$ and $E(per, +1)$ are the total energies of the supercell with an electron or a hole. Table 4 shows a comparison of the values calculated using these methods and experimental results. Although there is some variation in the experimental results, it is generally clear that the band gap is severely underestimated in the DFT calculations. It is possible to improve this by applying a many-body perturbation theory with PW DFT results as a reference. For zirconia this technique predicts a much improved band gap of 5.4 eV [118]. However, these calculations are still prohibitively expensive for the systems larger than a few atoms.

System	Method (i)	Method (ii)	Exp.
Zirconia	3.41	3.19	4.2 [111] 5.4 [136] 5.83 [110]
Hafnia	4.12	3.92	5.68 [112]

Table 4
Comparison of calculated and experimental band gaps in eV for zirconia and hafnia.

Although we are limited in the accuracy of absolute values, especially for excited and conducting states, the accuracy of the structures means that the occupied electron states must be reliable. Hence qualitatively the electron density of states (DOS) is reliable, and gives a good idea of how different orbitals lie in the energy spectra. The electron DOS for ideal monoclinic hafnia is shown in Fig. 6a. Note that the DOS for zirconia is almost exactly the same, and the following discussion applies equally well to both oxides. For better presentation, each of the discrete one-electron energies

forming the spectrum was broadened by a Gaussian with a smearing factor equal to 0.3 eV. The tails at the band edges are determined by this factor and do not have a quantitative meaning, but the raw DOS at the valence band edge can be seen in Fig. 6b. The DOS for the monoclinic phase is very similar to the DOS obtained for the cubic and tetragonal phases (not shown here), and has three clear bands. A valence band of oxygen $2s$ character at around -15 eV, a valence band of oxygen $2p$ character at around 0 eV and a conduction band of hafnium $5d$ character at around 7 eV. There is a small number of states of Hf $5d$ character in the middle band, but it is dominated by the O $2p$ states. This is consistent with the picture of hafnia as an ionic insulator, with some degree of covalent bonding between Hf and O.

3.2. Perfect crystals in the embedded cluster approach

As has been mentioned before (see section 2), there is a number of embedded cluster techniques ranging from fairly simple, where a quantum cluster is embedded into an electrostatic potential represented by point charges, to quite complicated, where the wave-function of the cluster is coupled with the wave-function of the ideal crystal. In this section we discuss the results of ideal crystal calculations obtained using an approach in which the quantum cluster is embedded in the shell model environment [43,44]. To illustrate the accuracy of the embedding technique we consider two generic examples: MgO, which has a tightly packed structure, and α-quartz, which is much more "open" and flexible. We show that the results depend not only on the quantum mechanical method and parameters of the shell model, but also on their consistency.

When two different methods are combined together in embedding technique, perturbations to the local geometry and the electronic structure usually occur due to two main reasons: i) approximate nature of embedding potential; ii) difference in the geometric parameters of the lattice given by both methods separately. Proper embedding aims at making these perturbations (negligibly) small.

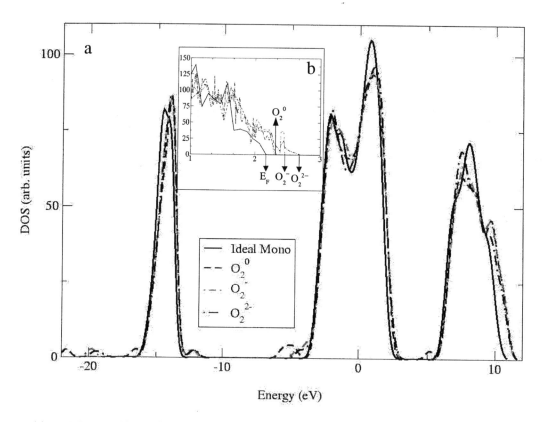

Figure 6. (a) Total Density of States (DOS) for ideal and defected monoclinic hafnia. Note that gaussian smearing has been applied to the DOS for clarity. (b) A blow-up of the unsmeared DOS around the Fermi level (E_F) of the ideal system, at 2.1 eV. The arrows show the highest occupied one-electron states of the defected systems, O_2^0 at 2.25 eV, O_2^- at 2.40 eV and O_2^{2-} at 2.70 eV. *Reproduced with permission [7].*

Consistency and accuracy of the embedding technique is checked with a so-called "perfect lattice test" in which a defect-free quantum cluster is embedded in the defect-free classical environment. The system is then fully relaxed and the displacements of atoms, particularly at the cluster-environment interface, and the electronic structure are monitored. To be informative, this test should be carried out for several clusters of different size treated by the same quantum mechanical method using the same basis set. To estimate the lattice distortions, changes of the interatomic distances and nearest-neighbour angles are calculated for atoms inside the QM cluster, in the classical environment and at the interface. The electronic structure of the QM cluster can be characterised by the atomic charges. This is a simple and conve-

nient approach which allows one to compare the charge density inside and at the boundary of the cluster. A more sensitive characteristic, such as atom-projected DOS, can also be used, although large clusters are needed to make this analysis feasible. An "ideal" embedding would correspond to the undistorted host lattice and homogeneous charge distribution within the cluster. Below we describe these tests for MgO and α-quartz.

MgO

The high symmetry of the rock-salt structure, and generally large ionic charges, make the MgO lattice rather rigid. The shell model is fitted to reproduce the experimental lattice parameters and dielectric constants, and the classical ions usually carry formal charges (+/- 2 in the case of MgO). Quantum mechanical calculations provide generally good, although not perfect, agreement with the experimental lattice parameters and ionic charges also close to formal. Therefore, one should expect only slight distortions at the cluster-environment interface in this case.

Indeed, we find that the spread of Mg-O distances in most perfect lattice calculations is within 1-3 % , with the largest values at the quantum/classical interface. Typical spread of ionic charges within the quantum cluster is about 0.05 e with the smallest charges at the centre of a cluster (see, for example, ref. [43]). The best results are obtained for stoichiometric clusters of cuboidal shapes.

α-quartz

α-quartz and SiO_2 polymorphs in general are more difficult to reproduce due to their structure: rigid SiO_4 tetrahedra connected by flexible Si-O-Si bridges. These Si-O-Si bridges act as "shock absorbers" and readily respond to the lattice size mismatch between the QM and classical regions. Consequently, the Si-O-Si angles, mostly at the interface, are perturbed by as much as 5% and the interatomic distances at the interface deviate from their ideal crystal values by about 4%. The distortions in the rest of the system are at least three times smaller. Calculated ionic charges in the quantum cluster are about 2.70 - 2.75 e for Si ions and -1.35 - -1.40 e for O ions. This is larger than the shell model charges of 2.4 and -1.2 e for Si and O respectively, which indicates a mismatch of the electrostatic potential.

This accuracy is generally sufficient for defect studies. However, it can still be improved if the parameters of the shell model are fitted to match the geometry obtained in quantum mechanical calculations. A possible approach to reducing this mismatch can be briefly described as follows: First, a crystal unit cell is relaxed using a QM method and a localized basis set in a periodic model calculation; second, parameters of the classical shell model are reoptimised to reproduce the relaxed unit cell structure. Then, the crystal geometry would be the same in both QM and classical calculation.

3.3. *Vibrational and dielectric properties of bulk materials*

The electronic characteristics of materials discussed above are tightly interconnected with their vibrational and dielectric properties. This fundamental physical property directly affects the performance of CMOS devices, which depends on a dielectric capacitance in a gate stack, band off-sets and heat dissipation parameters. One of the means for increasing capacitance is to employ for a gate dielectric a material with a dielectric constant higher than that of currently used silicon dioxide. However, in order to maintain sufficiently high band energy off-sets (see Fig. 3) with the channel layer and metallic contact, the material of choice must have a sufficiently large electron band gap. In general, band gaps tend to be smaller for materials with higher ϵ_0 (see for example discussion by Robertson [137]). How to satisfy these contradictory requirements is in the focus of many recent studies.

In order to identify key factors affecting dielectric response, let us recall a relation of the dielectric tensor to microscopic parameters of the system. Microscopically, the static dielectric tensor, ϵ_{ij}^0, has three separable contributions [138]:

$$\epsilon_{\alpha\beta}^0 = \epsilon_{\alpha\beta}^\infty + \epsilon_{\alpha\beta}^a + \epsilon_{\alpha\beta}^p. \qquad (3)$$

Here the first term is the electronic dielectric screening with atoms being fixed at the perfect lattice sites; the second term is due to atomic displacements with fixed lattice volume; and the last term is related to changes in lattice volume (and shape), that is, to the piezoelectric properties of the system.

Piezoelectrics usually have too narrow band gaps and are strongly susceptible to stress induced damage. Therefore, they cannot be effectively used as gate materials. The electronic contribution, ϵ_{ij}^∞, is mainly defined by the polarisability of ions in the perfect lattice. Typical values of ϵ^∞ in oxides range between 1.6 (Li_2O) and 6.5 (Cu_2O). Although values as high as 10 are observed in materials with anomolously high cation polarisability (e.g. Fe_xO_y or Ti_xO_y), these are narrow gap materials, and therefore, do not fit into the context of a gate stack in CMOS. Therefore, one should conclude that despite the importance of the electronic screening contribution, potentially practical oxide materials with ϵ^0 in a range between 20 and 40, would have a dominant contribution from atomic screening. In the harmonic approximation the static atomic dielectric tensor ϵ_{ij}^a can be decomposed into contributions of individual phonon modes [122]:

$$\epsilon_{\alpha\beta}^a = \frac{4\pi e^2}{V} \sum_\lambda \frac{Z_{\lambda\alpha}^* Z_{\lambda\beta}^*}{\omega_\lambda^2}, \tag{4}$$

where α and β denote Cartesian coordinates, and λ numerates phonon modes with the corresponding frequency ω_λ; e is the electron charge, and V is the volume of the unit cell. The mode effective charges, $Z_{\lambda\alpha}^*$, are defined by the expression:

$$Z_{\lambda\alpha}^* = \sum_{i\beta} Z_{\alpha\beta}^{*i} \frac{\xi_{i,\lambda\beta}}{M_i^{1/2}}, \tag{5}$$

where $\xi_{i,\lambda\alpha}$ corresponds to a normalized displacement of the atom i of mass M_i in the Cartesian direction α according to a phonon λ; $Z_{\alpha\beta}^{*i}$ is a Born dynamic charge tensor defined as the polarisation P_α induced in the crystal in the direction α by the infinitesimally small displacement of an atom i in the direction β:

$$Z_{\alpha\beta}^{*i} = \frac{V}{e} \frac{\partial P_\alpha}{\partial u_\beta^i}. \tag{6}$$

It follows from eq. (4) that the ionic response is defined by infrared active (IR) modes, with the main contribution coming from the low frequency modes with large mode effective charges. The understanding of the origin of such modes will bring a deeper insight and, therefore better control of dielectric properties of the materials of interest. In particular, one is interested in how the atomic dynamics is affected by the amorphisation or alloying of a material, and the role of internal stresses and strains induced in a dielectric film by the electric and elastic fields. Can a low level doping inhibit or enhance dielectric response of a material? Another important issue is that of anisotropy and its role in the long time reliability of the CMOS devices.

Recent developments in the quantum theory of polarisation [138,139,140,141] allowed a direct evaluation of all the ingredients nessessary for calculation of the dielectric tensor. The method is based on the Kohn-Sham implementation of the density functional theory, with the evaluation of electronic polarisation using the quantum Berry phase, and the phonon modes obtained form the direct diagonalization of the dynamical matrix. The latter can be constructed either numerically or using density functional perturbation theory [142]. This approach is already implemented in a few standard DFT codes (ABINIT, VASP, SIESTA), and rapidly brings an interesting insight into the physics of a dielectric response. First calulations by Bernardini et al. [138] validated the method by calculating the high frequency and static dielectric tensors for the III-V nitrides, where the ionic and electronic screening are approximately the same.

With respect to gate materials, very interesting studies have been recently presented on the dynamical and dielectric properties of low pressure polymorphs of zirconia and its silicates [121,122,143,144,145]. Detraux et al calculated phonon dispersion curves for the cubic fluorite [143] and tetragonal [121] phases of ZrO_2. They demonstrated softening of the $\Gamma - X$ TO branch in the cubic phase and its instability at the BZ boundary (X^{2-} phonon mode). This finding confirmed numerous earlier calculations (see e.g. previous sections, and refs. [116,146]) suggesting mechanical instability of the fluorite phase with respect to the cubic-tetragonal transformation. They also reported strong anisotropy of the static dielectric tensor in the tetragonal phase ($\epsilon_\parallel \sim 20.3$, $\epsilon_\perp \sim 48.1$) compared with $\epsilon \sim 35.1$ in the cubic phase. Very high $\epsilon_\perp$ was attributed to the low frequency phonon mode along the tetragonal axis - the phonon mode related to the tetragonal to cubic phase transformation. Interestingly, in both polymorphs the single phonon mode amounted to more than 70 % of the total ($\epsilon^\infty + \epsilon^a$) dielectric response. In view of this one might expect that monoclinic zirconia (the lowest energy phase) has substantially lower static dielectric constant than that of cubic or tetragonal phases, since it is intrinsically stable and, thus unlikely to possess very soft IR modes. This consideration was confirmed in a recent DFT study by Zhao and Vanderbilt [122], who calculated the ionic part of the static dielectric tensor in all three zirconia phases. Unlike in the tetragonal or in the cubic phase, no mode with dominating contribution to the dielectric tensor has been found in m-ZrO$_2$. The orientationally averaged static dielectric constant resulting from these calculations are 36.8, 46.6 and 19.7 for the cubic, tetragonal and monoclinic zirconia respectively. Not unexpectedly, the electronic screening ϵ^∞ does not display strong variations between the phases ($\sim$ 5 in the m-ZrO$_2$ and $\sim$ 5.75 the c-ZrO$_2$), or a large anisotropy.

The dielectric tensor was also calculated for zircon (ZrSiO$_4$) [144] by Rignanese and co-authors. The phonon mode associated with the anti-phase motion of the Zr ions and strongly deforming SiO$_4$ tetrahedra was found to make more than 60 % of the ionic contribution to ϵ^0. Once again, the role of a local atomic environment and of a selected few phonon modes was emphasised. Based on this argument, the same authors recently attempted to rationalise dielectric properties of pseudo-amorphous Zr silicates as a function of Zr composition [145]. They considered a model system in which silicate network consists of structural units (SU) comprised of cations (Si or Zr) with various oxygen coordinations ranging from 4 (as in a typical SiO$_4$ tetrahedra) to 8 (as in ZrO$_8$ octahedra in cubic zirconia). The fraction of particular SUs depends on the Zr/Si ratio. Assuming that only strictly local IR phonon modes associated with SUs are important, the authors then calculated contributions of these modes to the dielectric constant. The important conclusion of this analysis is that the dielectric constant of silicates varies sub-linearly with respect to Zr concentration. This is in contrast with earlier semiquantative study by Lucovsky and Rayner [147], who predicted a supra-linear dependence supported by experimental evidence presented by Wilk and co-authors ([148] and references therein). The argument of Lucovsky and Rayner was based on the assumption of the increase of the effective charge of transverse optical Zr-O mode with an increase of Zr coordination.

The *ab initio* calculations of dielectric properties of zirconia and Zr silicates provide an important insight into the origin of high dielectric permittivity of these materials. If the dielectric response is dominated by the atomic contribution, it is often coupled to very few specific low frequency IR active phonon modes. If, due to a crystal symmetry, these modes propagate only in specific directions, the static dielectric tensor will also exhibit strong anisotropy. Another consideration is that low frequency modes can be particularly susceptible to a change in a local environment, as in amorphous films or in the presence of defects (especially charged defects), interfaces, grain boundaries and internal and external stresses. This would inevitably lead to a decrease of the dielectric constant. This effect was indeed observed in the perovskite titanates used for the DRAM devices, where a significant reduction of the static dielectric constant has been observed in the films of SrTiO$_3$ [149]. This suggests a stiffening of the specific TO soft modes in the films as compared to the bulk.

A different situation may occur when many modes contribute comparably into a dielectric tensor. Although the bulk dielectric constant in this case is usually not as large (as in the m-ZrO$_2$), it may be strongly enhanced near the interface. Recent calculations of the Si/ZrO$_2$ and Si/HfO$_2$ interfaces by Fiorentini and Gullery [150] seem to confirm this argument. One should note however, that the enhanced dielectric screening at the interface may become damaging for CMOS perfomance. The modes associated with strong polarisation will also couple strongly to electrons. Therefore the mobility of carriers in the channel layer may be significantly eroded by a strong scattering by the soft polar modes at the interface.

Clearly, calculations of a dielectric response at the interfaces and in the presence of defects is the next logical and desirable step. This however, envolves modelling of very large systems, for which the phonon calculations become prohibitively expensive. In this situation, a coarse molecular dynamics approach may prove a fruitful alternative. The eigenvectors can be recovered from the direct diagonalisation of the resulting atomic covariance matrix, while the eigenfrequencies can be obtained from the relevant velocity autocorrelation functions. Then a dielectric constant can be evaluated using, say, the generalised Lyddane-Sachs-Teller (LST) relation [151]:

$$\frac{\epsilon^0}{\epsilon^\infty} = \prod_j^N \frac{\omega_{LOj}^2}{\omega_{TOj}^2}, \quad (7)$$

where the product is taken over all infrared active modes, and ϵ^∞ is assumed to take its perfect crystal value. The advantage of such an approach is that it would allow one to calculate dielectric response of larger disordered systems without making explicit harmonic approximations: the phonons obtained in MD are in essence effective normal modes accounting for the phonon-phonon interaction.

4. Oxygen vacancies

Vacancies are the defects common to nearly all materials, and oxygen vacancies in particular are likely to be the most prevalent defects in many oxides. Oxygen vacancies in all oxides may exist in several charge states. They can be characterised by formation energies, position of defect levels with respect to the valence and conduction bands, structural models, spectroscopic properties (including optical absorption and luminescence energies, parameters of EPR spectra and IR and Raman spectra), mechanisms of diffusion, electron transfer and recombination with other defects, and other characteristics. The study of vacancies in oxides has a long history (e.g. see the review in ref. [11]), and here we will try to only highlight those aspects which are of specific interest. First, we will introduce some general models and will illustrate them by several examples. Then we will consider in more detail a case study: oxygen vacancies in hafnia.

Dependent on material and particular application, vacancies can be classified as electron donors or hole traps. Their most stable charge state depends on chemical potential, i.e. whether the oxide is in contact with Si or other electron source, and on applied voltage. Another type of classification is based on structural models. It provides a rough but very simple cross-section through different oxides in terms of two common types: F - centres and E' - centres. We will use these two distinct models to discuss further properties of these defects in oxides. In all cases we will start from "neutral" vacancies, i.e. vacancies obtained by removal of one oxygen atom from its original lattice site. The main difference between the two structural models discussed below is in the character of localization of two electrons left by removed oxygen and of the lattice relaxation.

F-like centres are best known in cubic alkali halides and alkali earth oxides, such as MgO, CaO and BaO. In these ionic materials, the oxygen site is surrounded by six equivalent metal ions. A "classical" F-centre has one (alkali halides) or two (cubic oxides) electrons in an s-type deep state in the band gap. Formation of this centre is accompanied by small and fully symmetric (O_h point group) relaxation of the surrounding ions. A significant amount of the electron density is localized in the vacancy with considerable and equal contributions of six neighbouring metal ions (see Fig. 7(a)). Ionisation of the F-centre will lead to characteristic strong and, again, symmetric distortion of surrounding ions, where the nearest metal ions move outwards from the vacancy and twelve next nearest neighbour anions move inwards.

Another characteristic defect type is represented by what may be called a E'-like model. We should note that this notation is not fully consistent, as E' centre notifies a special type of paramagnetic defect in SiO_2 associated with an oxygen vacancy. However, these defects have been studied very extensively and therefore we believe that using their name to represent a class of models will not cause any problem. They are formed in different forms of SiO_2, GeO_2 and in silicates, where each oxygen ion is surrounded by only two nearest neighbour positive ions. Formation of a neutral oxygen vacancy in these materials is accompanied by the creation of a Me – Me bond and by strong displacements of these two Me ions towards each other (see Fig. 7(b)). The best-known example is a neutral vacancy in SiO_2, which is discussed in more detail below. A paramagnetic positively charged vacancy has a different structure to that of the neutral vacancy and originally was called a E' - centre. When one electron is removed from the neutral vacancy, the Me – Me bond becomes much weaker and can relax into two distinct configurations. One is still very similar to that of the neutral vacancy, where the remaining electron is delocalized over the two

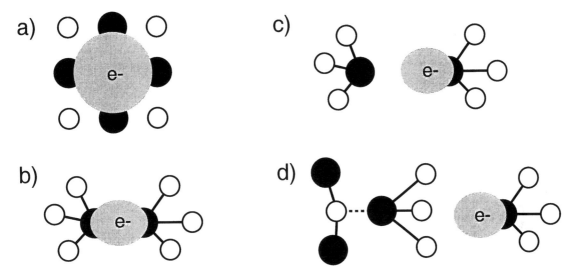

Figure 7. Schematic of different types of oxygen vacancies in insulators. Cations are shown in black and anions in white. a) An F-like center defect characterised by symmetric electron distribution centered in the vacancy (shown in grey); b-d) various types of E' - centre models (see text for discussion). Again electron localization is shown schematically in grey. Note that vacancy models (a) and (b) may contain two as well as one electron. The oxygen ion in (d) forming a dashed bond with the puckered Si ion is referred to as a "back oxygen" in the text.

Me ions, but with a much longer Me – Me bond distance. Another one is asymmetric with an hole almost fully localized on one Me ion, which is accompanied by strong and asymmetric lattice distortion [152,153] (see Fig. 7(c,d)). We should note that this description is valid mainly for crystalline silicates and germinates, and becomes more complicated in amorphous materials. However, it provides a generic model useful for discussion and classification in other materials.

We will now briefly review these two models in MgO and SiO_2 before considering "intermediate" models of oxygen vacancies in several other oxides, and finally moving to the case study.

Oxygen vacancies in MgO

MgO represents perhaps the most studied of oxides, and therefore we will use it as a reference for other oxide materials. The properties of neutral and charged oxygen vacancies in MgO have been studied experimentally in great detail. It is well established that vacancies in MgO can be described by an F centre model (see also Fig. 7(a)).

As discussed previously, a neutral vacancy possessing two electrons is called an F centre and a positively charged vacancy, which keeps only one electron, is called an F^+ centre, respectively. The paramagnetic F^+ centre has been well characterized using ESR [154] and ENDOR [155]. The unpaired electron of the F^+ centre is shown to be well-localized in the vacancy with only small hyperfine interaction with the surrounding Mg nuclei [154]. Both centres have close optical absorption energies - 4.92 and 5.01 eV for the F and F^+ centre, respectively [156]. According to the model suggested in ref. [157], the optically excited states of both centres are located very close to the CBM and can be easily thermally ionized. They undergo complex photo-conversion, which involves hydrogen-related defects and hole release under excitation [157,158]. Surface analogues of bulk F centres have long been considered to be part of various surface processes (see, for example, the recent review [159]).

The abundance of experimental data [9,10], the wide physical and the chemical applications of MgO and simplicity of its structure facilitated extensive theoretical studies. There is practically no computational technique which has not been tested on F centres in the bulk or at the surface of MgO. Relatively recent studies include extensive semi-empirical calculations in an embedded cluster model [11,160], *ab initio* embedded cluster calculations using different approaches [39,40,161], *ab initio* molecular cluster calculations including the crystal electrostatic potential [41,162,163,164], and periodic calculations using various approximations of Density Functional Theory [165,166,167]. This list is in the best case representative rather than exhaustive.

Practically all calculations agree that both one- and two-electron vacancy defects belong to the F-centre type in the sense that a significant part of the electron density attributed to the extra electron(s) is localized in the vacancy and decays with the distance from its centre. The exact amount of the electron density localized inside the vacancy strongly depends on the method of calculation, basis set and the method of analysis [162,165]. In particular, the often used Mulliken population analysis gives misleading results for extended basis sets [162]. A more consistent Bader analysis [168] demonstrates that two electrons are strongly localized inside the vacancy. For the neutral vacancy the calculations predict very small displacements of surrounding ions and small lattice relaxation energies. For the charged vacancy, the F^+ centre, the calculations predict outward displacements of the nearest neighbour Mg ions between 2 and 5 percent of the lattice constant [39,160,161]. We should note that comparison of different studies can only be qualitative as they predict different lattice parameters for the perfect lattice. The analysis of the ENDOR data [155] predicts the outward displacements of Mg ions by about 6% of the lattice constant.

A related characteristic is the hyperfine interaction with the nuclei of Mg ions surrounding the vacancy. The experimental value for the isotropic part of the hyperfine coupling for the bulk F^+ centre is quite small and equal to 3.94 G [154]. Again theoretical results strongly depend on the method and basis set, as discussed in detail in ref. [162]. Good embedding is crucial and provides very good agreement with the experimental result [161,162].

As we saw in Sec. 3, most of the calculations do not reproduce the band gaps of perfect crystals.

MgO is not an exception: Hartree-Fock calculations usually overestimate the experimental value of the gap (7.833 eV [169]) by about factor of two [170,171,172], whereas PW DFT calculations give a much smaller gap (e.g. 4.8 eV in ref. [167]). Various semi-empirical corrections to improve the agreement of DFT calculations have been discussed in ref. [165]. Therefore the position of the F-centre state in the gap can be determined relatively reliably only with respect to the valence band edge. Assuming that the first optically excited defect state is very close to the conduction band [157], one can estimate that the ground state for both defects should be at about 2.7 eV above the valence band edge [167]. In calculations, this energy can be estimated as a difference between the corresponding one-electron energies. Due to the defect-defect interaction the result depends on the size of periodic cell. Nevertheless, similar results have been obtained for the F-centre by different methods, e.g. 2.7 eV in ref. [167], 2.34 eV in ref. [166], 2.7 eV in ref. [161]. A critical analysis of the accuracy of determination of this parameter is given in ref. [165].

Optical absorption energy calculations are particularly challenging. They have been performed only in cluster techniques and recently have been critically reviewed in ref. [41]. The authors used an embedded cluster approach and several configuration interaction techniques based on the Hartree-Fock method for the calculation of the defect ground state. They have compared calculations for the F and F^+ centres and concluded that, although it is relatively easy to reproduce that both centres have very similar absorption energies, the absolute energies exceed the experimental values by about 1 eV, i.e. by 20%. Semi-empirical calculations [11,160] give good agreement of optical absorption energies with the experimental values, but the parameters of the method have been specially fitted to reproduce the defect properties. By calculating the adiabatic relaxation of excited states for both centres they achieve fair agreement with the experimental luminescence energies. An attempt to calculate the Franck-Condon absorption and luminescence energies has been made also in ref. [166], however agreement with experiment was not impressive.

Defect formation energies strongly depend on the extent of defect-induced lattice distortion accounted for in calculations and on their accuracy. In particular, many studies take into account the relaxation of just nearest neighbours. As we will see below, for some defects this is clearly inadequate. However, in the case of the F-centre in MgO all studies agree that the relaxation energy of the neutral vacancy is very small (of the order of 0.1 eV). Therefore the main discrepancies between the results again come from different models and computational techniques used in calculations. The neutral vacancy formation energies calculated using different techniques agree within 20%. For example, embedded cluster Hartree-Fock calculations [161] give 8.7 eV, in good agreement with the periodic Hartree-Fock calculations [173]. The value of 10.5 eV was obtained in a more simple cluster embedding scheme [163] and in periodic plane wave LDA calculations [167].

In the next section we will consider in more detail models and properties of oxygen vacancies in silica. As in the case of MgO, silica has been extensively studied both experimentally and theoretically over the last 40 years. With a complex structure built from corner-sharing tetrahedra and very strong ionic-covalent Si – O bonding, it represents another characteristic type of oxygen vacancy, the E' - centre.

4.1. Oxygen vacancies in silica

SiO_2 has several crystalline modifications, but most applications of this versatile dielectric concern amorphous silica (a-SiO_2). A traditional approach is, however, to regard crystalline forms of silicon dioxide as useful mimics of a-SiO_2. Although there is growing body of evidence that such an approach may lead to misleading conclusions, studies of defects in crystalline SiO_2 is always a useful starting point. Recent reviews of extensive experimental and theoretical studies of oxygen vacancy and other defects in silica can be found in ref. [4]. As we are mainly concerned with generic defect models and theoretical simulations as means for developing these models, we will confine ourselves to a brief survey of the representative results obtained for neutral and positively charged oxygen vacancies in α-quartz.

The neutral oxygen vacancy in silica is a diamagnetic defect and is usually characterized theoretically in terms of its geometric structure, formation energy, optical properties and its relation to the positively charged E' - centre. All existing calculations do not make much distinction between amorphous silica and α-quartz and indeed a model of neutral vacancy seems to be robust enough to withstand this negligence. The main reason is that it is basically a strong Si – Si bond made by two electrons left in a vacancy by oxygen atom. Indeed, all calculations of the neutral oxygen vacancy in α- quartz predict a significant displacement of the two Si atoms neighbouring the vacancy to one another, and the formation of a chemical bond between them (see Fig. 7(b)). The equilibrium distance between these atoms predicted in most ab initio and semi-empirical cluster calculations [162,174,175,176,177,178,?,180] is in the range of 2.3 - 2.5 Å. The dependence of this parameter on the cluster size has recently been studied in ref. [181] using the mechanical embedding method ONIOM. A larger Si – Si distance of 2.68 Å has been obtained by the quantum embedded cluster method [50]. PW DFT calculations reported in refs. [182,183] predict 2.5 Å and about 2.7 Å, correspondingly. PW DFT calculations using gradient corrections and the projected augmented wave method [78] predict 2.44 Å. The Si – Si distance at the vacancy site is similar to the Si – Si spacing in elemental Si (2.35 Å) and much shorter than the equilibrium distance of 3.08 Å between the two Si atoms in ideal α-quartz. It is interesting to note that there is almost no discussion in the literature of the relaxation of the atoms surrounding the neutral vacancy in silica. As demonstrated in ref. [44], the extent and character of this relaxation strongly alters the existing model of the neutral vacancy and affects the calculated defect properties.

Optical absorption energy is crucial for experimental identification of this diamagnetic defect. In early calculations using non-selfconsistent tight-binding methods (see references in ref. [178]) and in later, more rigorous semi-empirical MNDO studies [177,178,179], the absorption and luminescence bands of the neutral vacancy were related to the one-electron transition between two states in the band gap. These were formed mainly by the bonding and anti- bonding combinations of the sp orbitals of the two silicon atoms adjacent to the vacant site. The energy of the lowest singlet-to-singlet excitation calculated by the MNDO method in [177,178,179] was 5.0 eV. Further *ab initio* calculations in isolated clusters by Stefanov and Raghavachari [174] did not confirm these results. Instead, it was suggested that the lowest excitation has a much higher energy of 7 eV and corresponds to a transition of different character: from the Si – Si bonding orbital to a diffuse state (i.e., a Rydberg-type excitation, not the bonding – anti-bonding transition between strongly localized orbitals). On the other hand, the cluster calculations by Pacchioni *et al.* [162,176,102], which also included diffuse atomic orbitals, confirmed the bonding-antibonding nature of the lowest one-electron allowed transition between the two vacancy levels in the gap. The energy of this transition was also 7-8 eV, similar to the result of Stefanov and Raghavachari [174], and much higher than estimated with the semi-empirical MNDO method in [177,178,179]. Recent *ab initio* calculations of the oxygen vacancy in β-cristobalite, using the quantum embedded cluster method implemented in the computer code EMBED [50], again suggested the bonding – antibonding transition model for the lowest vacancy excitation proposed earlier. The value of 8.3 eV for the lowest singlet-to-singlet excitation energy calculated in ref. [50] corresponds better to the *ab initio* results obtained in refs. [162,176,102] with an isolated cluster approach. Overall, according to *ab initio* calculations, the neutral oxygen vacancy is a good candidate for the oxygen deficient centre ODC(I) with an absorption band at 7.6 eV [184].

Calculations also give the vacancy formation energy, with respect to the free oxygen atom in its ground triplet state. As for the optical energies, yje calculated values vary significantly. *Ab initio* calculations using the isolated cluster model give estimates of 5.5 eV and 6.7 eV at the Hartree-Fock (HF) level with and without d-functions, respectively. The electron correlation correction calculated at the MP2 level for the basis set containing

d- functions brings the vacancy formation energy to 8.5 eV [162,174]. Quantum embedded cluster calculations at the Hartree-Fock level [50] give about 6.6 eV for the vacancy formation energy in quartz and about 7.9 eV in β-cristobalite, respectively. PW DFT calculations give 6.97 eV in α-quartz [185] (with respect to the half of the energy of the oxygen molecule). Other PW DFT plane-wave calculations predicted somewhat larger values of 7.85 eV [186] and 8.64 eV [187] for α-quartz, and 9.3 eV [188] and 8.92 eV [187] for β-cristobalite. Thus most of the *ab initio* results lie between 6.5 and 9.5 eV. The thermodynamic estimate [189] of the vacancy formation energy is larger than 7.3 eV.

We can see again some discrepancies between quantitative values of basic defect parameters predicted by different techniques. As shown in ref. [44], part of these discrepancies stem from the fact that most of the cluster and periodic calculations do not fully account for the defect-induced distortion of silica lattice around the neutral vacancy. The calculations [44] using the embedded cluster method show that the neutral vacancy induces very strong and anisotropic lattice distortion, which extends further than about 13 Å from the vacant site. These results suggest that the neutral oxygen vacancy in SiO_2 cannot be viewed as merely a symmetric displacement of the two silicon atoms towards each other due to formation of a chemical bond between them. They demonstrate that in spite of the rigidness of Si – O bonds the character of defect-induced lattice distortion in SiO_2 can be much more long-range than intuitively anticipated and predicted in previous calculations. This should hold also for the charged vacancy, peroxy linkage and other defects in SiO_2. The calculations [44] demonstrate direct dependence of the defect formation energies and optical absorption and luminescence energies on whether the lattice distortion is included in the calculations.

The model of a neutral vacancy in silica is in marked contrast with cubic oxides, such as MgO and the more-complex HfO_2 discussed below. In those materials, despite the substantial ionic polarisation, the formation of a neutral oxygen vacancy leads to only very small displacements of the surrounding metal atoms. In the case of MgO, ionic displacements are isotropic and at a distance of 8.0

Å from the neutral vacancy are less than 0.002 Å. A double positively charged anion vacancy in MgO produces a much stronger perturbation, but again, the displacements of the lattice ions decrease relatively rapidly and do not exceed 0.02 Å at a distance 8.0 Å from the vacancy. This results from the different structure and chemical bonding in these materials. In particular, in MgO, unlike α-quartz, the oxygen site is the centre of inversion. The peculiar structure of α-quartz is also manifested in the strong asymmetry of the relaxation of the two Si atoms neighbouring the vacancy (see ref. [44]).

The positively charged oxygen vacancy in SiO_2, the E'-centre, is perhaps the most famous defect related to oxygen deficiency in silica, which also plays a very important role in many applications of silica in microelectronics and fibre optics. Feigl, Fowler and Yip [152] first suggested that the E'-centre in a-SiO_2 is a positively charged oxygen vacancy. Ionization of the neutral vacancy leaves a relatively weak bond between two silicon atoms. Flexibility of the amorphous structure gives room for the formation of several centres. In particular, the E'_γ-centre identified by EPR in a-SiO_2 [190,191] has an unpaired spin localized on a dangling bond of a silicon atom (see Fig. 7(c,d)). This defect has an optical absorption band at 5.85 eV with linewidth (FWHM) of 0.8 eV and an oscillator strength of f=0.14 ± 0.1. This absorption yields no luminescence. The EPR spectrum of E'_1-centre in α-quartz is similar to that of the E'_γ, as shown in refs. [192,193,194]. The optical absorption observed in irradiated samples and attributed to E'_1-centres (see ref. [184] for discussion) has the maximum at 6.2 eV. Another analogue of the E'_γ-centre which exists at a silica surface [195] has a very similar EPR signal and an absorption band at 6.3 eV.

In the perfect α-quartz lattice every oxygen is coordinated by two inequivalent silicon atoms. One Si – O distance is equal to 1.611 Å and the corresponding Si atom is called "long bond silicon", Si_L, whereas the other ("short bond" Si atom, Si_S) is at 1.604 Å from the oxygen site. According to the initial proposal [152], in the E'_1 centre the unpaired electron is localized on a dangling bond on short bond silicon and Si_L, moves from the vacant position into the near-planar configuration with its

remaining three oxygen neighbours. The important feature of the experimental hyperfine parameters is a very small hyperfine interaction of the unpaired electron with the Si_L. To explain that, in the refined model [153], the long-bond Si_L was displaced even further from the original site through the plane of its oxygens. In α-quartz it is stabilized in the so-called "puckered configuration" (PC) by one of lattice oxygens called "back oxygen", becoming effectively four-coordinated again. This model, shown in Fig. 7(d), is currently accepted also as a model of the E'_γ centre in a-SiO_2. The fact that Si_L does not interact with the unpaired electron may explain the noted similarity between the EPR and optical parameters of the bulk and surface E'-centres (where Si_L is missing altogether). On the other hand, it allows one to present this configuration in a generic form $O_3 \equiv Si\bullet$, where the $\bullet$ represents an unpaired electron and $\equiv$ bonds with three oxygen ions.

Besides the puckered configuration, there is also the second configuration of a positively charged vacancy where the positions of two silicon atoms Si_L and Si_S are close to those in the perfect lattice with an unpaired electron shared almost equally between them. This "normal" configuration (NC) has been identified with the E'_δ centre in a-SiO_2 [196]. Thermal conversion of E'_δ to E'_γ was observed in ref. [196] during isochronal annealing within a temperature range of 200 − 400°C. According to this study, the E'_γ centre in a-SiO_2 is much more stable than E'_δ. This implies that the puckered configuration of the positively charged vacancy has the lower energy, something what theory should be able to prove.

The original semi-empirical Hartree-Fock calculations in a cluster model [153] suggested that NC is higher in energy than PC by about 0.1 - 0.2 eV, with a 0.4 eV potential barrier separating them. Later several more sophisticated first principles studies validated this model. PW LDA calculations with plane waves basis set [183] led to the energy difference between NC and PC of ΔE=0.3 eV and the barrier E_B=0.4 eV. The hyperfine parameters obtained were in reasonable agreement with experiment. *Ab initio* Hartree-Fock calculations in a cluster model by Pacchioni *et al.* [197,198] resulted in values of ΔE=0.64 eV and E_B=0.32 eV. Recent PW DFT augmented plane wave calculations [78] give a similar barrier $E_B=0.38$ eV, but a much smaller energy difference between the two configurations $\Delta E=0.04$ eV. Thus again there is no quantitative agreement between the results by different methods and the relative stability of the two configurations is still an open issue (see discussion in ref. [78]).

Perhaps a more sensitive test of the defect model is whether it gives correct EPR parameters. An application of core pseudo-potentials in periodic DFT calculations has long been a problem for calculating the hyperfine parameters. On the other hand, local basis set molecular codes employed in cluster studies perform such calculations routinely. Therefore it has been an important achievement that techniques have been developed for calculating the hyperfine parameters also within periodic DFT methods (see, for example, refs. [78,183,199]). Recent refined calculations [78,199] gave very good agreement with the experimental hyperfine parameters for both "normal" and "puckered" configurations of the positively charged vacancy. The results in cluster models depend on the cluster size and basis set, as discussed in ref. [162]. However, for large clusters and extended basis sets the agreement with experiment is again fair.

E'_γ centres play an important role in applications of amorphous silica as a gate dielectric in microelectronics. They are also thought to be responsible for Bragg grating formation in silica glass [4,200]. In the first case defect charge state levels are of crucial importance, in the second case – their optical properties. Perhaps the most comprehensive study of charge state levels of oxygen vacancies in silica is presented in ref. [78]. The nature of the optical absorption of E'_γ centres with a maximum at 5.85 eV is still controversial. The present status of the experimental and theoretical data on this issue has been critically discussed in detail in the recent reviews [184,201] as well as in ref. [202].

Finally, we would like to stress that in spite of some analogies between models of vacancy defects in α-quartz and amorphous silica, it is becoming increasingly clear that there are significant differences too. Some of them stem from the fact that, in contrast with crystalline α-quartz, there are various types of a-SiO$_2$, which differ by preparation,

treatment, water content and other parameters. Although the mechanisms of their formation are still unclear, some of the defects exhibit different properties. For example, neutral E'-type defects have been found by EPR in SiO$_2$ films prepared by plasma-enhanced chemical vapour deposition on silicon followed by electron transport across the deposited film [203]. There is other evidence which suggests that there is no direct correlation between positive charge and E'-type paramagnetic defects in thermal SiO$_2$ layers on Si [204] (see also recent discussion and calculations in refs. [199,205]).

On the other hand, some of the differences are induced by the structure and disorder of a-SiO$_2$. In particular, there is not always a "back oxygen" conveniently located to stabilise the puckered configuration of E'_γ-centre in a-SiO$_2$. Therefore other types of generic $\equiv Si\bullet$-type centres can be formed [206]. For example, recent molecular dynamics simulations [207] demonstrate the transformation of an optically excited neutral oxygen vacancy in a-SiO$_2$ into two E' centres, which do not have the puckered Si_L "half" at all (see also Fig. 7(d)). The analysis of several configurations of E' centres in amorphous silica found in ref. [206] suggests that the vast majority of oxygen vacancies do not pucker after capture of a hole, but rather form E'_δ centres (see Fig. 7(b)). The remaining vacancies exhibit at least two types of stable puckered configurations. Although the number of such configurations can be even larger, determining their relative concentrations is a challenging problem. Solving this problem requires a statistical analysis of large amorphous samples, taking into account the mechanisms of formation of neutral and charged oxygen vacancies. Some of the examples of this approach in relation to other defects and defect properties are discussed in the following sections.

To summarise, recent state of the art calculations are able to make semi-quantitative predictions of experimental defect parameters with systematic discrepancies within about 20% of the experimental values. As in experiment, each method has its advantages and disadvantages, which will be further highlighted in the following discussion.

Having discussed the "classical" F-centre and E'-centre case, we will now consider several "intermediate" cases where the structural model of oxygen vacancy is in between the two models outline above.

Intermediate cases

Sapphire or α-Al_2O_3 is a representative of a wide class of ceramic materials important for applications in microelectronics, laser optics, as a substrate and in many other areas. Each Al atom in this material is bonded to six oxygen ions, and each oxygen ion is surrounded by four nearest neighbour Al ions in the form of a distorted tetrahedron. The semi-empirical calculations of F and F^+ centres in α-Al_2O_3 have been reviewed in ref. [11]. Recent *ab initio* calculations [208] investigated the properties of neutral and charged oxygen vacancies in this material using a periodic orthogonalized linear combination of atomic orbitals method (OLCAO) and the Local Density Approximation. They found a formation energy of 5.83 eV for the neutral vacancy, and demonstrated that it formes an F centre, trapping two electrons. However, the low symmetry of the vacancy and mixed ionic-covalent bonding of alumina meant that there was significant relaxation (up to 16% of the original bond-lengths) of the Al and O ions around the vacancy. The neutral vacancy, F-centre, introduces a doubly occupied deep level of about 3.4 eV above the valence band edge whereas the position of the F^+-centre state is predicted to be only 0.25 eV lower. Analysis of the electron density shows that the ground state of the F-centre is not very strongly localized inside the vacancy with large contributions on the surrounding Al and O ions. The semi-empirical calculations [11] predict stronger electron localization, but qualitatively very similar models for both neutral and charged vacancies. They also suggest a model for photo-stimulated conversion of F^+ centres into F centres.

Ferroelectric materials have been in the focus of intense study because of applications in making dynamical and non-volatile memory components. Again, among various other defects, oxygen vacancies are thought to be the most mobile and abundant in perovskite ferroelectrics, such as $PbTiO_3$, $BaTiO_3$ and $KNbO_3$. They have been a subject of

recent theoretical studies and demonstrate an interesting intermediate character between the two types discussed above. In particular, recent PW LDA calculations of oxygen vacancies in the tetragonal phase of $PbTiO_3$, have examined two types of oxygen vacancies [209]. These vacancies belong to Ti–O–Ti chains along the c axis (parallel to the direction of polarisation), and to Ti–O–Ti chains along a and b axes normal to the c axis. In this material, the oxygen sites of the first type are surrounded by two Ti ions and four Pb ions. They can be attributed to the F-centre type with small displacements of the nearest metal ions. However, the electron density is now localized both in the vacancy and to a large extent on the two nearest Ti ions. The positively charged vacancy is shown to have the same symmetry as the neutral one with larger outward displacements of surrounding metal ions. An interesting result of this work is that it indicates that oxygen vacancies in this material are one possible source of domain pinning and polarisation fatigue.

An *ab initio* embedded cluster study of oxygen vacancies in $BaTiO_3$ [42] examined both symmetric and asymmetric types of relaxation of the paramagnetic charged oxygen vacancy. The authors attribute this defect to the F-centre type, however, it demonstrates clear features of the classic E' centre. Again, in the cubic phase, the vacancy is embedded in a Ti–O–Ti chain and is surrounded by for Ba ions in the perpendicular plane. The calculations demonstrate a very similar pattern of the spin density distribution and lattice relaxation to that found in $PbTiO_3$: the electron is delocalized in the centre of the vacancy and by the two nearest Ti ions with strong outward displacements of these ions. However, there is also a broken symmetry configuration, with the unpaired electron localized just on one Ti ion and asymmetric relaxation. This configuration is thought to be higher in energy by about 1 eV. It is interesting to note that a similar configuration could perhaps also exist in $PbTiO_3$.

Yet another anion vacancy type is predicted in CeO_2. In stoichiometric CeO_2, oxygen ions are surrounded by four Ce ions. According to full-potential linear muffin-tin orbitals (FP-LMTO) periodic DFT calculations [210], the f states of Ce play an important role in neutral oxygen vacancy formation in this material. It is accompanied by localization of two electrons in f states of two nearest Ce ions [210]. The calculations predict 4.55 eV oxygen vacancy formation energy in perfectly stoichiometric material and a much smaller value of 0.26 eV if electrons localize in the f states of surrounding Ce atoms. Thus the process of oxygen vacancy formation is coupled with localization/delocalization of $4f$ electrons on Ce. This property is related to fast rate of oxygen ion diffusion and the oxygen storage capability of cerium oxide.

There has been recently a surge of modelling and simulation activity in the area of high-k oxides. However, only three have been subjected to any systematic studies of vacancies: zircon ($ZrSiO_4$) [75], zirconia (ZrO_2) [6,117,118] and hafnia (HfO_2) [7]. In each of these materials the defect properties are quite similar and correspond more to the F-centre type. To further illustrate the possibilities and problems of periodic DFT methods as well as to discuss in more detail the defect models, we will consider oxygen vacancies in hafnia in more detail.

Oxygen vacancies in hafnia

In this section we turn to a case study of oxygen vacancies in hafnia. All defect calculations in these systems, as well as most calculations of more complex zircon, have been made so far using periodic models and DFT. We should note that zirconia is an important refractive oxide, which has many technological applications. Addition of impurities like Ca^{2+} or Y^{3+} stabilizes the cubic phase (see section 3) of zirconia and improves its thermochemical properties allowing the production of extremely durable materials. These impurities are compensated in the lattice by oxygen vacancies. There is a very extensive literature concerning this subject (see, for example, refs. [110,116,117,211]) and we will not consider it in this section, but rather focus on the properties of single oxygen vacancies in monoclinic hafnia calculated in ref. [7].

All the calculations were made using a 96 atom unit cell, which is generated by extending the 12 atom monoclinic unit cell by two in three dimen-

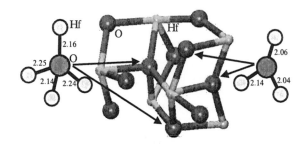

Figure 8. Diagram showing the 4-coordinated tetragonal (left) and 3-coordinated trigonal (right) bonding of the oxygen ions in the monoclinic phase of hafnia as calculated in this work. The numbers show interatomic distances in Å. *Reproduced with permission* [7].

sions. The monoclinic structure is used in calculations since it is the most stable phase, even for some thin films [112,212,213]. Details of the bonding in the monoclinic structure are given in Fig. 8.

We start our discussion from neutral oxygen vacancies. They can form at either a 4-coordinated or 3-coordinated lattice oxygen site (see Fig. 5). For ease of reference all values in the following sections associated with 3-coordinated defects will be labelled X_3 and all associated with a 4-coordinated oxygen will be labelled X_4, where X is the defect species. Hence, there are two types of oxygen vacancies in monoclinic hafnia: 3- and 4-coordinated (V_3 and V_4, respectively). The lattice relaxation around them involves small displacements of the nearest neighbour Hf ions of 0.01 - 0.02 Å, or 0.5 - 1.0 % of the Hf-O bond length. Such small displacements are characteristic for the F-centre type defects discussed above. They correspond to almost full screening of the anion vacancy by the two remaining electrons, which are strongly localized around the vacancy site. The relaxation energies with respect to the perfect lattice are 0.09 eV and 0.06 eV, for V_3 and V_4, respectively. Using Eq. (1), we obtain vacancy formation energies of $E_{for}(V_3) = 9.36$ eV and $E_{for}(V_4) = 9.34$ eV, correspondingly. The formation of a vacancy introduces a new double-occupied level in the band-gap situated at 2.8 eV and 2.3 eV with respect to VBM for V_3 and V_4, respectively.

Removing an electron from the relaxed neutral vacancies results in formation of the positively charged defects, V_3^+ and V_4^+. The Hf ions surrounding the vacancy displace outwards by about 0.1 - 0.2 Å (5-10% of the Hf-O distance). This is much bigger than for the neutral vacancy. The relaxation energies are 0.65 eV and 0.61 eV for V_3^+ and V_4^+, respectively. The strong relaxation is caused by the fact that the neighbouring Hf ions lose part of the screening effect provided by the two electrons in the neutral vacancy case. The remaining electron is strongly localized inside the vacancy site, as can be seen in the spin density map shown in Fig. 9. The vertical ionization energies for both types of neutral vacancies are given in Table 6. The single-occupied one-electron state of this defect lies lower in the band-gap than that for the neutral vacancy.

Removal of yet another electron from the system leads to the formation of doubly positively charged vacancies, V_3^{2+} and V_4^{2+}. The Hf ions neighbouring the vacancy site are displaced outwards by about 0.1 - 0.2 Åwith reference to the singly charged case, and the relaxation energy is about 0.8 eV. Note that for both the V^+ and V^{2+} defects, the system total energy is much lower (0.44 eV and 0.76 eV respectively) for the 3-coordinated site. This implies that although formation of an initial neutral vacancy is energetically balanced between sites, once electrons are removed the V_3^X species is strongly favoured and vacancies are likely to diffuse to these sites.

As discussed in Sec. 3, the band gap of hafnia obtained in DFT is strongly underestimated. This precludes studies of the possibility of trapping of an extra electron by a neutral vacancy. Optical excitation energies also cannot be calculated with this technique and calculations of the hyperfine interaction with the nuclei of atoms surrounding the defect, although possible [78,199], have not been reported so far.

5. Transparent conductors

Transparent conductive oxide films present yet another type of system where oxygen vacancies play a very prominent role and in some cases determine electrical properties. They are widely used in various types of displays, solar cells, electrochromic

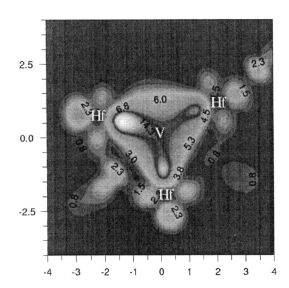

Figure 9. Spin density in a plane containing three Hf ions neighbouring the 3-coordinated oxygen vacancy (V) with one electron removed from the system. Charge density is in 0.1 $e/Å^3$ and all distances are in Å. *Reproduced with permission* [7].

devices, window heaters and other applications (see, for example, refs. [215,216,217,218,219]). The best-known examples are doped oxides, such as indium tin oxide (ITO) and SnO_2:F, which exhibit some of the best electrical properties [219,220]. These films have high n-type conductivity, which is achieved mainly in two ways: i) creation of intrinsic donors by lattice defects, such as oxygen vacancies or metal atoms on interstitial lattice sites; and ii) intentional doping with metals with higher oxidation number (e.g. Sn in In_2O_3) or halogen with oxidation number minus one on oxygen lattice sites (e.g. F in SnO_2). This donor-doping involves the production of shallow donor states which effectively dope the conduction band with electrons.

The design of p-type TCOs is more difficult to achieve [221] as the valence band should have low effective mass to provide shallow acceptor levels and good hole mobility [222]. However, the top of valence band of most oxides is due to non-bonding 2p oxygen states, which have high effective mass and give deep hole levels. A way forward has been proposed in ref. [221] where it has been suggested to use a cation with closed shell levels, which are degenerate with O 2p states. It has been demonstrated [221] that thin films of $CuAlO_2$ (with Cu^+ ions in $d^{10}s^0$ configuration) designed according to this prescription combine good optical transparency with high electrical conductivity.

An interesting new candidate TCO material has recently been reported in ref. [223]. It has been demonstrated that a main-group metal oxide $12CaO \cdot 7Al_2O_3$ (or C12A7) can be converted from an insulating state to a conductive state and still retain high transparency. Crystalline and thin film C12A7 samples are optically transparent. However, loaded with hydrogen and irradiated with UV light they acquire high electrical conductivity. This opens an exciting possibility for writing conductive wires in transparent insulating films. The mechanism of conductivity in this material is different from that in traditional n- and p-type TCOs and is determined by electrons localized in vacant lattice sites, as discussed in detail below.

The reasons for co-existence of optical transparency and electrical conductivity of n-type TCOs, such as ITO and SnO_2 have been studied in several theoretical papers. Most of them draw conclusions on the basis of band structure calculations of particular materials. In particular, Mryasov and Freeman recently performed local density full-potential linear muffin-tin orbital electronic structure calculations of indium tin oxide [224]. The calculated band structure of the parent material In_2O_3 has a single highly dispersed band of In 5s character at the very bottom of the conduction band. Mryasov and Freeman argue that the s-character and high dispersion of this band due to the strong overlap between neighbouring In sites may explain the high mobility of electrons in the doped material. Another important issue is that this band is separated by 4 eV in the Gamma point from the higher band. This ensures high plasmon frequency of conducting electrons occupying this band. Together with a 3.6 eV band gap with the top valence band of In_2O_3 this provides optical transparency of the doped highly conductive material. Further analysis of the band structure of Sn doped In_2O_3 supports these conclusions [224]. Thus the electronic structure of the host material determine its conducting behaviour with electron doping. Mryasov and Freeman generalized these

results and the results of electronic structure calculations of ZnO and SnO_2 by formulating conditions favouring transparent conducting behaviour: (i) a highly dispersed and single character s-type band at the bottom of the conduction band, (ii) this band is separated from the valence band by a large enough fundamental gap to exclude interband transitions in the visible range; (iii) the properties of this band are such that the plasma frequency is below the visible range.

Another typical TCO is SnO_2. It has a band gap of 3.6 eV and conduction band with a broad minimum of empty 5s Sn states, similar to In_2O_3 [222]. Since the conductivity of thin tin oxide films directly depends on oxygen partial pressure [225], it has been assumed that the transparent conductivity of this material is caused by electron donation from shallow levels induced by oxygen vacancies. Recent theoretical modelling [226] supports the general features of an oxygen vacancy model, but suggests that tin interstitials can be more important for electron donation. The calculations [226] predict very low formation energies for tin interstitial atoms and attribute that to the multi-valence of tin. They argue that a tin interstitial produces a donor level inside the conduction band leading to its instant ionisation. Thus high conductivity of undoped SnO_2 can be attributed to high concentration of interstitial Sn atoms, which donate their electrons to the conduction band. Again similar to ITO, the optical transparency is explained by a special feature of the band structure of SnO_2: a large gap between the Fermi level and first unoccupied states in the conduction band.

The s-type conduction band with a broad minimum is characteristic of other n-type TCO materials [227] too. It determines the maximum mobility and conductivity of these systems, which is further affected by Coulomb scattering by ionised dopants [228].

The nature of transparent conductivity of p-type TCOs is, however, less understood. Recently Robertson et al. [222] have performed DFT band structure calculations of prototype p-type TCOs of $CuAlO_2$ family. They broadly confirmed the general design principles proposed by Kawazoe et al. [221]: holes should be in the filled p^6 valence band, broadened by interaction with filled cation

d^{10} states at a similar energy. Cu_2O is actually one of the simplest p-type conductors exhibiting these characteristics, but its band gap of 2.2 eV is too small. It has a three-dimensional network of inter-penetrating $O - Cu - O$ chains on a tetrahedral lattice, and a $Cu - Cu$ coordination equal to six. Robertson *et al.* [222] argue that by reducing the $Cu - Cu$ coordination one should be able to increase the band gap and thus to make the conductor transparent. This is what occurs in $CuAlO_2$ and similar materials, although the experimental data on direct and indirect band gaps are still controversial [222].

As the search for new electro-active insulators continues, it is important to understand and predict other mechanisms of this interesting phenomenon. In the next section we will consider a different type of TCO, $12CaO\cdot7Al_2O_3$, and will discuss a theory of its conductance and optical properties. This case study is also used to illustrate further the applications of the embedded cluster approach discussed above.

Electronic structure and optical properties of C12A7

As discussed above, the existing mechanism in the best-known n-type material - indium tin oxide (ITO) is based on electron donation into the highly dispersed s-type conduction band [224,226]. The mechanism of conductivity in C12A7 is very different primarily due to its unusual structure [230]. C12A7 is a cubic crystal, which belongs to the $\bar{I}43d$ space group and has a lattice constant of 11.99 Å. The unit cell contains two $12CaO\cdot7Al_2O_3$ molecules. It is composed of a positively charged framework built from 12 cages and two extra-framework O^{2-} ions occupying two different cages, and can be represented by a chemical formula $[Ca_{24}Al_{28}O_{64}]^{4+}\cdot2O^{2-}$. The cages are approximately 5 Å wide and connected to 8 other cages via 4 Å wide openings.

The effective positive charge of each cage ($+1/3$ e) is a peculiar property of this system, which strongly differs from other TCOs, and cage sodalites (see, for example, ref. [231]). C12A7 samples are optically transparent up to 5.0 eV. How-

ever, loaded with hydrogen and irradiated with UV light, single crystals, as well as sintered powders and thin films, acquire high electrical conductivity. This is due to formation of a significant number of unpaired electrons, which correlates with the creation of two broad optical absorption bands with maxima at about 0.4 eV and 2.8 eV. It has been suggested [223] that similar to positively charged oxygen vacancies in many oxides (e.g. F^+-centers in MgO and HfO_2 discussed above), positive cages may serve as trapping centres for at least one electron. The main difference is that these cages constitute the crystal lattice, whereas vacancies are located a long distance apart. The results of *ab initio* calculations described in this section provide an explanation for conductivity and optical transparency in this system, and suggest criteria for choosing new potentially electro-active insulators of this type.

The geometric and electronic structures of an ideal C12A7 crystal was calculated using the periodic DFT and the SIESTA code [24]. These calculations were carried out for a supercell of 118 atoms (a single unit cell) where the extra-framework O^{2-} ions were arranged in such a way as to minimise the system dipole moment. The optimised crystal geometry is very similar to that reported in ref. [230]. Analysis of the electron density demonstrated that C12A7 is an ionic crystal. The dispersion of the lowest conduction band was several times smaller than that obtained in other cubic TCOs [227]. This suggests that the mobility of electrons in this band should be smaller than in other TCOs. However, to determine whether extra electrons in C12A7 would form shallow or deep states in the gap one requires that calculations predict the right band gap value. However, as in most periodic DFT calculations (see Sec. 2), the band gap turns out to be much narrower than the experimental value. Besides, extensive defect calculations for a unit cell of 118 atoms are very time-consuming.

One way forward is to use an embedded cluster technique based on a local basis set. This allows one to use DFT with non-local functionals and calculate optical, paramagnetic and transport properties of individual defects. In this study [229] an embedded cluster method which takes into account the full extent of lattice polarisation has been em-

184

ployed. It has been discussed in section 2 and thoroughly described in refs. [43,44]. Briefly, a crystal with a defect is represented by a large finite cluster, divided into regions I and II (see Fig. 10). A spherical Region I is centred on a defect site and includes a quantum-mechanically treated cluster (QM cluster) surrounded by interface ions and classical polarizable ions represented using the shell model. The remaining part of the cluster, region II, is represented by classical non-polarisable ions. The interface between the QM cluster and classical ions is needed in order to prevent an artificial spreading of electronic states outside the QM cluster. All the classical ions interact among themselves via inter-atomic potentials. The interface atoms interact quantum mechanically with atoms in the QM cluster and classically with other interface atoms and with classical atoms in regions I and II. The interaction between the QM atoms and classical atoms in regions I and II is also included and is described using short-range classical potentials. All the centres in region I are allowed to relax simultaneously during the geometry optimisation. Ions in region II remain fixed and provide an accurate electrostatic potential within region I. This method allows one to treat nano-clusters, infinite crystals and amorphous solids.

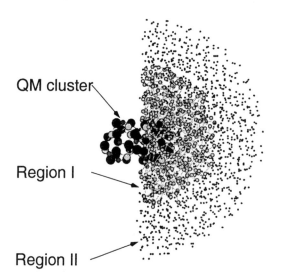

Figure 10. Set-up of embedded cluster calculations of C12A7:e^- system. Half of the two-cage quantum cluster is hidden behind atoms of Region I.

To model the electronic states of C12A7, a finite nano-cluster of nearly spherical shape was built from 64 unit cells (see Fig. 10). Two adjacent cages at the centre of the nano-cluster are treated quantum-mechanically (QM cluster) and surrounded by a spherical region I with the radius of about 13 Å. Ions outside the QM cluster are treated using the classical shell model and interatomic potentials [232]. All oxygen ions in the QM cluster are represented as all-electron atoms with the standard 6-31G basis set. The Ca and Al ions are treated using the lanl2 effective core pseudopotential (ECP) and the lanl2dz basis set [35]. To provide a better flexibility an additional basis, obtained by extending the standard 6-311G* basis set for oxygen atoms by two sp functions, was positioned at the centre of each of the two QM cages. Ca and Al ions in the interface region are represented using ECPs and interact quantum-mechanically with atoms of the QM cluster and classically with the classical environment. Calculations were made using DFT with a hybrid density functional B3LYP and the optical transition energies were calculated using the Time-Dependent DFT method (TD-DFT) as implemented in Gaussian98 [233].

Extra-framework oxygen ions compensating the positive framework charge tend to minimise their electrostatic repulsion, but do not form an ordered sub-lattice. The oxygen configuration used in the calculations corresponds to the smallest dipole moment per crystal unit cell. An extra-framework oxygen ion in the QM cluster was substituted by one or two electrons. The two cages of the QM cluster are not equivalent: one of them (cage 1) has a classical neighbouring cage with an O^{2-} ion in it, whereas the other one (cage 2) does not have an extra-framework oxygen ion in its nearest neighbours. First we consider two electrons in a singlet state. Calculations demonstrate that they prefer to delocalize over the two cages so that each cage contains one electron and interact only weakly. Therefore in the following discussion we focus on the one electron case.

The results of calculations demonstrate that the extra electron can be localized in either of the cages (see Fig. 11). The electron density has predominantly s-character and is localized at the centre of a cage in the same fashion as a single electron is localized in an anion vacancy of, e.g. MgO [43]. The iso-surface of the one-electron function associated with the cage electron in Fig. 11 and the integration of the electron density show that most of the electron density (about 85%) is confined within one cage, while the other one remains almost empty. Similar electronic structure was obtained for the electron localized in the other cage. Since all cages are almost equivalent, the unpaired electron can be treated as a small radius polaron. The calculated hyperfine interaction of the unpaired electron with the nuclei of the neighbouring Al ions is quite small and does not exceed 0.5 mT. This is mainly due to the fact that Al ions are screened by the surrounding oxygen ions, which repel the extra electron (see Fig. 11).

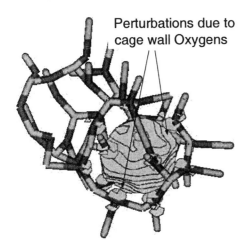

Perturbations due to cage wall Oxygens

Figure 11. The electron localized in one of the two C12A7 cages. The iso-surface of the single-occupied one-electron function is shown.

Comparison of the local atomic structures obtained for the electron localized in two different cages reveals a characteristic lattice distortion, with the strongest relaxation of two Ca ions in each cage located on the opposite sides of the wall with the centre of the cage approximately between them. Unlike other Ca and Al ions, they are exposed to the electron density localized in the cage and displace towards its centre by about 0.4

Å. Most of the other ions displace away from the centre and their relaxation propagates by up to 10 Å from the cage centre. These large displacements reflect the strong electron-phonon interaction due to flexible cage walls.

The number of variables involved in the geometry optimisation in these calculations exceeds two thousand. Therefore to estimate an adiabatic barrier for the electron hopping from cage 1 to cage 2 the authors [229] used a linear interpolation approach. Let R_1 and R_2 be coordinates of all atoms for the two configurations of the electron localized in cage 1 and cage 2 respectively. Then the coordinates of all the lattice atoms along the hopping path can be approximated using $R = tR_1 + (1 - t)R_2$, where t is a parameter varied in the range from 0 to 1. This allows one to calculate all points along the adiabatic path with the same accuracy, but the calculated barrier represents the upper limit for the "true" hopping barrier. The calculated barrier is about 0.1 eV for the hopping from cage 1 to 2, and 0.17 eV for the hopping from cage 2 to cage 1.

The optical absorption energies and the corresponding transition matrix elements for the lowest excited state, E_1, along the hopping path were calculated using TD-DFT. The vertical transition energy in the minima is about 1 eV with a relatively small transition matrix element. If the system is displaced from an equilibrium position in either cage towards the barrier point, $E_1(t)$ decreases to about 0.6 eV. At the same time the transition matrix element increases, indicating a higher transition probability. The highest occupied and lowest unoccupied one-electron molecular orbitals (HOMO and LUMO, respectively) calculated using B3LYP are related to each other in the following way. If the system is in one of its two fully relaxed ground state configurations, say $t = 1$, then the HOMO is fully localized in cage 1 and has s-character, as discussed previously. The corresponding LUMO has two components: its predominantly s-character maximum is localized in cage 2 with a smaller p-character contribution extended to cage 1. The opposite applies to the HOMO and LUMO calculated for the parameter $t = 0$. Thus the lowest optical excitation corresponds to electron transfer between cages (see Fig. 12). When the system is displaced from the minimum towards the barrier, the HOMO attains two s-type components localized in cages 1 and 2 with the ratio between their maxima determined by t; the LUMO has two s-type components of different sign localized in the two cages with the ratio between their maxima reverse with respect to that found for HOMO. This leads to a strong increase in the transition matrix element. At the barrier point the HOMO and LUMO can be qualitatively described as symmetric and anti-symmetric linear combination of s-type functions localized within the two cages: $s_1 + s_2$ and $s_1 - s_2$ respectively.

The optical transitions into three higher excited states calculated using the TD-DFT approach for $t = 0$ have energies of 2.4 eV, 2.6 eV and 2.7 eV. These transitions correspond to $s - p$ excitations within the cage and depend only slightly on the parameter t and thus on details of the cage geometry. The matrix elements of these transitions are similar and about twice as large as that for the inter-cage transition E_1. Taking into account that each cage is surrounded by eight other cages, the probability of the inter-cage transitions should be in fact about four times higher than that of the intra-cage ones. This is indeed observed experimentally [223].

In Fig. 12 the calculated levels of the (C12A7:e^-) system are shown with respect to the valence and conduction bands. The occupied state of the electron localized within a cage is about 4 eV above VBM and about 2.5 eV below CBM. This is a deep level, which agrees well with its localized character. The lowest unoccupied state is about 0.6 – 1.1 eV higher and is localized in a neighbouring cage. The shallow energy minima and the strong dependence of the first transition energy on the parameter t suggest that the corresponding optical absorption band should be broadened with a maximum at around 0.6 eV and a tail towards larger energies. Other factors, such as disorder of the extra-framework oxygen ions, can also contribute to additional broadening. These results agree very well with the experimental optical absorption and conductivity data [223]. The second and several higher transitions have predominantly intra-cage character and smaller probability. The corresponding electronic states are very close to

the CBM. This suggests that excitation in this band can lead to electron ionisation into the conduction band and thus to photo-conductivity.

Finally we note that the unusually deep position of the electron level in the gap resulting in large intra-cage optical transitions is the direct consequence of positive framework charge. One of the obvious directions of improvement of this system therefore would be cation substitution leading to smaller inter-cage distances and larger optical gap.

6. Interstitial oxygen and nitrogen defects

Studies of oxygen excess species in oxides also has a long history. These defects can be created radiolitically, i.e. under neutron or ionising radiation, in the process of metal or silicon oxidation or annealing of grown oxide films, or thermally, as in ionic conductors. In particular, ceramics based on MgO- Al_2O_3 systems are known as prospective materials for fusion reactors. To be successfully used in these applications they should maintain structural and electric integrity under irradiation with fast neutrons, γ-rays and high energy ionising particles, such as electrons [234]. The same criteria apply to gate dielectrics to be used in microelectronic devices, first of all SiO_2, and more recently to perspective high-k materials based on zirconia and hafnia. Irradiation of these materials produces a number of interstitial oxygen atoms and ions and vacancies in different charge states. Further reactions between these defects can lead to the formation of defect clusters, dislocation loops, and voids affecting the mechanical and electrical properties of oxides and ultimately leading to their failure. Another common example, which always involves oxygen incorporation and diffusion through oxide, is oxidation. It can be desirable as in oxide growth and annealing, and harmful, as in corrosion. Again understanding of the structure and stability of oxygen species and control over their diffusion is crucial. Oxide materials that exhibit high ionic conductivity provide another example. Some of them, such as perovskites, e.g. $La_{1-x}Sr_xMnO_3$ and $LaGaO_3$, are used as oxide electrodes for solid oxide fuel cells and as ceramic membranes designed

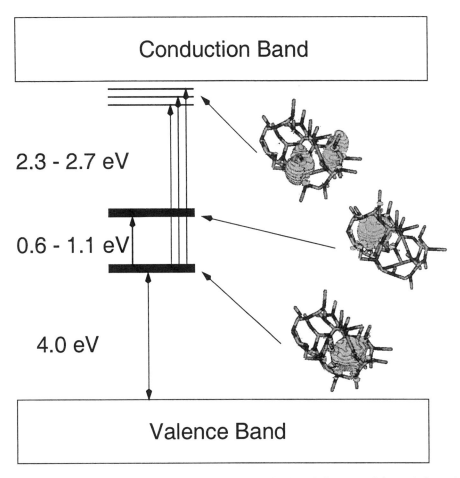

Figure 12. Schematic diagram of the energy levels and character of the optical transitions in C12A7.

to generate oxygen by separation from air. However, in this case interstitial oxygen formation and migration is expected to be highly unfavourable and ionic conductivity is governed by migration of oxygen vacancies (see, for example, ref. [16]). Recent reviews of oxygen excess defects in silica glass derived from optical and EPR studies can be found in refs. [191,201].

6.1. *Atomic and molecular interstitial oxygen in oxides*

As before, we will focus mostly on the models relevant for silica and perspective high-k oxides, especially on those derived from recent modelling

of silicon oxidation and annealing of perspective high-k oxides grown on silicon. An extensive review of oxygen-related defects in irradiated silica, such as the non-bridging oxygen hole center and peroxy radical, is presented in refs. [191,201] and references therein.

From the previous studies of other materials, some general ideas of the properties of interstitial defects can be established. Interstitial anion defects are well known in alkali and alkali-earth halides and also in classic oxides such as MgO. In MgO, interstitial oxygen forms a dumbbell-like bond with a lattice oxygen (see Fig. 13(a)) drawing charge density to itself [11,235]. This behaviour is due to the high electron affinity of oxygen, as it seeks to fill its outer electronic shell. As we will see below, it is quite common for other oxides too.

Excess atomic and molecular oxygen species in silica have been simulated mainly in relation to the mechanism of Si oxidation and annealing of oxide films. The presence of silicon as a source of electrons affects most of the generic questions pertaining to these systems: the energies of incorporation; whether atomic or molecular defect species are favoured; whether these defects act as electron and hole traps. As has been pointed out in refs. [236,237], the polarization of Si and oxide by charged species at the interface may affect their incorporation energies and diffusion. However, only oxide polarization is taken into account as the interface is rarely considered explicitly. Yet another issue concerns disorder in the oxide structure. In this case one should talk about distributions of incorporation energies and diffusion barriers, as will be shown latter. However, structural models remain unaffected and we will look at them first. Incorporated species can be ionised by irradiation or by hot electrons and holes from electrodes or due to electron transfer reactions inside the oxide. To consider these processes it is useful to know thermodynamic and switching charge state defect levels, which are discussed in Sec. 2.

In discussing the static models of oxygen species in the bulk of α-quartz we will follow the PW DFT calculations discussed in refs. [77,238,239]. However, other PW DFT and cluster calculations give very similar results (see, for example, refs. [240,241,242]). The PW DFT calculations [76,238] demonstrate that an oxygen atom inserted into the Si–O–Si bond forms a peroxy linkage (POL). There are two stable POL forms, related by rotation of the Si–O–O–Si complex about the line joining the two Si atoms, as shown in Fig. 13(b). In α-quartz, the energy difference between the configurations is about 0.2 eV [238]. The center of the O–O component of POL is located in the oxygen site. Thus qualitatively the structure of this defect is similar to the dumbbell configuration of an interstitial oxygen atom in MgO.

The atomic oxygen incorporation energy from periodic DFT calculations is usually calculated relative to the non-defective quartz structure plus half of the energy of an isolated O_2 molecule in the triplet state [238]. In the lowest energy configuration the incorporation energy is +2.0 eV; a positive value indicates an endothermic process due to dissociation of the molecule. The dissociation energy of the O_2 molecule is evaluated in this approach to be 5.88 eV (with respect to the broken-symmetry triplet state solution for free oxygen atom). Then a single free O atom would be incorporated in the SiO_2 lattice with an energy gain of 0.91 eV (i.e., a negative incorporation energy of -0.91 eV). This value should be compared with those reported in the literature for the same process in α-quartz. Using a PW DFT at GGA level, Hamann found an incorporation energy of -0.86 eV [240]; a similar value, -0.70 eV has been reported from correlated cluster calculations [242].

Incorporation of an oxygen molecule into α-quartz is more favourable than that of atomic oxygen, if the latter includes dissociation of gas phase molecules [76,238]. In particular, the incorporation energy of two interstitial atomic oxygens is about 4.1 eV, whereas the incorporation energy of molecular oxygen is only 2.1 eV [76,238]. As similar value has been found in cluster calculations [243]. This is due to the fact that the oxygen molecule does not form a chemical bond with the lattice ions and stays in the centre of a big helix of the α-quartz structure, as shown in Fig. 13(c). It incorporates in the triplet state, which is also the most stable state in the gas phase, and causes a significant lattice relaxation with some ions displaced by about 0.5 Å [76,238]. Dissociation of the molecule inside the lattice requires only 2.0 eV, much less energy than in the gas phase due to formation of POL. In the context of radiolitic processes, interstitial atomic oxygen species, if they encounter each other, will form molecular oxygen with an energy gain of 2.0 eV. Since atomic oxygen incorporates into the SiO_2 network forming POLs, diffusion of atomic species will lead inevitably to oxygen exchange with the network. However, diffusion of more independent molecular species cannot easily lead to oxygen exchange, unless they interact with structural defects, such as oxygen vacancies.

The calculations [76,238] also demonstrate that charged species are energetically more favourable

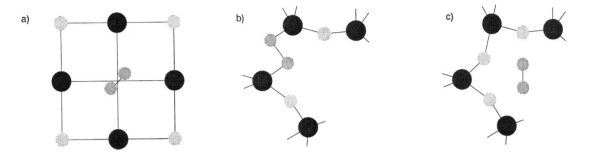

Figure 13. Models of interstitial oxygen species in oxides: a) dumbbell configuration of interstitial oxygen atom in MgO; b) peroxy linkage in α-quartz; c) molecular oxygen in α-quartz .

than the neutral species, if electrons are available from the bottom of the conduction band of silicon or above. They form a variety of configurations [76,238]. The incorporation energy decreases with the increase of negative charge. Upon elastic capture of an electron, the molecular species can either relax in molecular form or dissociate into two atomic species in the appropriate charge states. The former process leads to a lower energy configuration, but will require releasing more energy. The dissociation therefore can be preferred in some cases from kinetic point of view.

These observations may help in identifying species dominating in different processes. In particular, most of the as-manufactured silicas are either oxygen-deficient or contain negligible concentrations of interstitial oxygen species ($< 5 \cdot 10^{13}$ cm^{-3}). Hence, the experimental information on the oxygen-excess defects in SiO$_2$ materials is quite limited compared with the amount of data on oxygen-deficient centres. For instance, methods for identification of interstitial molecular oxygen have been only recently developed [244]. In oxygen-rich silica glass, Suprasil W, the natural concentration of the interstitial molecules (O$_{2i}$) may reach 10^{18} per cubic centimeter, which corresponds to an average distance of 100 Å between the incorporated molecules. In the other types of amorphous silica, interstitial molecules are observed after neutron or proton irradiation [191]. Interstitial molecules are believed to be the dominant species diffusing through silica materials up to a temperature of about 1300°C, at which appreciable exchange of oxygen between the molecules and the SiO$_2$ network was observed. Isotope exchange experiments during dry oxidation [245] show significant exchange only close to both interfaces. This isotope exchange can be due to creation and motion of charged atomic species, in particular, near to Si/SiO$_2$ interface where the electron transfer from Si can encourage the dissociation of molecular oxygen species and formation of O$^-$ and O^{2-} species [238,246].

Similar defect structures were seen in studies of interstitial oxygen in zircon [75] and zirconia [6], where atomic oxygen again forms a dumbbell-like bond with network oxygen ions with a formation energy of about 8.5 eV and 8.9 eV, respectively. As in silica, charged oxygen species are more favourable than neutral for experimental values of band off-set with silicon at the interface. We will illustrate this trend in oxygen interstitial properties in the next section by discussing in more detail the results of calculations for oxygen excess defects in hafnia [7]. However, before that we would like to use the example of POL in a-SiO$_2$ to highlight again some of the issues pertaining to the effect of structural disorder in amorphous materials on defect properties.

Properties of peroxy linkage in amorphous silica

As discussed above, the POL defect in α-quartz induces a strong local distortion into the surrounding network. One can expect that this distortion will depend on the particular site in the amorphous structure. A systematic study of the site-to-site variations of oxygen atom incorporation energies in a-SiO$_2$ was performed in refs. [76,77] using PW

DFT.

In α-quartz there are only two stable POL forms, related by rotation of the Si-O–O-Si complex about the line joining the two Si atoms (see Fig. 13b). The atomic oxygen incorporation energy from plane-wave DFT in the lowest energy configuration is about 2.0 eV (relative to the non-defective α- quartz structure plus half of the energy of an isolated O_2 molecule in the triplet state). In amorphous structures there are again two stable configurations of POL per site, but their incorporation energies have characteristic distributions with a big spread around the average value. Calculations have been made for both configurations of the POL incorporated at all 48 oxygen sites in the 72 atom periodic cell. The distribution of incorporation energies is centred on 1.9 eV, varying from about 1.1 eV to 2.7 eV. Most POL defects have incorporation energies between 1.5 and 2.3 eV. The formation of a peroxy linkage causes a distortion of the surrounding network, which decays with the distance from the defect. An important conclusion reached in ref. [77] is that O-atom incorporation can either create or release strain in the embedding network. This appears to be true even without the changes in topology (connectivity), which can occur at higher temperatures, perhaps by diffusion. A statistical analysis [77] shows that the variation of the incorporation energy does not originate solely from the local geometry of particular oxygen sites, but rather from the contribution from the more distant amorphous network.

These results raise a more general question: to what extent do local and long- range disorder affect other defect properties, such as formation energy, optical, IR and XPS spectra, *etc.* This question is closely related to more technical issues: which are the most appropriate methods to simulate amorphous structures and to which extent cluster models are applicable to study a-SiO_2. Some of these issues have been addressed in ref. [239]. This work considered the applicability of cluster models to studying the local and medium-range disorder around the defect, such as the POL, which induces relaxation in an area of the amorphous structure extending beyond the first and second neighbours. It demonstrates that the geometric structures

and the distribution of incorporation energies of such a defect cannot be adequately described in a simple molecular cluster model. On the other hand, optical absorption through electronic transitions localized on the -O–O- bridge itself are not strongly affected by disorder. For this reason, these optical absorptions are much less sensitive to the model used (i.e. molecular versus embedded cluster). However, the transitions involving delocalized states are likely to be affected by the size of the quantum cluster described on the higher level of theory, and optical emission is very likely to be sensitive to the method chosen. Other properties considered in ref. [239], core level photoemission and local vibrational spectra, do not seem to depend markedly on the model used. The authors were not able to find a systematic dependence of the site-to-site variations on any of characteristics of disorder.

Oxygen in hafnia

The initial question with regard to interstitial oxygen incorporation, is in which state oxygen defects form: atomic or molecular. Atomic interstitials could form due to the creation of vacancy interstitial pairs, dissociation of molecules within the crystal or due to direct incorporation via dissociated oxygen gas. Molecular oxygen species may incorporate into the hafnia lattice from the gas phase or form from atomic oxygen species already existing inside the lattice.

As discussed previously, the formation energy of oxygen interstitials in hafnia depends on the reference for the oxygen source. Assuming a molecular oxygen source, we use half of the energy of an isolated O_2 molecule (4.91 eV) as the reference energy, E_O, (see Eq. 1). We then obtain atomic defect formation energies $+1.6$ eV for O_3^0 and $+2.3$ eV for O_4^0. A positive value indicates an endothermic process, which involves dissociation of the O_2 molecule. Using the atomic E_O as a reference, we find formation energies for single O atom incorporation into the monoclinic hafnia lattice of -1.3 eV (O_3^0) and -0.6 eV (O_4^0).

For molecular oxygen, with E_O equal to 9.81 eV, we find incorporation energies of $+4.2$ eV for $(O_2^0)_3$

and $+5.8$ eV for $(O_2^0)_4$. As for the atomic case, incorporation near the less cramped 3-coordinated oxygen lattice site is more favoured. These molecular incorporation energies are double that found in silica [247], as to be expected considering the greater interstitial space available in the silica lattice.

Again, comparison of atomic and molecular incorporation depends on the process by which the oxygen enters the system. If the process includes dissociation of the molecule, at a cost of 5.88 eV, we can compare the incorporation energy of a molecule in hafnia (4.2 eV) with the incorporation of two oxygen atoms (3.2 eV), implying that atomic incorporation is favoured by only 1 eV. Again this contrasts with the favouring of molecular adsorption in silica by 2 eV [247]. However, if incorporation may occur effectively from an atomic gas, without dissociation, this reduces the incorporation energy of two separate atoms to -2.6 eV and favours atomic vs. molecular incorporation by almost 7 eV.

Although we performed full calculations for the O_4^x and $(O_2^x)_4$ series of defects, in every case the O_3^x and $(O_2^x)_3$ equivalents were lower in energy, so we will focus in detail only on the latter case. Note also that studies in zirconia have also shown that there is no stable minimum for charged defects at the O_4 sites [6].

Atomic oxygen

For the neutral interstitial, a charge density map in Fig. 14 shows that the oxygen defect forms a clear covalent bond with the 3-coordinated lattice oxygen. The interstitial and lattice oxygen form a "dumbbell" defect pair structure, which is characteristic of other oxides, such as zirconia [6] and zircon (ZrSiO$_4$) [75]. To accommodate the interstitial, the lattice oxygen displaces from its planar position between three Hf ions to form a shallow pyramid. In this structure the two oxygens are effectively identical structurally and electronically, although small differences in bond lengths evidence slightly different environments, and the interstitial is bound to only two of the three hafnium ions in the pyramid. The interstitial incorporates in the lowest singlet state - the triplet state is 0.9 eV higher in energy.

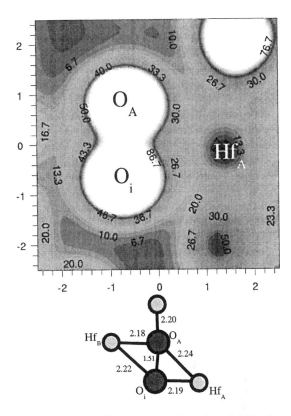

Figure 14. Charge density in the plane through Hf$_A$, O$_A$ and O$_i$, and a schematic diagram of neutral oxygen interstitial (O$_i$) near a 3-coordinated oxygen (O$_A$) in hafnia. Charge density is in 0.1 e/Å^3 and all distances are in Å. *Reproduced with permission [7].*

To study electron trapping by the defect, an extra electron was added into the system in the atomic configuration of the neutral interstitial. The electron does not initially localize at the defect site, but rather remains delocalized with its energy very close to the CBM. Therefore the vertical electron affinity is close to zero. We should note that this can be an artifact of our DFT calculation caused by the wrong position of the bottom of the conduction band (see section 3).

System relaxation leads, however, to complete localization of the electron on the oxygen pair. The increased charge on the defect oxygen pair causes the ions to separate, also increasing the depth of the triple oxygen pyramid. Figure 15 demonstrates that the covalent bond between the two has almost disappeared, although they remain effectively still a similar species within the crystal. The relaxation energy from the initial O_3^0 configuration is equal to 2.3 eV. The electron is fully localized in a doublet state on the defect pair.

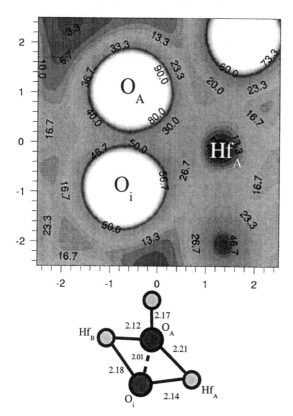

Figure 15. Charge density in the plane through Hf_A, O_A and O_i, and schematic diagram of singly charged oxygen interstitial (O_i) near a 3-coordinated oxygen (O_A) in hafnia. Charge density is in $0.1\ e/\text{Å}^3$ and all distances are in Å. *Reproduced with permission* [7].

Addition of the second electron from the conduction band to the negatively charged defect is again accompanied by strong lattice relaxation. Introduction of the second electron effectively produces two independent lattice oxygen ions. The separation between them increases even further, to about 2.4 Å. The corresponding relaxation energy is about 1.53 eV. The original 3-coordinated lattice oxygen is pushed into a deeper pyramid structure with the three Hf ions, but the interstitial now

creates a new 3-coordinated site, bonding with a third, independent, hafnium ion at 2.2 Å. In the previous defect structures, this hafnium ion was over 2.6 Å from the interstitial (and over 4 Å from the lattice oxygen) and no bond could be seen in the charge density. The new electron is fully localized on the defect oxygen pair, which is in the singlet state with equal spin components.

The changing nature of the defect pair can also be seen in the evolution of the total DOS of the systems. On addition of the neutral oxygen interstitial, the main band structure remains the same, but new states can be seen. These are bonding and anti-bonding states of the O_i-O_A defect pair due to the extensive charge transfer and formation of a strong covalent bond. The highest occupied defect states are in the band gap at 2.3 eV above VBM. For the singly charged oxygen interstitial, the O_i-O_A bond is weaker and the DOS is even closer to the ideal bulk DOS. The fact that the defect pair separates and becomes much more ionic means that there are no other clear defect states in the gap. Defect states appear in the energy gap at about 2.8 eV above VBM. The DOS for the doubly charged interstitial is, again, very similar to that for the perfect lattice, with O_i^{2-} related defect states at about 2.7 eV above VBM.

Molecular oxygen

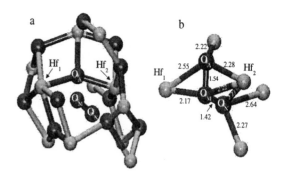

Figure 16. (a) Atomic structure of ideal monoclinic hafnia with neutral interstitial oxygen molecule (O_N -O_M) introduced near to 3-coordinated lattice oxygen site (O_A). (b) Close-up of defect structure after relaxation. Equivalent atoms in the ideal and defected structure are labelled. All distances are in Å. *Reproduced with permission* [7].

Figure 16 shows the atomic structure of a neutral oxygen molecule incorporated near to a 3-coordinated lattice oxygen site. The associated charge density plot is shown in Fig. 17. It is immediately evident that the molecule shares an electron density with the lattice oxygen: there is a high electron density between O_A and O_M, and a bond length very close to the molecule, O_M - O_N, bond length. Atoms around the molecule displace away from the site to create more space in the crystal, but these relaxations decay rapidly to less than 0.1 Å beyond the immediate neighbours. As a consequence of these displacements all the Hf–O bond lengths are significantly longer than in the ideal bulk crystal (see Fig. 8). In this configuration the molecule incorporates in the singlet state, and the triplet state is 0.5 eV higher in energy.

To study whether there are other stable configurations of a neutral molecule inside the hafnia lattice, we started the geometry optimization from initial configurations far from any oxygen sites. We have found that a stable minimum exists with bonding of the molecule only to hafnium ions, however, this state was 0.23 eV higher in energy. This result suggests that the molecule has a high electron affinity in the oxide, as it seeks the most abundant source of electrons in the system - lattice oxygen sites.

Adding an electron to the molecular defect produces a similar story to that for the atomic oxygen defect. The electron localizes fully on the defect only after atomic relaxation. The molecule displaces away from the lattice oxygen and is now effectively bonded only to Hf ions. The intramolecular bond length remains the same as in the neutral case. The large relaxation energy of 2.67 eV from the neutral geometry for this defect reflects the strong reconstruction. Addition of a further electron continues the development of the oxygen molecule as an ionic species in the crystal, returning to an almost ideal 3-coordinated oxygen structure. The molecular bond length increases very slightly, but overall displacements are not very large. The relaxation energy from the singly charged geometry is 1.05 eV.

Figure 6a shows how the introduction of molecular defects to hafnia changes the total DOS. The incorporation of the neutral molecule produces sev-

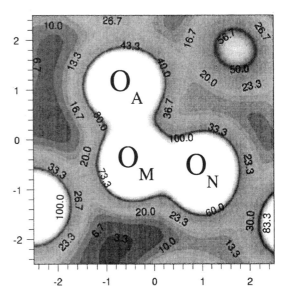

Figure 17. Charge density of neutral interstitial oxygen molecule (O_N–O_M) bonding with 3-coordinated lattice oxygen site (O_A). Charge density is taken in a plane through O_M, O_N and O_A in 0.1 $e/Å^3$ and all distances are in Å. *Reproduced with permission* [7].

eral defect states due to the covalent bonds between the oxygens in the molecule and the lattice oxygen. Clear states due to charge transfer and the formation of covalent bonds between oxygen atoms can be seen at about -22 eV, -17 eV, -13 eV, -5 eV. As electrons are added to the system it tends towards the spectrum of the ideal bulk crystal. After an electron is added, the state at -17 eV disappears, and the other states decrease in energy and density, mirroring the dissipation of the covalent bond between O_A and O_M. A state now appears in the gap at 2.4 eV (see Fig. 6b). When a second electron is added only very small changes in the DOS can be observed, in agreement with the similar bonding seen around the molecule in both charged states.

Nitrogen in hafnia

Several recent experiments [248,249,250] have demonstrated that many characteristics of high-k oxides, and in particular hafnia, improve significantly as a result of post-deposition annealing in nitrogen. Nitrogen clearly incorporates into oxide and passivates some of the pre-existing defects,

194

however, the form of incorporation and the way nitrogen interacts with oxide defects remain largely unknown. This paucity of data extends to theory, where very few studies of nitrogen in oxides have been performed (see, for example, refs. [251,252]. Hence, here we present our own systematic study of nitrogen defects in hafnia. The methodology is the same as that discussed for oxygen incorporation, with the focus on discovering in what form nitrogen favours incorporation, and also how nitrogen interacts with oxygen vacancies in the oxide.

The formation energies of nitrogen defects also depend on the reference source of nitrogen used. If we consider incorporation from a molecular source, where the large N_2 dissociation energy must be paid, then atomic nitrogen incorporates at a cost of +4.60 eV (+5.5 eV at 4-coordinated site) and molecular oxygen +6.06 eV (+7.8 eV at 4-coordinated site). Hence molecular incorporation is favoured by 3.14 eV. However, if we consider adsorption from an atomic source the atomic formation energy reduces to -0.38 eV and atomic incorporation is favoured by 6.44 eV. All the defect formation energies are summarized in Tab. 5.

D	$E_{form}(E_N)$	$E_{form}(E_{N_2})$
N	-0.38	4.60
N_2	-	6.06
$(V+N)$	3.37	8.14
$(V+N_2)$	-	9.15

Table 5
Formation energies (E_{form}) of nitrogen defect species in hafnia with respect to an atomic (E_N) and molecular (E_{N_2}) reference states.

Atomic nitrogen

As in the case of oxygen, Fig. 18 shows that the nitrogen interstitial defect forms a clear covalent bond with the lattice oxygen site. The defect forms the characteristic 'dumbell' shape discussed previously in the context of oxygen in zircon, zirconia and hafnia. The defect forms in the doublet state, with a single unpaired electron - the quad state being over 3 eV higher in energy. If we add an electron to system it localizes fully on the defect pair, increasing the separation between O and N by about 0.1 Å. The bond between them remains,

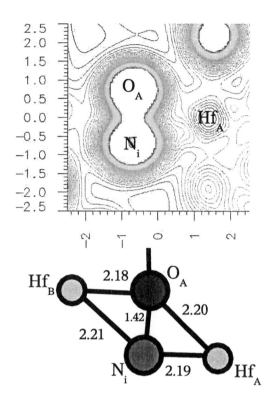

Figure 18. Charge density in the plane through Hf_A, O_A and N_i, and a schematic diagram of neutral nitrogen interstitial (N_i) near a 3-coordinated oxygen (O_A) in hafnia. Charge density is in 0.1 $e/Å^3$ and all distances are in Å.

and the surrounding geometry does not change significantly - the relaxation energy is about 1.2 eV, compared to over 2 eV for oxygen.

If we remove an electron (i.e. create a hole) then Fig. 19 shows that the defect model changes more dramatically. The nitrogen-oxygen pair is now very electron deficient, and in fact nitrogen forms a bond with a second lattice oxygen, O_B, although the relaxation energy is again about 1.2 eV.

We can also consider the incorporation of nitrogen at oxygen vacancy sites. Tab. 5 shows that if we remove the cost of the creation of a vacancy (about 9.4 eV), then it is very energetically favourable for nitrogen to be trapped there. The bond lengths between nitrogen and hafnium are slightly larger than that shown for the 3-coordinated site in Fig. 8, but the structure is the same. Fig. 20 demonstrates that when an electron is added, it localizes on the nitrogen and the bond lengths become much

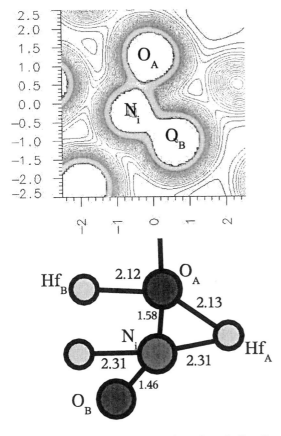

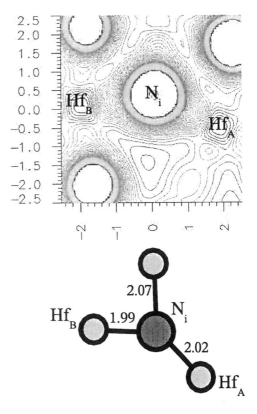

Figure 19. Charge density in the plane through O_A, O_B and N_i, and a schematic diagram of a positively charged nitrogen interstitial (N_i) near a 3-coordinated oxygen (O_A) in hafnia. Charge density is in 0.1 $e/\text{Å}^3$ and all distances are in Å.

Figure 20. Charge density in the plane through Hf_A, Hf_B and N_i, and a schematic diagram of a negatively charged nitrogen interstitial (N_i) at a 3-coordinated oxygen vacancy site in hafnia. Charge density is in 0.1 $e/\text{Å}^3$ and all distances are in Å.

smaller. The relaxation energy is about 0.6 eV. If we remove an electron, the reverse process occurs with the bond lengths increasing, but with a similar relaxation energy.

Molecular nitrogen

Fig. 21 shows that, unlike for an oxygen molecule, a nitrogen molecule does not form a strong covalent bond with the lattice oxygen, but rather incorporates in the interstitial space near to the site. Introduction of an electron to the system increases the independence of the N_2 species, increasing the O–N distance from 1.96 to 2.20 Å, with a relaxation energy of about 1.2 eV. In contrast to the atomic case, a second electron will

localize on the molecular defect, increasing the O–N distance to 2.34 Å at a relaxation cost of 0.6 eV. Adding a third electron or removing 1 or 2 electrons from the neutral system produces a de-localized result.

Once again, we can consider the case where the molecule is incorporated near a vacancy site - Tab. 5 shows that the incorporation energy when the vacancy is already present is practically zero. The similarity of Fig. 22 and Fig. 18 demonstrates that the molecule forms a defect pair which is almost identical to the O–N defect complex. Adding/removing an electron just increases/decreases the N–N bond length and decreases/increases the surrounding Hf–N bonds.

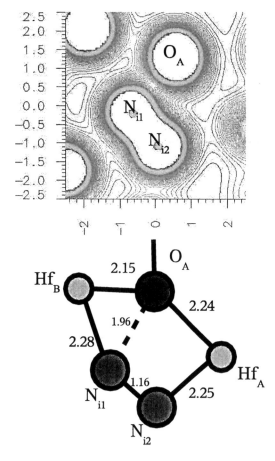

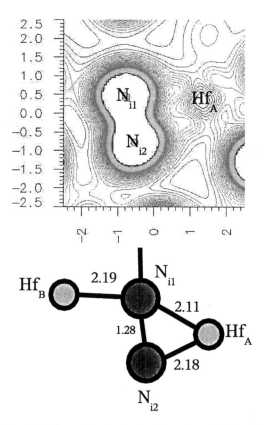

Figure 21. Charge density in the plane through Hf_A, N_{i1} and N_{i2}, and a schematic diagram of a neutral nitrogen molecule (N_{i1}-N_{i2}) at a 3-coordinated oxygen site in hafnia. Charge density is in $0.1\ e/\text{Å}^3$ and all distances are in Å.

Figure 22. Charge density in the plane through Hf_A, N_{i1} and N_{i2}, and a schematic diagram of a neutral nitrogen molecule (N_{i1}-N_{i2}) at a 3-coordinated oxygen vacancy site in hafnia. Charge density is in $0.1\ e/\text{Å}^3$ and all distances are in Å.

Nitrogen-oxygen exchange

Finally, one can consider the energetics of an exchange of a nitrogen interstitial and a lattice oxygen (O⇔N). We find that if you exchange O_A and N_i in the neutral nitrogen interstitial case (see Fig. 18), and re-relax, the total energy is in fact 0.02 eV lower. This means that if the barrier to the exchange can be overcome, then it is energetically favourable for nitrogen to exchange with the lattice oxygens. Upon adding an electron to the system we observe very similar behaviour to that for the neutral nitrogen interstitial, the O–N distance increases, although the relaxation energy (about 1.0 eV) is slightly smaller.

7. Hydrogen defects

The lightest element, hydrogen, is probably also the most abundant, as well as the least detectable impurity in the dielectric oxides. It is also believed to be most damaging for the dielectric properties of the gate stack in MOSFET, leading to interface charging [253], stress induced leakage currents, and ultimately to an electrical breakdown of the dielectric film [254,255,256]. Also, in traditionally used Si/SiO_2 systems protons may play a positive role in passivating Si dangling bonds at the interface.

Abundance of hydrogen in a dielectric film arises from the fact that, being present throughout the manufacturing process (e.g. hydration step in the ALCVD techniques), hydrogen is easily incorporated into a film. Although hydrogen can be subsequently partly removed using post deposition annealing, some quantity remains trapped in the lattice in various stable forms, not all of which are easily detectable. For example, molecular hydrogen does not usually interact strongly with the dielectric lattice and may be identified by Raman spectroscopy. Neutral hydrogen atoms can in principle be detected in the EPR measurements, but the complications arise from a high mobility of hydrogen atoms. In contrast, various hydroxide ions (OH)$^-$ and water molecules can be probed using the Fourier Transform Infrared Spectroscopy (FTIR) techniques, although an ambiguity often arises as to where from in the sample (device) an IR signal originates.

Due to difficulties in experimental detection, most of the evidence of the involvement of hydrogen in the electric degradation of CMOS devices is indirect. The experiments indicating hydrogen's role include the following:

1. Effect of hydrolytic weakening of silica glass - generation of oxygen vacancies by irradiation is significantly higher in wet silica. [257,258]. (In fact, hydrolytic weakening is common for all natural oxides. Its mechanism has been addressed e.g. using ab initio calculations [259,260].

2. The positive charging of the Si/SiO$_2$ interface, observed at certain conditions, is generally associated with building up the proton concentration at the interface [253,254,255].

3. Hydrogen has also the effect of reducing a material's dielectric constant [148].

It is understood that some forms of hydrogen in the dielectric are more damaging than others. Therefore, at least in principle, adverse effects can be eliminated, reduced or delayed to a degree by carefully tuned film deposition conditions and post deposition treatment [255]. This, however, requires a comprehensive understanding of the forms and charge states in which hydrogen is trapped in a dielectric. Such understanding includes the energetic hierarchy of various hydrogen defects, their ability to accept or release carriers as well as details of kinetics of their transformations in free and electrically loaded conditions. *Ab initio* modelling once again emerges here as indispensable tool used in conjunction with the experimental data

for gaining this understanding. One should note that the high reactivity of hydrogen and its ability to make and break bonds in polar crystals make modelling with various atomic force-field approximations highly unreliable. So, a quantum mechanical treatment of electrons in hydrogen species is essential. More specifically, quantum mechanical calculations have been particularly instructive in understanding the role of hydrogen related defects in silica [78,260,261,262,?,264,265,266,267,268]. Atomistic studies of hydrogen may be divided into two interrelated groups. The first group of papers addresses the question of what defects are likely to occur, given the specific manufacturing conditions? The second is concerned mostly with the mechanisms of the dielectric film degradation. Although most of the calculations related to hydrogen consider it to be embedded in α-quartz or β-crystabolite bulk materials, an influence of structural disorder was recently addressed by Bunson *et al.* [264] and Bakos *et al.* [266], and hydrogen at the Si/SiO$_2$ interface was studied by Tuttle [265], Chadi [268] and Bakos *et al* [266].

H in silica

The main question to be addressed first, as for oxygen and nitrogen, is in what form hydrogen is present in silica. Calculations suggest a variety of possibilities. Neutral hydrogen is found not to make bonds with the SiO$_2$ framework, while interstitial protons can make an ionic bond with a bridging oxygen [253,269] or make a hydrogen bond between two bridging oxygens [268] (Fig. 23a,b). The latter configuration is found to be more stable. Hydrogen may also be present as an acceptor when bonded to a silicon [269]. However, H$^-$ ions are very reactive and are likely to evolve into more stable defects. The following reactions were proposed by Chadi [268]:

$$H^- + O_{int} \rightarrow OH^-; \tag{8}$$

$$H^- + O_{int} \rightarrow H^+ + O_{int}^{2-}. \tag{9}$$

Both reactions were found exothermic by 3.4 eV and 2 eV respectively. H$^-$ may further recombine with the interstitial proton resulting in the formation of a neutral H$_2$ molecule (exothermic by

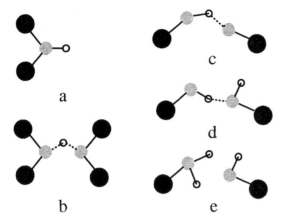

Figure 23. Hydrogen defects in the SiO$_2$ (after Chadi [268]. a) Proton bound to bridging oxygen; b) proton bound between two oxygens c) (OH)$^-$ in the peroxide bridge configuration; d) H$_2$O in the peroxide bridge; e) H$_3$O$^+$ in the peroxide bridge. The elongated hydrogen bonds are indicated by dashed lines.

2 eV). An interstitial hydroxide ion, (OH)$^-$, can accept few forms, but most stable is considered to be a peroxide bridge [268,269] (Fig. 23c). This model suggests a very low mobility of the OH$^-$ which is consistent with the experimental data in quartz and rhyolite glass [270]. Being negatively charged with respect to the lattice, (OH)$^-$ may trap highly mobile protons. The result is an H$_2$O molecule with one elongated bond and an energy gain of 0.9 eV (Fig. 23d). Water in this form also has a small affinity to an additional proton (0.2 eV), so a H$_3$O$^+$ complex can be formed (Fig. 23e). All the interstitial defects discussed in Ref. [268] (except H$_2$ and H^0) assume a strong interaction with the silica framework. However, recent studies by Bakos *et al* [266] suggest that H$_2$O and H$_3$O$^+$ also exist as almost free molecules within large (6-7 membered) SiO$_2$ rings, and any accommodation of these species into a framework is a subject to an energy barrier of the order of a few electron volts.

It must be noted that the relative stability of different species obtained just from the total energy calculations may not reflect thermodynamic reality, due to an essentially unspecified chemical potential (both ionic and electronic). A good example is a study of atomic hydrogen. As discussed before, calculations predict that it does not make chemical bonds with the perfect crystalline SiO$_2$ framework.

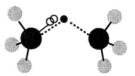

Figure 24. Hydrogen bridge defect (after ref. [78]). Small open circles indicate the displacement of hydrogen toward one of the silicon ions upon addition of 1 and 2 electrons.

Tuttle [267] presented a DFT MD study of H^0 and reported an activation energy for diffusion of $\sim$ 0.2 eV, indicating extremely high hydrogen mobility. However, as first suggested by Yokozawa and Miyamoto [269], a neutral hydrogen is a negative U defect and is thermodynamically unstable at any position of Fermi energy. For example, an electron transfer reaction $2H^0 \rightarrow H^+ + H^-$ is exothermic [78]. Of course, the possibilities are not exhausted by this reaction. As SiO$_2$ films in the CMOS device are subjected to elastic and electric stress, a significant concentration of vacancies is expected. Also oxygen vacancy generation is known to be enhanced in wet oxides [257,258,259].

Oxygen vacancies can effectively trap hydrogen resulting in e.g. a hydrogen bridge centre where hydrogen replaces the bridging oxygen. As discussed earlier, anion vacancies have a large electron affinity, and so does the hydrogen bridge centre. Furthermore, charging leads to a substantial defect relaxation - addition of electrons causes an off-centre shift of hydrogen towards one of the silicon sites (see fig. 24).

This relaxation is accompanied by a substantial shift of electron energy levels. This finding led to a suggestion that a hydrogen bridge is responsible for the SILC, as discussed by Blöchl and Stathis [78,262]. Oxygen vacancies and atomic disorder also affect the lattice interaction with H$_2$O and H$_3$O$^+$ defects. The latter have been the subject of a detailed density functional study by Bakos and co-authors [266], who also discussed various models for the effect of proton release upon hole capture in the irradiated SiO$_2$. There were also numerous DFT studies of hydrogen interaction with a Si/SiO$_2$ interface. Hydrogen has been observed to passivate P_b centers at the interface [271]. Using a DFT cluster approach Tuttle [265] modelled the P_b centre at the (111)Si/SiO$_2$ interface as a simple dangling bond. Calculated parameters for hyperfine splitting qualitatively agreed with the ESR experiments supporting this model. He then considered interaction of atomic and molecular hydrogen with this defect and suggested a mechanism for P_b passivation by hydrogen involving H_2 dissociation. Note, that the use of a cluster approach and a localized basis set allowed calculations of the EPR parameters and hence a direct comparison of the model with the experimental data. A similar approach based on the Hartree-Fock method has been used by Pacchioni and Vitiello [261] to calculate hyperfine interaction parameters and infrared frequencies and relative intensities for various hydrogen defects in the bulk and on the surface of crystalline α-quartz.

Finally, silicon vacancies nominally have a large formation energy. However, due to the kinetics of the oxide formation they may also be present in appreciable concentrations. In the presence of hydrogen this will lead to the formation of hydrogarnet type defects, where up to 4 hydrogen atoms make bonds with the oxygens closest to a vacancy cite. The hydrogarnet defect is present in most of the natural minerals and has been addressed by DFT calculations [259,260].

All the studies presented here give only a semiquantitative picture of the hydrogen related processes. Nor do they exhaust all the processes and defects which may be important. However, being used in conjunction with experimental data they provide a general insight into these processes and give an invaluable experience for the technology when this knowledge is extrapolated to systems less studied. In the next section we discuss our own modelling on the effect of hydrogen in high-k dielectrics.

H in zirconia

As the applicability of zirconia and hafnia films as gate dielectrics in CMOS devices is vigorously examined, the role of hydrogen in these materials needs to be investigated. It is paramount for technology to understand to what extent its role and behaviour is similar to that in silica films.

It is observed that unlike silica, zirconia and even more so hafnia, have a strong tendency for crystallization. In the absence of impurities deposited and subsequently annealed films, as well as natural minerals, are polycrystalline. It is a matter of a current discussion whether the annealed thin films consist of a pure monoclinic, tetragonal, cubic phase or a mixture of all three. It is believed that the stabilization of tetragonal and cubic zirconia is due to the presence of anion vacancies, whose role is to decrease an effective coordination of the cation sites, and thus to increase the bond strength (each anion vacancy results in four seven-fold coordinated cations in contrast to normal eight-fold coordination in cubic and tetragonal phases). One of the methods to create such vacancies and thus to stabilize the tetragonal phase, is doping by trivalent impurities such as yttrium. However, there is strong experimental evidence that water may significantly affect the relative stability of zirconia polymorphs. Sato and Shimata [272] first demonstrated that zirconia is subject to a substantial water uptake at elevated temperatures. They further observed that water uptake leads to a degradation of yttria stabilized tetragonal zirconia into the monoclinic phase. Recently Guo [273,274] suggested that water molecules (or $(OH)^-$ ions) passivate anion vacancies, and in doing so, reduce the stability of tetragonal phase. As we shall see, there is an alternative explanation of the influence of water, which does not require anion vacancies. A more recent FTIR study of deuterium substitution of hydroxyls in zirconia by Merle-Méjean *et al* also confirmed the substantial amount of various types of $(OH)^-$ centres trapped in the bulk of tetragonal and monoclinic zirconia after 24 hours of annealing in an O_2/H_2O atmosphere at 1170 K.

A further question is in what form hydrogen is stabilized in bulk zirconia and what is its role in relative phase stability. Here we address this problem using a plane wave DFT approach. Details of the calculations are essentially the same as were used in the study of vacancies and interstitials discussed previously.

Atomic hydrogen or proton?
PW DFT calculations predict two possible configurations for the hydrogen atom: i) an H^0 inter-

stitial whose electron occupies the band gap level located 1.2 eV above the VBM; and ii) a hydrogen donating its electron into the conduction band and making the OH$^-$ bond with an oxygen ([(OH)$^-$ + e]). These configurations for the tetragonal lattice are depicted in Fig. 25a,b. H^0 does not interact strongly with the lattice and hence produces very small relaxation. Electronically this defect is very similar to H^0 in SiO$_2$ cavities discussed previously. However, further analysis of the adiabatic surface using static and MD DFT simulations shows that the H^0 configuration in ZrO$_2$ is metastable and relaxes spontaneously into [(OH)$^-$ + e]. The latter is by 0.95 eV lower than H^0 in the monoclinic phase and by 0.7 eV in the tetragonal phase of zirconia.

Although hydrogen is predicted to be a donor in monoclinic, tetragonal and cubic zirconia, there is a substantial difference between the [(OH)$^-$ + e] defect in these phases. In the former case the proton makes a bond explicitly with the 3-coordinated oxygens and it is unstable near the 4-coordinated oxygen sites. Moreover, the oxygen ion in OH$^-$ is only slightly displaced from its site. On the contrary, in tetragonal and cubic phases the formation of the (OH)$^-$ ion is associated with large oxygen and hydrogen displacements towards each other along the (111) direction: oxygen moves out of the perfect site by 0.7 Å while the proton's displacement from its cubo-interstitial position is only 0.5 Å (fig 25b). As seen in the figure, the oxygen ion involved in the OH$^-$, brakes the bond with one of the nn zirconium ions and moves into the plane of its other three Zr neighbours (the sum of the three Zr-O-Zr angles is 359°). It is easy to notice that the local coordination of this oxygen becomes very similar to a 3-coordinated anion in the monoclinic phase. Furthermore, recalling that the tetragonal (or cubic) to monoclinic transformation is of martensitic type [109], one can consider a continuous path from the cubic to monoclinic phase as a superposition of an optical mode related to a displacement of half of the anions from the 4-coordinated to 3- coordinated environment and a sheer deformation tilting the c-axis by approximately 10°. The resulting displacement for the relevant lattice fragment is depicted in Fig. 26a,b. One can see, that the oxygen related optical mode

consists of a displacement of the relevant oxygen ions towards their nearest cubo-interstitial positions and into the plane formed by the three nn zirconium ions. This is similar to the relaxation imposed by the proton in the lattice. In other words, the proton in the cubic (tetragonal) lattice instigates locally the phonon mode responsible for the transformation into the monoclinic phase, thus creating locally a monoclinic precursor.

We note that the monoclinic phase has quite low local symmetry, and it is not clear whether the discussed interstitial hydrogen centres exhaust all possibilities. A qualitative analysis suggests that stability of the ([(OH)$^-$ + e]) complex results primarily from the interplay between an (OH)$^-$ bond formation and the electrostatic repulsion of the proton from the nearest neighbour zirconium ions as illustrated in Fig. 27a.

Indeed, the 3-coordinated oxygen ion in the monoclinic phase is located approximately within the plane of three nearest neighbour Zr ions. Thus to minimize the repulsion from these cations the proton sticks out of the plane at approximately the right angle (Fig. 27a). There are only two possible ways of proton attachment in this arrangement. At the 4-coordinated site oxygen is approximately tetrahedrally coordinated (Fig. 27b), and therefore, a strong coulomb repulsion of the proton from the nn cations is likely to exceed the energy gained from OH$^-$ bond formation. Interestingly, an interstitial OH$^-$ in the m-ZrO$_2$ has also three nearest neighbour cations, but these form a pyramidal rather than in-plane arrangement (Fig. 27c). In this situation the only feasible OH$^-$ orientation is the one where the OH bond is directed primarily along the pyramid axes with a proton facing away from the nn cations. The above simplified arguments should hold also for tetrahedral zirconia. All oxygen ions in this phase are tetrahedrally coordinated, thus the formation of the (OH)$^-$ bond in the lattice site is expected to be unfavourable unless significant oxygen relaxation changing its coordination takes place. Our calculations confirm this argument.

Finally, we emphasize once again that the PW DFT approach at the GGA level predicts atomic hydrogen to be a donor in all zero pressure phases of zirconia, as discussed above. However, this contra-

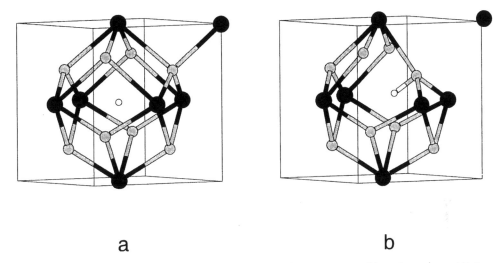

Figure 25. H^0 interstitial configuration (a) and ($[(OH)^- + e]$) configuration (b) in the tetragonal ZrO$_2$.

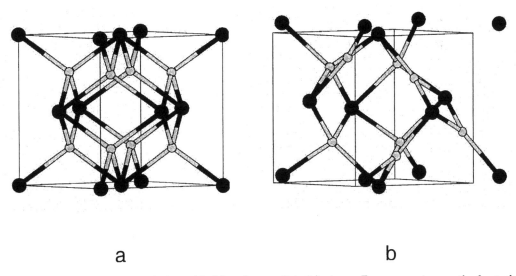

Figure 26. Continuous transformation relating cubic (a) and monoclinic (b) phases. Four oxygen ions on the front plane change their coordination from four to three by relaxing towards their nearest cubo-interstitial positions.

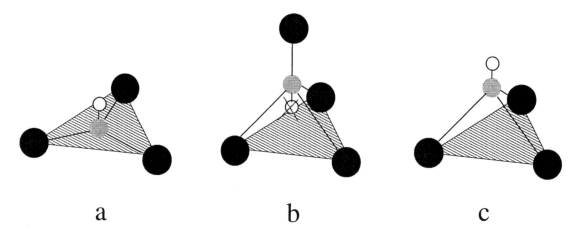

a   b   c

Figure 27. Possible configurations of the proton in the monoclinic lattice: near the 3-coordinated oxygen site (a); near the 4-coordinated oxygen site is unfavourable due to a strong Zr-H repulsion (b), near the oxygen interstitial ion ((OH)$^-$) (c);

dicts earlier results by Nishizaki et al. [275] whose X_α cluster calculations predicted a neutral interstitial H^0 as the only stable atomic hydrogen form in t-ZrO_2. Also, in their recent work Kiliç and Zunger [276] predicted that H^0 in ZrO_2 does not donate an electron. This discrepancy is most likely to originate from the fact that the energy balance between H^0 and $[(OH)^- + e]$ involves a band gap energy predicted differently by different methods. Our calculations of H^0 in t-HfO_2 support this argument. The single particle band gap in hafnia is approximately 0.9 eV larger than in zirconia. As a result, an interstitial H^0 is predicted to be a global minimum configuration in hafnia, in contrast with zirconia. It follows from the calculations that zirconia and hafnia present near threshold systems, were the character of hydrogen atomic species change from ambivalent to donor- like. Thus, more refined approaches must be developed to pinpoint an exact position of this threshold. However, the question of stability of atomic hydrogen in static calculations is largely academic. It is very likely, that similarly to SiO_2, hydrogen in zirconia and hafnia is a negative U defect and it is charged (negatively or positively) at all values of chemical potential. Also even if it is stable, an atomic hydrogen is likely to donate its electron to more strong electron traps, e.g. anion vacancies, or atomic oxygen interstitials as discussed in sections 4 and 6.

H_2O in monoclinic zirconia

Our PW DFT calculations of H_2O molecule in m-ZrO_2 can be summarized as follows:

1. A water molecule is unstable in bulk m-ZrO_2 relative to the dissociation reaction:

$$H_2O + O^{2-}(3) \to O(3)H^- + (OH)^-, \qquad (10)$$

where O(3) stands for the three coordinated oxygen site, and (OH)$^-$ is located in the interstitial position. The other channel of dissociation resulting in the formation of a H_2^0 molecule and the O_2^{2-} defect, as suggested by Freund in MgO crystal [277], is energetically unfavourable.

2. The immersion energy for the water molecule in m-ZrO_2 (calculated with respect to the H_2O in vacuum and perfect m-ZrO_2) is of the order of 2.3 eV, which explains the efficient water uptake by zirconia.

3. The H^+–(OH)$^-$ separation energy (in the approximation of fully screened monopole-monopole correction) is 0.13 eV. However, even at close distances we have not found any direct correlation between the energy and H^+–(OH)$^-$ distance, which suggests that an electrostatic interaction between the proton and the hydroxide group is effectively screened.

4. The proton produced in the reaction (10), bonds explicitly to the three-coordinated oxygen sites, in agreement with our findings for an isolated hydrogen atom/ion. Due to a large electrostatic

potential in the lattice site, the $O(3)H^-$ bond is substantially weaker than that of the interstitial OH^- resulting in a large red shift of the frequency of the $O(3)H^-$ stretch mode.

The weak $O(3)H^-$ bond and hydroxide-proton interaction results in an extremely fast proton migration, which will be discussed in Sec. 9.

We note here that, although protons in the bulk zirconia are extremely mobile, they cannot be completely evacuated from the sample unless some effective neutralizing mechanism is available. Therefore, they are likely to be trapped at the interface or in the lattice. Our simulations suggest that protons can be effectively trapped by neutral anion vacancies (exothermic by 2.7 eV), creating H^- centers. They also may form the hydrogarnet type defects, provided cation vacancies are present. Although the formation energy for the cation vacancies in zirconia is rather high, they still can be present in appreciable concentration if the material is not fully annealed.

The presented study of hydrogen in zirconia suggests a new mechanism for the water assisted tetragonal-to-monoclinic phase transformation in zirconia: relatively low immersion energy of water into the tetragonal material results in high concentration of protons, who strongly interact with the lattice and instigate a reduction of the coordination of some oxygen ions from four to three, which ultimately leads to the formation of the monoclinic precursors (Fig 25,26). In contrast with the mechanism earlier proposed by Guo [273,274], our model does not require anion vacancies, and thus a necessary step of diffusion of $(OH)^-$ ions towards them. Calculations reveal that similarly to SiO_2, hydroxide ions are highly immobile, thus making the vacancy passivation process inefficient.

8. Defect charge states and reactions

In order to determine the stability of charged defect states and the possible role of defects in photo- and thermo-stimulated processes, and in electronic devices one needs to know the "thermodynamic" and "switching" charge state levels (see Sec. 2). These can be alternatively defined as

electron affinities and ionization energies of defect states with respect to the CBM of oxide or with respect to other electron or hole sources, such as silicon bands. Recent calculations of defect levels for several types of oxygen vacancy defects in α-quartz are presented, for example, in ref. [78]. Defect levels for interstitial oxygen species in α-quartz have been calculated in ref. [246]. Hydrogen-related defect states in SnO_2 and in MgO have been calculated in ref. [276].

In this section we will summarize the results of similar calculations for oxygen vacancies, and interstitial oxygen and nitrogen defects in hafnia.

Electron and hole trapping

To calculate the charge state levels of defects in hafnia, we compare total energies of the initial and final systems with the same number of electrons. The main inaccuracy of this approach, when one is using PW DFT methods, is related to the underestimated band-gap in DFT calculations (see discussion in Sec. 3). The corresponding error can be corrected using the experimental value for the energy gap.

Defining the absolute value of the defect ionization energy $I_p(D^q)$ as the vertical excitation energy of an electron from the defect with charge q to the CBM, we have:

$$I_p(D^q) = E_0^- + E_D^{q+1} - E_0^0 - E_D^q + \kappa 1, \quad (11)$$

where E_0^- and E_0^0 are the calculated energies of the perfect supercell with charge -1 and 0, respectively, and E_D^q is the energy of the defect with charge q (i.e. number of protons minus number of electrons). In Eq. (11) the value E_D^{q+1} is calculated for the geometry of the defect relaxed with charge q, and $\kappa 1$ is a correction for the position of the CBM. Similarly we can define the electron affinity of the defect $\chi_e(D^q)$ (i.e. the energy gain when the electron from the CBM is trapped at the defect) as follows:

$$\chi_e(D^q) = E_0^- + E_D^q - E_0^0 - E_D^{q-1} + \kappa 2. \quad (12)$$

Here the correction $\kappa 2$ can be generally different from $\kappa 1$. One can consider both "vertical" and "relaxed" or "thermodynamic" electron affinities. In the latter case the lattice relaxation after electron trapping is included in E_D^{q-1}. We can also define the hole affinity of the defect $\chi_h(D^q)$, i.e. the energy gain when the a free hole is trapped from VBM to the defect state as follows:

$$\chi_h(D^q) = E_0^+ + E_D^q - E_0^0 - E_D^{q+1} + \kappa 3. \quad (13)$$

Again, dependent on whether the lattice relaxation in the final state is included or not, one will obtain either thermodynamic, i.e. fully relaxed, or vertical hole affinities. The vertical hole affinity provides a useful estimate of the position of the defect state with respect to VBM. To define the corrections $\kappa 1$, $\kappa 2$, $\kappa 3$ we use the following considerations. i) We assume that the main inaccuracy in defining the relative positions of defect states with respect to the band-gap edges is due to unoccupied Kohn-Sham states, and that the underestimated band gap is mainly due to the too low position of the CBM. Therefore we use an approximation that $\kappa 1 = \kappa 2 = \kappa$ and $\kappa 3 = 0$. These conditions are difficult to fully justify without comparison with experiment. As we will show below, the vertical hole affinities calculated as the difference between one-electron states and using Eq. 13 agree within 0.5 eV, which gives an indication of an error made by this assumption. ii) Using these conditions and definitions (12) and (13) it is easy to obtain:

$$\chi_h(D^q) + \chi_e(D^{q+1}) = E_g(exp), \quad (14)$$

where both affinities correspond to relaxed final defect states. This condition holds in all calculations, which insures the consistency of our approach. iii) In the case of hafnia, we use the experimental value of $E_g(exp) = 5.68$ eV [112] to define the difference

$$\kappa = E_g(exp) - E_g(theor), \quad (15)$$

and correct the defect excitation energies, the ionization potentials and electron affinities. This gives $\kappa = 5.68 - 3.92 = 1.76$ eV, which is used in all further calculations.

Although this method is approximate, fixing the value of κ allows us to present the results of our calculations in one scale. Another advantage is that, in order to find defect affinities with respect to electrons at the CBM of silicon or holes at the VBM of silicon within the same method, one can use the experimental value of band offset with Si. This scale can be changed if a more "accurate" or relevant value for κ will be found. This will require only a shift of our predicted values by a constant.

D	$I_p(D)$	$\chi_e(D)$	$\chi_h(D)$
V_4^0	3.88 (3.80)	-	2.42 (2.07)
V_4^+	4.10 (3.98)	3.27 (3.33)	2.42 (1.86)
V_4^{2+}	-	3.26 (3.54)	-
V_3^0	3.41	-	2.92
V_3^+	3.75	2.76	2.75
V_3^{2+}	-	2.93	-
O_3^0	5.55 (5.39)	3.95 (3.73)	0.19 (0.07)
O_3^-	5.38 (5.19)	4.75 (4.62)	1.73 (1.67)
O_3^{2-}	5.35 (5.19)	-	0.92 (0.78)
$(O_2^0)_3$	5.53	4.67	0.20
$(O_2^-)_3$	5.46	5.06	1.01
$(O_2^{2-})_3$	5.32	-	0.62
$(V_3+S_{Zr})^0$	3.46	2.03	2.82
$(V_3+S_{Zr})^+$	3.86	2.86	2.63
$(V_3+S_{Zr})^{2+}$	-	3.05	-

Table 6

Ionizational potential $I_p(D)$, relaxed electron $\chi_e(D)$ and hole $\chi_h(D)$ affinities (in eV) of defects in different charge states in hafnia. Values in brackets are for similar defects in zirconia. Here S_{Zr} represents substitution of a lattice hafnium by zirconium.

The calculated ionization energies and relaxed electron affinities of various defects are summarized in Tables 6 and 7. Large electron affinities clearly indicate that interstitial species and charged vacancies may serve as traps for electrons from the hafnia conduction band. The calculated absolute values of relaxed hole affinities for charged interstitial species are large due to the strong defect relaxation. The "vertical" values for these affinities are about 0.3 eV, in line with what one would expect from the DOS and vertical ionization energies.

To facilitate further discussion, some of the electron affinities are also shown in a schematic energy diagram in Fig. 28. These can be used to estimate the electron affinities of these defects with respect

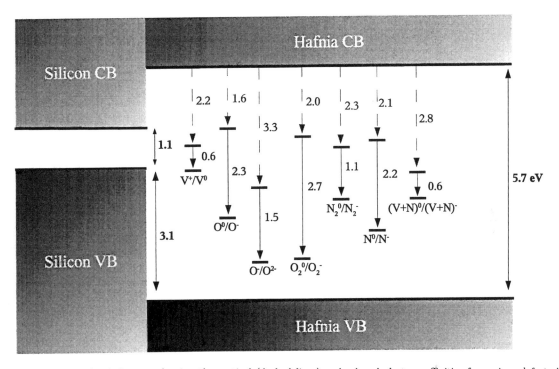

Figure 28. Energy level diagram showing the vertical (dashed lines) and relaxed electron affinities for various defects in monoclinic hafnia (see discussion in chapter 2). All energies are in eV.

to electrons at the CBM of silicon at the Si/HfO$_2$ interface. This is particularly relevant for thin oxide films where electrons can tunnel from the interface into defect states [278]. Counting from the VBM of HfO$_2$ (see Fig. 28), we can use a theoretical estimate [279] of the valence band offset at the interface and the band gap of Si (1.1 eV) to estimate the energy of an electron at the CBM of Si with respect to the defect levels. In all cases the relaxed electron states of defects are lower than the CBM of the bulk silicon, suggesting the possibility of electron tunneling from Si into oxide. We should stress, however, that these results do not provide information regarding cross-sections for electron/hole trapping on these defects, they only predict the energies gained by the process. Calculation of the probability of trapping requires a further study to determine the kinetics of electron/hole trapping.

One can estimate the lowest excitation energy of the vacancy by calculating the singlet to triplet transition, $S_0 \rightarrow T_1$. Comparing the total energies of the supercells we obtain $E_{(S_0 \rightarrow T_1)}(V_3) = 1.30$ eV and $E_{(S_0 \rightarrow T_1)}(V_4) = 1.59$ eV, which are very similar to the values obtained by directly subtracting the energies of the one-electron states (1.22 eV for V_3 and 1.66 eV for V_4). We should note, however, that these excited states are delocalized at the bottom of the crystal conduction band. This again may be due to the fact that the position of the CBM is too low in our DFT calculations. Localized defect excited states can be located below a proper conduction band, as is the case in e.g. MgO [157,166]. This problem is also highlighted in several other atomic and molecular charged defects which were studied in this work, specifically O_3^+, $(O_2^+)_3$, V_3^- and V_4^-. The charge density maps of the relaxed structure show that the hole/electron is delocalized over the whole cell and the one-electron energy spectrum is typical for a metallic state. The relaxation of the ions is also much smaller than for the equivalent oppositely charged defect. This type of phenomenon in DFT is discussed in more detail in Sec. 2.

D	$I_p(D)$	$\chi_e(D)$	$\chi_h(D)$
N^-	5.08	-	0.60
N^0	5.24	4.27	1.66
N^+	5.41	4.01	0.26
N^{2+}	-	4.83	-
N_2^{2-}	4.39	-	1.29
N_2^-	4.44	3.64	1.24
N_2^0	-	3.36	-
$(V+N)^-$	5.28	-	0.39
$(V+N)^0$	5.48	4.66	0.20
$(V+N)^+$	-	5.10	-
$(V+N_2)^-$	4.04	-	1.64
$(V+N_2)^0$	4.51	3.42	1.17
$(V+N_2)^+$	-	3.75	-
$O \Leftrightarrow N$	-	4.24	-

Table 7

Vertical ionizational potential $I_p(D)$, relaxed electron $\chi_e(D)$ and vertical hole $\chi_h(D)$ affinities (in eV) of defects in different charge states in hafnia. Here all energies are with respect to a three-coordinated lattice oxygen site. The last value in the table refers to values for a nitrogen atom which has exchanged with a lattice oxygen.

Defect reactions

Various reactions and their energies are presented in Table 8. These energies have been calculated as differences in total energies of pairs of individual defects and each pair has the same total charge state and number of atoms. Positive energies indicate that a reaction in the direction of the arrow is energetically favourable. Note that we do not consider any reactions which include total energies with delocalized states, e.g. O^+ and V^-. The energies presented in Table 8 also do not include the interaction between defects, which can be strong especially in close charged defect pairs.

Reactions 1 and 2 in table 8 indicate that charge transfer between oxygen vacancies and interstitials is favourable. A separated pair of doubly charged defects has 1.5 eV lower energy than the neutral pair. The associated Frenkel pair energies are 8.0 eV for formation of the neutral pair, 7.3 eV for the singly charged pair and 5.8 eV for the doubly

charged pair. This confirms the intuition that the combination of the high electron affinity of oxygen and the highly ionic crystalline potential of the oxides would favour defects with higher charge state.

No.	Reaction	Energy (eV)
1	$O^0 + V^0 \Rightarrow O^- + V^+$	0.7 (0.5)
2	$O^- + V^+ \Rightarrow O^{2-} + V^{2+}$	1.5 (1.4)
3	$O^{2-} + O^0 \Rightarrow 2O^-$	-0.8 (-0.9)
4	$V_4^{2+} + V_4^0 \Rightarrow 2V_4^+$	0.0 (0.2)
5	$V_3^{2+} + V_3^0 \Rightarrow 2V_3^+$	0.2
6	$O_2^0 + V^0 \Rightarrow O_2^- + V^+$	1.4
7	$O_2^- + V^+ \Rightarrow O_2^{2-} + V^{2+}$	1.8
8	$O_2^{2-} + O_2^0 \Rightarrow 2O_2^-$	-0.4
9	$2O^0 \Rightarrow O_2^0 + E_0^0$	-1.0
10	$O^0 + O^- \Rightarrow O_2^- + E_0^0$	-0.3
11	$2O^- \Rightarrow O_2^{2-} + E_0^0$	0.8
12	$O_2^0 + V^0 \Rightarrow O^0 + E_0^0$	9.0
13	$O_2^- + V^+ \Rightarrow O^0 + E_0^0$	7.6
14	$O_2^{2-} + V^{2+} \Rightarrow O^0 + E_0^0$	5.7

Table 8

Defect reactions and associated energies for hafnia. Values in brackets are for similar reactions in zirconia. Positive energies indicate exothermic reactions.

As for calculations of zirconia [6], reaction 3 predicts "negative U" behaviour of an oxygen ion in hafnia, i.e. that two isolated O^- species would decay into O^{2-} and O^0. The same is true to molecular species but with a much smaller energy gain: reaction 8 shows that the oxygen molecule also has "negative U", and two O_2^- species would decay forming a doubly charged and a neutral molecule. This terminology comes from Anderson's model for semiconductors where this effect is much more common than in insulators [280,281]. Experimentally this means that atomic and molecular oxygen species in hafnia prefer to stay diamagnetic and will be difficult or impossible to detect using paramagnetic resonance. A similar process for vacancies in reaction 4 shows a balance between the different pairs for 4-coordinated and a slight energy gain of 0.2 eV for 3-coordinated to form two singly charged vacancies. Again this is very similar to the results found in zirconia. However, in this case concentration of paramagnetic V^+ vacancies will strongly depend on temperature. Molecular species also demonstrate a tendency for charge transfer between the vacancies and interstitial molecules. The doubly charged molecule and vacancy pair is 3.2 eV lower in energy than the neutral pair (reactions 6 and 7).

Finally we study some reactions with respect to changes in the type of defect, rather than just the charge state. In these reactions we consider two independent defects combining to form a different defect, where the energy of the perfect lattice, E_0^0 (as in Eq. 1), is included to conserve the number of particles. Here we see in reaction 9, as discussed previously, that it is energetically favourable for an oxygen molecule to separate into two interstitial atoms. However, two singly charged interstitial oxygen ions would like to recombine and form a doubly charged interstitial molecule (reaction 11). Also, we see that the calculations predict that all combinations of molecule and vacancy defect pairs (reactions 12-14) would "annihilate", leaving only a neutral oxygen interstitial. The energy gain is less for the more favourable doubly charged defect pair. These reactions are likely to take place during annealing of oxide in an oxygen atmosphere.

9. Diffusion

Diffusion affects most of the defect processes discussed in this paper. A model of silicon oxidation assumes that oxygen diffuses through the already grown oxide before reacting at the oxide/substrate interface [282]. Growth and annealing of high-k oxides on silicon is again accompanied by diffusion of oxygen, nitrogen or hydrogen species through the network. For example, one of the roles of nitrogen treatment of hafnia films is to prevent boron diffusion from the poly-Si electrode into the CMOS channel layer through the gate oxide (as discussed at the [283]). The mechanisms of defect diffusion are important for their aggregation, especially in radiation-induced processes. They determine the performance of conducting oxides in fuel cells and long-term stability of electronic devices. However, atomistic models of diffusion mechanisms are ex-

tremely difficult to interpret from experiment. The role of modelling is therefore to develop such models and explain the kinetics of experimentally observed processes.

Later in this section we will discuss in detail the models of hydrogen and oxygen diffusion in zirconia and hafnia. But before turning to these examples, it is instructive to survey what has been learned about the mechanisms of diffusion of similar species in other oxides. Of particular relevance to further discussion are the results on oxygen vacancy and interstitial diffusion.

As usual, MgO provides some archetype models. In particular, Kotomin *et al.* [11] calculated adiabatic barriers for diffusion of oxygen vacancies in MgO in different charge states. They used a static approach and the semi- empirical INDO method in an embedded cluster model. Due to high lattice symmetry, the saddle point for diffusion can be easily identified. It is located in the middle between two regular oxygen lattice sites along the (110) crystal axis. So in effect it is an oxygen ion which is moving and the electron trapped in the vacancy exchanges its place with the ion at the saddle point. This quantum effect is not easy to model and experimental data are also limited. Nevertheless the barrier energies calculated in ref. [11] are in good agreement with available experimental results. They predict a trend, which is later seen in other systems: the barrier for the neutral vacancy (F-center) jump is larger than that for single charge vacancy (F^+-center) and for bare vacancy, 3.1 eV, 2.7 eV and 2.5 eV, respectively. The reason for that is the repulsion between the trapped electron(s) and the oxygen ion.

A very similar mechanism has been predicted for oxygen vacancy diffusion in pure tetragonal zirconia using plane wave DFT calculations [211]. Again a saddle point is located between two oxygen sites, however, due to the lower symmetry different directions are inequivalent (see discussion in section 3 and Fig. 5). Therefore there is a small difference in the distance between neighbouring oxygen sites, which results in slightly different barrier energies. However, the same trend prevails: the barrier for the neutral vacancy (about 1.4 eV) is much higher than that for the double charged bare va-

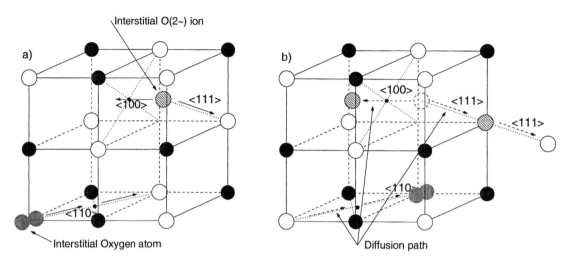

Figure 29. Diffusion models of interstitial oxygen atom and Q^{2-} ion in MgO. a) Inital state: Oxygen atom is at the lattice site forming O_2^{2-} ion oriented along <110>, Oxygen ion is at the centre of the Mg_4O_4 cube. b) Final state: Oxygen atms hops to the nearest site along <110> or other equivalent direction (1.45 eV), O^{2-} hops to the centre of a nearest Mg_4O_4 cube along <100> (1.2 eV) or substitutes a lattice O^{2-} along <111> direction (0.54 eV).

cancy (about 0.2 eV) [211].

The mechanism of diffusion of oxygen interstitials in oxides (and many materials in general) can be classified as occurring via either an exchange (interstitialcy [284]) or interstitial process. The exchange mechanism involves the continuous replacement of a lattice site by the diffusing defect, and the lattice site then becoming the diffusing species. It is characteristic of diffusion of anions, for example, in oxides such as MgO [235] and fluorides such as CaF_2 [285]. In the interstitial mechanism, the defect diffuses through empty space between the lattice sites. This mechanism is characteristic of diffusion in oxides such as silica [77,266]. To highlight the difference between these mechanism we review in detail oxygen diffusion in MgO and silica.

As discussed in Sec. 6, in MgO the interstitial oxygen atom forms a dumbbell configuration with the regular oxygen ion with the center of mass in the regular oxygen site (see Fig. 29). However, unlike similar defects in silica and zirconia, there is only weak chemical bonding between the two dumbbell atoms, which are mainly kept together by a balance of lattice distortion and Coulomb interaction with the surrounding ions [235]. Therefore the interstitial doubly charged O^{2-} ion prefers to accommodate itself in the cube center. The mech-

anisms of diffusion of the two species are markedly different too. The dumbbell rotates easily in the lattice site and can change its orientation at a cost of only 0.15 eV [235]. As stated previously, diffusion of the interstitial O atom takes place by the so-called "Interstitialcy" mechanism, where one of the oxygen ions making the dumbbell breaks a bond with its partner and switches to another nearest neighbour oxygen ion (see Fig. 29). The adiabatic barrier calculated for this process is 1.45 eV [235]. The diffusion of interstitial O^{2-} ion requires overcoming a much smaller barrier. It may happen either by direct jumps between to cube sites (see Fig. 29), which requires about 1.2 eV, or by collinear interstitial mechanism (see Fig. 29). In the latter case, the interstitial ion substitutes a regular site ion and kicks it out into the interstitial site. The dumbbell configuration oriented along the <111> crystal axis is the saddle point for this process, which requires a much smaller energy of 0.54 eV [235]. As we will see later, many features of the discussed mechanisms persist in other systems too.

The fact that the negative oxygen interstitial has a much smaller diffusion barrier than the neutral atom in MgO, has also been found in α-quartz [238,246]. The chemical bonding in peroxy linkage (POL) in SiO_2 is much stronger that in the dumbbell configuration of O_2^{2-} in MgO. Nevertheless the barrier for atomic oxygen diffusion is found to be lower, about 1.3 eV [238,240,246]. Because the bonding configuration of O interstitial depends on the charge state, the diffusion mechanism of O ions is quite different from that of the neutral interstitial [240]. In this case, the interstitial oxygen ion is almost identical to the host oxygen ion. Therefore the most effective mechanism, again similar to MgO, is to kick out the host oxygen ion and substitute it into a lattice site. However, in α- quartz this costs much less energy - 0.27 eV and 0.11 eV for O^- and O^{2-} interstitial ions, respectively [246]. Very similar energies have been predicted in ref. [238].

The activation barriers for interstitial oxygen molecule diffusion in α-quartz show an opposite tendency as a function of charge state [238]. The barrier for the neutral molecule has been predicted to be only 0.09 eV, whereas for charged molecules it increases to about 0.6 eV for a single negative charge and to about 2.1 eV for a doubly charged molecule. This is not surprising as molecules are shown to diffuse via interstitial sites without cost-saving exchange with host ions. Hence one can expect a strong effect from disorder on molecular diffusion too. Indeed, as has been shown in ref. [286], the formation energy of interstitial molecular oxygen shows a strong correlation with the size of the interstitial void, strongly decreasing for increasing cage size. Consequently there is a wide distribution of transition barriers for diffusion between neighbouring voids. The statistically meaningful model of the long-range diffusion of O_2 molecules in a-SiO$_2$ has been developed in ref. [286]. This pioneering study highlights the percolative nature of the diffusion process and the critical dependence of the diffusion rate on the connectivity of the interstitial network. It identifies the interstitial O_2 molecule as the transported species during silicon oxidation.

Although very complex, these static calculations exploit a simple idea that diffusion takes place via adiabatic displacements of atoms over a saddle point on a potential energy surface. This means that all other atoms adjust their position as diffusing species moves from one site to another. The adiabaticity does not always work, as light species can move much faster than other atoms are able to respond. A classical example is

hydrogen diffusion (see, for example, ref. [287]). In particular, most calculations of hydrogen atom diffusion in silica assume that it is moving in a frozen lattice [267]. Searching for a saddle point in a multi-coordinate space can be very difficult too and requires some special techniques (see, for example, refs. [288,289,290,291]). Another method of studying diffusion is based on Molecular Dynamics (MD). In this case, the diffusion coefficient can be calculated directly without making any assumptions regarding adiabaticity and positions of saddle points. Applications of both static and MD methods are given below.

Proton diffusion in zirconia

In Sec. 7 we have established that the $(OH)^-$ stretching mode at the 3- coordinated lattice site is significantly softer than that of the $(OH)^-$ in the interstitial. This indicates weaker bonding for the proton at the lattice site and suggests a high mobility of protons in the monoclinic lattice. Indeed, MD simulations show that at temperatures around 600 K the proton undertakes frequent jumps between the adjacent 3-coordinated ions. In figure 30 the proton diffusion is illustrated by plots of the time evolution of the distances between the proton and the oxygen ions involved.

The formation of an OH bond is seen in this figure as rapid oscillations of the curve around a distance of ~ 1 Å ($(OH)^-$ stretching mode) . One can also see that the proton hops between the oxygen sites, that is, there is no stable position for the proton in the interstitial position. The hops occur within 0.1 ps and may be viewed as an $(OH)^-$ bond switching from one oxygen to another initiated by the lattice vibrations. The average delay between the jumps is ~ 0.5 ps. One should note that the bond switching occurs between the ions only $2.3 - 2.4$ Å apart. Therefore, it involves less then ~ 0.4 Å displacement of the proton itself. We estimate that the proton activation energy is ~ 0.1 eV. One can also see that the elementary jump may result in a significant (up to ~ 0.2 eV) accumulation of energy in the $(OH)^-$ stretching mode. The last two observations indicate that the overlap between the excited vibrational states of the $(OH)^-$ bond on adjacent sites may be strong, and

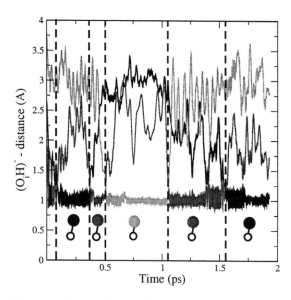

Figure 30. Proton diffusion illustrated as a time evolution of a distance between the proton and the 3-coordinated lattice oxygens envolved (black, red and green).

the effects of quantum tunneling not considered here may significantly enhance already fast proton diffusion further still.

$(OH)^-$ diffusion in zirconia

In contrast to proton interstitial, $(OH)^-$ diffusion occurs at much lower rate. In our simulations we have not observed diffusion events, although the activation energy for diffusion can be as low as 0.8 eV. As evidenced in MD simulations, interstitial hydroxide ion is involved in a soft motion whereby the oxygen ion swaps the bond between one of the nearest neighbour and next nearest neighbour zirconium ions (Fig. 31).

Although an overall displacement of the OH^- ion in such a jump is about 0.9 Å , it does not constitute a diffusion event, since it remains confined within the same cage. Therefore, no mass transfer occurs. $(OH)^-$ hops back and forth by swapping with one of the Zr neighbours, which apparently serve as an efficient energy dissipation channel, so an accumulation of energy neccessary to overcome a diffusion barrier becomes kinetically unlikely. We recall that the OH^- in a SiO_2 glass also has low mobility [270], although its origin there could be of purely energetic (i.e. high diffusion activation bar-

rier).

Atomic oxygen diffusion in hafnia

The previous sections established the confined interstitial space in the monoclinic structure favours incorporation of oxygen in atomic form, and that the crystalline potential of the oxide favours the O^{2-} charge state where an electron source is available (such as a silicon substrate). In this section we will look at how atomic oxygen can diffuse through the crystal, and how its charge state affects the barriers. Again, we focus on hafnia, although the results are relevant to zirconia too. The diffusion paths presented here are calculated in the same way as previous defects, using a static approximation and the Nudged Elastic Band (NEB) method [288,289].

D	$E_{ex}(D)$	$E_{in}(D)$
O^0	0.8	1.3
O^-	0.3	1.1
O^{2-}	0.6	1.8

Table 9
Exchange activation barriers (E_{ex}) and interstitial activation barriers (E_{in}) for different charge states of oxygen interstitial defects in hafnia. All values are in eV.

As discussed previously, there are two general classes for the mechanism of oxygen diffusion in oxides: exchange diffusion in simple cubic oxides; and interstitial diffusion in more structurally open oxides. The structure of hafnia is more complex than most of these classical cubic oxides, yet retains the same lack of interstitial space, therefore it is especially interesting to see which mechanism is energetically favoured. Also in some materials, such as MgO [235], the mechanism of diffusion is very dependent on the oxygen charge state. The specific barrier for each oxygen species is given in Table 9, and in the following sections we discuss in detail the mechanisms themselves.

As discussed previously, the neutral oxygen interstitial in hafnia is characterized by its strong bond with a lattice oxygen site (O_A in Fig. 33), and we find that it is this need to find an electron source which also dominates the diffusion process. For the exchange mechanism, the transition

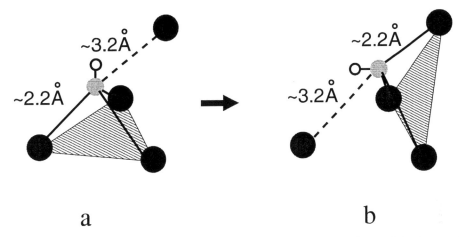

Figure 31. Schematics of the OH⁻ activation jump occuring via the bond swapping with one of the three nn Zr ions. The overall displacement is 0.9 Å .

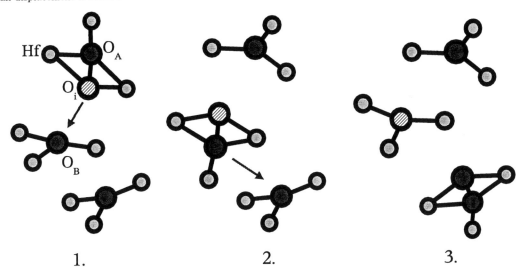

Figure 32. Schematic diagrams for the exchange mechanism appropriate to diffusion in hafnia: 1. initial diffusion of interstitial to nearest lattice oxygen. 2. new defect pair formed, but now lattice oxygen continues diffusion. 3. interstitial now effectively becomes a lattice site and O_B diffuses to another lattice oxygen.

point occurs when the oxygen atom is approximately equi-distant between the initial (O_A) and final (O_B) lattice oxygen, and the defect is furthest from a source of electrons (the fact that the distances are not exactly equal reflects the asymmetry of the bonding environment around the defect complex in hafnia). Fig. 33(b) clearly shows that the defect at this point has much smaller bonding with the lattice sites than in the equilibrium po-

sition. The diffusion of the oxygen begins with a re-orientation of the oxygen pair, such that the defect moves closer to the final lattice oxygen (O_B), but the O-O bond length does not change significantly. After this, the defect moves to the transition state (see Fig. 33(b)), breaking the bond and then moves almost linearly to bond with the final lattice oxygen. Finally, the new O-O bond reorients to reach the equilibrium. The energy barrier

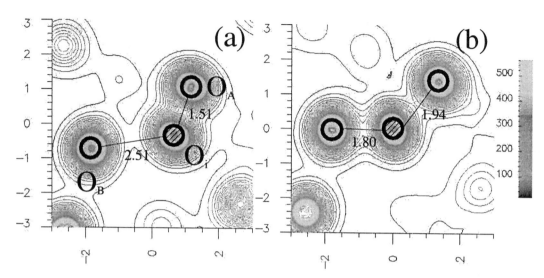

Figure 33. Charge density plots for neutral interstitial oxygen (O_i) in a plane through O_A, O_i and O_B for (a) equilibrium geometry near to a lattice oxygen (O_A), and (b) transition state during exchange diffusion from O_A to equivalent lattice oxygen site (O_B). Charge density is in 0.1 $e/\text{Å}^3$ and all distances are in Å. *Reproduced with permission [125].*

at the transition state is 0.8 eV.

Diffusion by the interstitial mechanism is in principle governed by similar effects, but travelling between the lattice sites produces a longer path, causes much greater relaxation in the crystal and it is more difficult for the defect to bond with lattice sites along the diffusion path. In the exchange mechanism the maximum atom displacement (aside from the diffusing defect itself) is 0.2 Å with only neighbouring oxygen atoms affected. However, for the interstitial mechanism displacements rise to a maximum of 0.4 Å with both oxygen and hafnium atoms displacing. This is reflected by the increased barrier of 1.3 eV for the interstitial mechanism. Note that due to the increased complexity of the interstitial diffusion path, and the closeness of the exchange and interstitial barriers, the interstitial barrier was checked with double the number of points along the path. This raised the barrier by about 0.1 eV, but did not qualitatively affect the diffusion mechanism.

Introduction of an electron to the system allows the oxygen interstitial to exist more independently from the lattice site, since the electron localizes fully onto the defect. However, the defect is still somewhat coupled to the lattice site with some small covalent bonding evident in Fig. 34(a). How-

ever, the diffusion mechanism is simpler than for the neutral case. At first the defect moves linearly in a plane with the initial and final lattice oxygen sites, and there is no re-orientation. At the transition point (see Fig. 34(b)), the defect is already almost at its equilibrium bond distance to the final lattice oxygen. However, the final stages of the diffusion involve displacement and re-orientation of the new O-O pair to their equilibrium position. The barrier at the transition point is 0.3 eV. Displacement of atoms during the exchange diffusion are of similar magnitude to that for the neutral case, although now more atoms are involved - as to be expected for the increased Coulomb interaction from the charged defect.

The interstitial barrier for the singly charged defect is 1.1 eV, similar to the neutral species, but the comparatively bigger difference to the exchange barrier is due to large displacements along the path. The maximum oxygen displacement is again 0.4 Å, but, as in the exchange case, the increased Coulomb interaction means that many more atoms are being displaced and the disruption is much more delocalized than for the neutral species.

Adding a second electron to the system effectively creates an independent oxygen ion with a

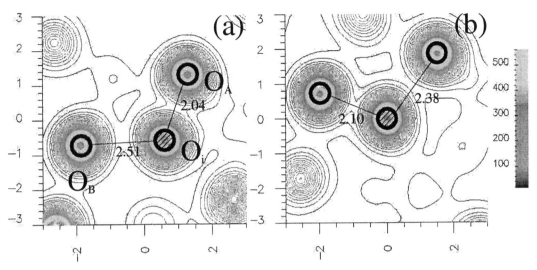

Figure 34. Charge density plots for singly charged interstitial oxygen (O_i) in a plane through O_A, O_i and O_B for (a) equilibrium geometry near to a lattice oxygen (O_A), and (b) transition state during exchange diffusion from O_A to equivalent lattice oxygen site (O_B). Charge density is in $0.1e/Å^3$ and all distances are in Å. *Reproduced with permission [125].*

full outer shell. Fig. 35(a) shows that there is now enough extra charge available for the oxygen interstitial to be stable in the crystal without any bonding to the original lattice oxygen sites. This is reflected in both diffusion mechanisms, where the barriers are totally dominated by crystal relaxations due to increased Coulomb interaction. The mechanism is very similar to that for the singly charged defect, with an initial planar diffusion followed by shift of the final oxygen site. However, we see displacements of oxygen atoms slightly larger to that seen for the singly charged defect, and now we also see for the first time significant displacements (over 0.1 Å) of hafnium atoms during exchange diffusion. It is these displacements which give a larger barrier of 0.6 eV for doubly charged diffusion, even though bonding with lattice sites is no longer an issue, and the mechanism is similar to the singly charged case. Fig. 35(b) shows that the density configuration of the O^{2-} defect complex changes little during diffusion, and it changes in the surrounding atoms that are responsible for the barrier.

The doubly charged interstitial diffusion produces the largest relaxations of all the processes, with maximum displacements of over 0.5 Å for surrounding oxygen atoms and up to 0.2 Å for hafnium

atoms. This produces a correspondingly large barrier of 1.8 eV.

In summary, oxygen diffusion in hafnia is governed by two competing processes: (i) the crystalline potential in hafnia means that oxygen defects are only stable as ions, and (ii) relaxation of atoms along the diffusion path. The neutral oxygen interstitial causes the least disruption of the surrounding crystal during diffusion, but its need to form an 'ion-pair' with a lattice oxygen produces a large barrier. The doubly charged oxygen is the most stable defect in hafnia film on silicon, yet its large Coulomb interaction means it generates large displacements during diffusion. Hence, the singly charged defect proves to be the best balance - it is more independent of the lattice oxygens than the neutral species, but does not produce as large disruption of the crystal as the doubly charged defect. In terms of the general mechanism, the small space in between atoms in hafnia means that the exchange mechanism is favoured over the interstitial for all defect species due to the reduction in lattice disruption. Note, however, that in the neutral case, where displacements are in general quite small, the difference between the two mechanisms is much smaller than for the other defect species.

These results show that although hafnia has a

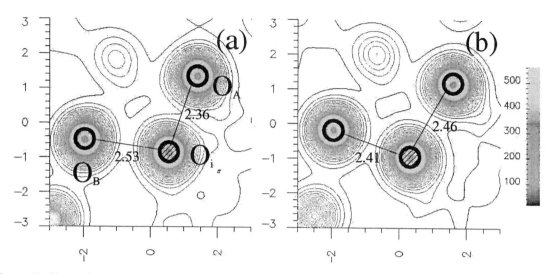

Figure 35. Charge density plots for doubly charged interstitial oxygen (O_i) in a plane through O_A, O_i and O_B for (a) equilibrium geometry near to a lattice oxygen (O_A), and (b) transition state during exchange diffusion. Charge density is in 0.1 $e/Å^3$ and all distances are in Å. *Reproduced with permission [125].*

much more complex atomic structure than other, simpler ionic materials, its geometry shares a similar lack of interstitial space and the lattice exchange (or interstitialcy) remains the favoured diffusion mechanism. In general, the barriers for interstitial oxygen diffusion in hafnia are small, and the defects will be very active, especially during the high temperature processing common in microelectronic processes. These barriers are much smaller than the measured activation energy of 2.3 eV for oxygen diffusion in m-zirconia [2]. However, this activation energy is dominated by the Schottky formation energies (about 2.2 eV [115]). The fact that oxygen is predicted to diffuse as a charged species suggests the possibility of using an applied electric field to influence the diffusion, and perhaps control defect concentrations. The large electron affinity of all the oxygen species means that they can all act as traps within a device, creating intrinsic electric fields and contributing to dielectric losses, so their control would be highly desirable for efficient device design.

10. Outlook

In this chapter we reviewed some of the main issues and recent calculations pertaining to point defects and defect processes in oxides for microelectronics. Several years of exploratory research and hundreds of publications have not lead so far to a final consensus about the materials to be used in DRAM and CMOS applications. The prevailing line of thought in gate oxide research at the moment seems to favour "something HfO_2 based". Although the situation remains fluid, HfO_2 and its silicates and aluminates emerged as strong contenders and provide good examples of typical defect related issues. Therefore we used them, where appropriate, to illustrate the present state of the art of calculations and main conclusions relevant to the performance of oxide materials. To some extent we were following empirical approaches, which rely on knowledge of other materials in similar states to those to be used. Therefore we have drawn some analogies with other oxides, such as silica and MgO. The properties we considered concern mainly non-stoichiometry and disorder of prospective oxides as well as related issues, such as calculation of structural and dielectric properties of oxides.

Theoretical modelling is a source of reliable models of defects and defect processes. In most cases they are impossible to develop fully from the experimental data alone. A comparison of models of similar species in different materials points to interesting trends, which are not obvious from disparate data available in different research areas. Particularly striking are similarities in models of structure and diffusion of interstitial atomic oxygen species in different oxides. They form peroxy linkage-like species symmetric with respect to the original oxygen site and usually diffuse by the exchange (or interstitialcy) type mechanism. On the other hand, oxygen vacancies, which are intrinsic structural defects, form a range of models depending on the ionicity of the material and coordination of oxygen sites. In ionic materials with high coordination, such as MgO, Al_2O_3 and HfO_2, they form F-centre like defects and deep states in the band gap. In more covalent materials, such as SnO_2 and CeO_2, electrons left in the vacancy tend to be more localized on surrounding metal atoms. In SiO_2 where oxygen sites are only two-coordinated, neutral vacancies can be viewed as a Me-Me chemical bond.

The direct relevance of this to technological applications is perhaps not immediately evident. Certainly defect models discussed above have been developed as a result of the analysis of a large amount of experimental data. In return, they helped to rationalize that data and predict models transferable between different materials. Still specific problems pertaining to microelectronics are different: defects initiate degradation; they cause charge trapping and provide fixed charge; dielectric performance will vary in time, as the defects or charge state populations evolve; thin film materials will often be inhomogeneous because of the different epitaxial constraints at the two interfaces; finally, there will be effects on interface properties, including interface dipoles. The mechanisms of these processes remain largely unresolved, with several encouraging examples trying to bridge the gap between defect modelling and real processes, for example [78,210,226]. Establishing a direct link between defect models and properties, and their role in processes leading to degradation of microelectronic devices remains one of the main challenges of both theoretical modelling and experiment.

Among other challenges one can note calculations of defect excited states: optical absorption and luminescence energies, exciton states and their relaxation. Computational studies of the effect of structural disorder on the structure and properties of defect in oxides have only started to appear, with the main emphasis on amorphous silica. They demonstrate the limited transferability of defect models developed in crystals within disordered amorphous materials. This issue is likely to become important for amorphous Hf silicates and aluminates too. One can also expect limited transferability of defect models onto thin films and interfaces. Direct modelling of defects in these systems will become extremely important with the ever decreasing thickness of oxide films and increasing computer resources. Linking the static properties of these defects with real processes in microelectronic devices requires more effort on the front of modelling dynamics of defect reactions. In particular, calculations of the cross-sections of defect states with respect to charge transfer processes, studies of effectiveness and rates of defect reactions and the effect of energy dissipation are bound to become extremely important. This should begin to bridge the size and timescale gap existing in most of the calculations discussed above.

Acknowledgements: The authors are grateful to A. M. Stoneham, F. Lopez Gejo, S. Mukhopadhyay, V. B. Sulimov, A. Korkin, G. Bersuker, V. Afanas'ev, K. Tanimura, H. Hosono, Y. J. Lee and R. M. Nieminen for many useful discussions. We wish to thank G. Pacchioni and E. Kotomin for critical reading of the manuscript. JLG is supported by EU Framework5 HIKE project, PVS is supported by Japan Science and Technology Agency. ASF has been supported by the Academy of Finland through its Centre of Excellence Program (2000-2005), and wishes to thank to the Centre of Scientific Computing (CSC), Helsinki for computational resources.

References

[1]   B. W. Busch, W. H. Schulte, E. Garfunkel, T. Gustafsson, W. Qi, R. Nieh, and J. Lee. *Phys.*

Rev. B, 62:R13290, 2000.

[2] U. Brossman, R. Würschum, U. Södervall, and H-E. Schaefer. *J. Appl. Phys.*, 85:7646, 1999.

[3] D. Martin and D. Duprez. *J. Phys. Chem.*, 100:9429, 1996.

[4] In G. Pacchioni, L. Skuja, and D. L. Griscom, editors, *Defects in SiO₂ and Related Dielectrics: Science and Technology, NATO Science Series*, pages 1–624. Kluwer, Dordrecht, 2000.

[5] J. R. Chavez, R. A. B. Devine, and L. Koltunski. *J. Appl. Phys.*, 90:4284, 2001.

[6] A. S. Foster, V. B. Sulimov, F. Lopez Gejo, A. L. Shluger, and R. M. Nieminen. *Phys. Rev. B*, 64:224108, 2001.

[7] A. S. Foster, F. Lopez Gejo, A. L. Shluger, and R. M. Nieminen. *Phys. Rev. B*, 65:174117, 2002.

[8] A. M. Stoneham. *J. Non-Cryst. Solids*, 303:114, 2002.

[9] B. Henderson. *CRC Critical Rev. Sol. St. Mater. Sci.*, 9:1–60, 1980.

[10] W. Hayes and A. M. Stoneham. *Defects and Defect processes in Non-Metallic Solids*. Wiley, New York, 1985.

[11] E. A. Kotomin and A. I. Popov. *Nucl. Instr. and Meth. in Phys B*, 141:1, 1998.

[12] P. M. Lenahan and Jr. J. F. Conley. *J. Vac. Sci. Technol.*, 16(4):2134–2153, 1998.

[13] A. H. Edwards and W. B. Fowler. *Microel. Reliab.*, 39:3–14, 1999.

[14] In H. Z. Massoud, I. J. R. Baumvol, M. Hirose, and E. H. Poindexter, editors, *The Physics and Chemistry of SiO₂ and the Si-SiO₂ Interface - 4*, pages 1–520. The Electrochemical Society, Pennington, NJ, 2000.

[15] G. F. Neumark. *Mater. Sci. Eng. Reports*, 21:1–46, 1997.

[16] M. S. Islam. *J. Materials Chem.*, 10:1027–1038, 2000.

[17] E. E. Kotomin, J. Maier, R. I. Eglitis, and G. Borstel. *Nuclear Instr. and Meth. Phys. Res. B*, 191:22, 2002.

[18] A. Zunger. *Current Opinion in Sol. St. and Mater. Sci*, 3:32–37, 1998.

[19] A. Kawamoto, J. Jameson, K. Cho, and R. W. Dutton. *IEEE Trans. Electr. Dev.*, 47(10):1787–1794, 2000.

[20] A. Kawamoto, K. Cho, and R. Dutton. *J. Comp.-Aided Mater. Design*, 8:39–57, 2002.

[21] P. J. Gellings and H. J. M. Bouwmeester. *Catal. Today*, 58:1–53, 2000.

[22] G. Makov and M. C. Payne. *Phys. Rev. B*, 51:4014, 1995.

[23] L. N. Kantorovich. *Phys. Rev. B*, 60(23):15476, 1999.

[24] J. M. Soler, E. Artacho, J. D. Gale, A. García, J. Junquera, P. Ordejón, and D. Sánchez-Portal. *J. Phys.: Condens. Matter*, 14:2745–2779, 2002.

[25] D. R. Bowler, T. Miyazaki, and M. J. Gillan. *J. Phys.: Condens. Matter*, 14:2781–2798, 2002.

[26] G. Onida, L. Reining, and A. Rubio. *Rep. Prog. Phys.*, 74:601-659, 2002.

[27] G. Pacchioni, P. S. Bagus, and F. Parmigiani. *Cluster models for surface and bulk phenomena*. Plenum Press, New York, 1992.

[28] J. Sauer. *Chem. Rev.*, 89:199–255, 1989.

[29] P. W. Fowler and P. A. Madden. *J. Phys. Chem.*, 89:2581–2585, 1985.

[30] N. C. Bacalis and A. B. Kunz. *Phys. Rev. B*, 32(8):4857–4865, 1985.

[31] Z. Barandiaran and L. Seijo. *J. Chem. Phys.*, 89(9):5739–5746, 1988.

[32] V. E. Puchin, E. V. Stefanovich, and T. N. Truong. *Phys. Rev. B*, 51:4014, 1995.

[33] K. Todnem, K. Børve, and M. Nygren. *Surf. Sci.*, 421:296–307, 1999.

[34] P. J. Hay and W. R. Wadt. *J. Chem. Phys.*, 82:270, 1985.

[35] W. R. Wadt and P. J. Hay. *J. Chem. Phys.*, 82:284, 1985.

[36] B. G. Dick and A. W. Overhauser. *Phys. Rev.*, 112:90, 1958.

[37] A. Aguado, L. Bernasconi, and P. A. Madden. *Chem. Phys. Lett.*, 356(5-6):437–444, 2002.

[38] C. Domene, P. W. Fowler, P. A. Madden, J. J. Xu, R. J. Wheatley, and M. Wilson. *J. Phys. Chem. A*, 105(16):4136–4142, 2001.

[39] R. Pandey and J. M. Vail. *J. Phys.: Condens. Matter*, 1:2801–2820, 1989.

[40] J. M. Vail. *J. Phys. Chem. Solids*, 51(7):589–607, 1990.

[41] C. Sousa and F. Illas. *J. Chem. Phys.*, 115(3):1435–1439, 2001.

[42] H. Donnerberg and A. Birkholz. *J. Phys.: Condens. Matter*, 12:8239–8247, 2000.

[43] P. V. Sushko, A. L. Shluger, and C. R. A. Catlow. *Surf. Sci.*, 450:153–170, 2000.

[44] V. B. Sulimov, P. V. Sushko, A. H. Edwards, A. L. Shluger, and A. M. Stoneham. *Phys. Rev. B*, 66:024108, 2002.

[45] C. Pisani, R. Dovesi, R. Nada, and L. N. Kantorovich. *J. Chem. Phys.*, 92:7448, 1990.

[46] C. Pisani, F. Còra, R. Nada, and R. Orlando. *Computer Physics Communications*, 82:139–156, 1994.

[47] N. Govind, Y. A. Wang, and E. Carter. *J. Chem. Phys.*, 110(16):7677, 1999.

[48] T. Klüner, N. Govind, Y. A. Wang, and E. Carter. *J. Chem. Phys.*, 116(1):42, 2002.

[49] L. N. Kantorovich. *Int. J. Quantum Chem.*, 78:306–330, 2000.

[50] V. B. Sulimov, S. Casassa, C. Pisani, J. Garapon, and B. Poumellec. *Modelling Simul. Mater. Sci. Eng.*, 8:763, 2000.

[51] P. V. Sushko, J. L. Gavartin, and A. L. Shluger. *J. Phys. Chem. B*, 106:2269–2276, 2002.

[52] M. A. Johnson, E. V. Stefanovich, and T. N. Truong. *J. Phys. Chem B*, 102(33):6391–6396, 1998.

[53] R. McWeeny. *Methods of molecular quantum mechanics*. London : Academic Press, 2 edition, 1989.

[54] R. G. Parr and W. Yang. *Density-functional theory of atoms and molecules*. New York : Oxford University Press ; Oxford : Clarendon, 1989.

[55] C. R. A. Catlow and W. C. Mackrodt, editors. *Computer simulations of solids*. Springer-Verlag, 1982.

[56] D. Vanderbilt. *Phys. Rev. B*, 41:7892, 1990.

[57] G. Kresse and J. Hafner. *J. Phys.: Condens. Matter*, 6:8245, 1994.

[58] P. E. Blöchl. *Phys. Rev. B*, 50:17953, 1994.

[59] G. Kresse and D. Joubert. *Phys. Rev. B*, 59:1758, 1999.

[60] G. Pacchioni, F. Frigoli, D. Ricci, and J. A. Weil. *Phys. Rev. B*, 63:054102, 2001.

[61] J. Lægsgaard and K. Stokbro. *Phys. Rev. Lett.*, 86:2834, 2001.

[62] J. L. Gavartin and A. L. Shluger. *Phys. Rev. B*, 64:245111, 2001.

[63] J. L. Gavartin, P. V. Sushko, and A. L. Shluger. *Phys. Rev. B*, 67:035108, 2003.

[64] H. B. Jansen and P. Ross. *Chem. Phys. Lett.*, 3:140, 1969.

[65] W. J. Hehre, R. F. Stewart, and J. A. Pople. *J. Chem. Phys.*, 51:2657, 1969.

[66] D. E. Woon and T. H. Dunning Jr. *J. Chem. Phys.*, 98:1358, 1993.

[67] E. Hernández, M. J. Gillan, and C. M. Goringe. *Phys. Rev. B*, 55:13485–13493, 1997.

[68] L. Füsti-Monlar and P. Pulay. *J. Chem. Phys.*, 116(18):7795–7805, 2002.

[69] L. Hedin. *Phys. Rev. A*, 139:796, 1965.

[70] D. Cepperley, G. Chester, and M. Kalos. *Phys. Rev. B*, 16:3081, 1971.

[71] M. Petersilka, U. J. Grossman, and E. K. U. Gross. *Phys. Rev. Lett.*, 76:1212, 1996.

[72] A. D. Becke. *J. Chem. Phys.*, 98:5648, 1993.

[73] J. Muscat, A. Wander, and N. M. Harrison. *Chem. Phys. Lett.*, 342:397, 2001.

[74] C. Lee, W. Yang, and R. G. Parr. *Phys. Rev. B*, 37:785, 1988.

[75] J. P. Crocombette. *Phys. Chem. Minerals*, 27:138, 1999.

[76] A. M. Stoneham, M. A. Szymanski, and A. L. Shluger. *Phys. Rev. B Rapid Commun.*, 63:241304, 2001.

[77] M. A. Szymanski, A. L. Shluger, and A. M. Stoneham. *Phys. Rev. B*, 63:224207, 2001.

[78] P.E. Blöchl. *Phys. Rev. B*, 62:6158–6179, 2000.

[79] W. B. Fowler, J. K. Rudra, M. E. Zvanut, and F. J. Feigl. *Phys. Rev. B*, 41:8313, 1990.

[80] J. L. Alay and M. Hirose. *J. Appl. Phys.*, 81:1606, 1997.

[81] J. G. Mihaychuk, N. Shamir, and H. M. van Driel. *Phys. Rev. B*, 59:2164, 1999.

[82] W. B. Fowler, J. K. Rudra, M. E. Zvanut, and F. J. Feigl. *Phys. Rev. B*, 41(12):8313–8317, 1990.

[83] Zvanut, F. J. Feigl, and J. D. Zook. *J. Appl. Phys.*, 64(4):2221 – 2223, 1988.

[84] P. Hohenberg and W. Kohn. *Phys. Rev.*, 136:B864, 1964.

[85] W. Kohn and L. J. Sham. *Phys. Rev*, 140:A1133, 1965.

[86] M. L. Cohen. *Phys. Rep.*, 110:293, 1984.

[87] W. Pickett. *Comput. Phys. Rep.*, 9:115, 1989.

[88] C. Pisani. *Quantum-mechanical ab-initio calculation of the properties of crystalline materials*. Springer, Berlin, 1996.

[89] N. A. W. Holzwarth, G. E. Matthews, R. B. Dunning, A. R. Tackett, and Y. Zeng. *Phys. Rev. B*, 55:2005, 1997.

[90] J. P. Perdew and S. Kurth. *Density Functionals: Theory and Applications*. Springer, Berlin, 1998.

[91] K. Ohno, K. Esfarjani, and Y. Kawazoe. *Computational Materials Science*. Springer, Berlin, 1999.

[92] W. Kohn. *Rev. Mod. Phys.*, 71:1253, 1999.

[93] J. Muscat, V. Swamy, and N. M. Harrison. *Phys. Rev. B*, 65:224112, 2002.

218

[94] K. J. Chang and M. L. Cohen. *Phys. Rev. B*, 30:4774, 1984.

[95] Y. Yourdshahyan, C. Ruberto, M. Halvarsson, L. Bengtsson, V. Langer, and B. Lundqvist. *J. Am. Ceram. Soc.*, 82:1365, 1999.

[96] N. V. Skorodumova, R. Ahuja, S. I. Simak, I. A. Abrikosov, B. Johansson, and B. I. Lundqvist. *Phys. Rev. B*, 64:115108, 2001.

[97] S. H. Cai, S. N. Rashkeev, S. T. Pantelides, and K. Sohlberg. *Phys. Rev. Lett.*, 89:235501, 2002.

[98] B. Civalleri, C. M. Zicovich-Wilson, P. Ugliengo, V. R. Saunders, and R. Dovesi. *Chem. Phys. Lett.*, 292:394, 1998.

[99] D. C. Allan and M. P. Teter. *Phys. Rev. Lett.*, 59:1136, 1987.

[100] N. Binggeli, N. Troullier, J. L. Martins, and J. R. Chelikowsky. *Phys. Rev. B*, 44:4771, 1991.

[101] F. Liu, S. H. Garofalini, R. D. King-Smith, and D. Vanderbilt. *Phys. Rev. Lett.*, 70:2750, 1993.

[102] G. Pacchioni and G. Ieranó. *Phys. Rev. B*, 57:818, 1998.

[103] L. Levien, C. T. Previtt, and D. J. Weidner. *Am. Mineral*, 65:920, 1980.

[104] J. Wang, H. P. Li, and R. Stevens. *J. Mater. Sci.*, 27:5397, 1992.

[105] P. Aldebert and J. P. Traverse. *J. Am. Ceram. Soc.*, 68:34, 1985.

[106] C. J. Howard, R. J. Hill, and B. E. Reichert. *Acta. Cystallogr. B*, 44:116, 1988.

[107] R. Ruh and P. W. R. Corfield. *J. Amer. Ceram. Soc.*, 53:126, 1970.

[108] D. M. Adams, S. Leonard, D. R. Russel, and R. J. Cernik. *J. Phys. Chem. Solids*, 52:1181, 1991.

[109] P. M. Kelly and L. R. F. Rose. *Prog. Mat. Sci.*, 47:463–557, 2002.

[110] R. H. French, S. J. Glass, F. S. Ohuchi, Y. N. Xu, and W. Y. Ching. *Phys. Rev. B*, 49:5133, 1994.

[111] D. W. McComb. *Phys. Rev. B*, 54:7094, 1996.

[112] M. Balog, M. Schieber, M. Michiman, and S. Patai. *Thin Solid Films*, 41:247, 1977.

[113] A. Lakhlifi, Ch. Leroux, P. Satre, B. Durand, M. Roubin, and G. Nihoul. *J. Sol. Stat. Chem.*, 119:289, 1995.

[114] W. C. Mackrodt and P. M. Woodrow. *J. Am. Ceram. Soc.*, 69:277, 1986.

[115] A. Dwivedi and A. N. Cormack. *Phil. Mag. A*, 61:1, 1990.

[116] E. V. Stefanovich, A. L. Shluger, and C. R. A. Catlow. *Phys. Rev. B*, 49:11560, 1994.

[117] G. Stapper, M. Bernasconi, N. Nicoloso, and M. Parrinello. *Phys. Rev. B*, 59:797, 1999.

[118] B. Králik, E. K. Chang, and S. G. Louie. *Phys. Rev. B*, 57:7027, 1998.

[119] G. Jomard, T. Petit, A. Pasturel, L. Magaud, G. Kresse, and J. Hafner. *Phys. Rev. B*, 59:4044, 1999.

[120] S. Fabris, A. T. Paxton, and M. W. Finnis. *Phys. Rev. B*, 61:6617, 2000.

[121] G. M. Rignanese, F. Detraux, X. Gonze, and A. Pasquarello. *Phys. Rev. B*, 64:134301, 2001.

[122] X. Zhao and D. Vanderbilt. *Phys. Rev. B*, 65:075105, 2002.

[123] A. A. Demkov. *Phys. Stat. Sol. B*, 226:57, 2001.

[124] X. Zhao and D. Vanderbilt. *Phys. Rev. B*, 65:233106, 2002.

[125] A. S. Foster, A. L. Shluger, and R. M. Nieminen. *Phys. Rev. Lett.*, 89:225901, 2002.

[126] S. Fabris, A. T. Paxton, and M. W. Finnis. *Phys. Rev. B*, 63:094101, 2001.

[127] M. W. Finnis, A. T. Paxton, M. Methfessel, and M. van Schilfgaarde. *Phys. Rev. Lett*, 81:5149, 1998.

[128] G. Kresse and J. Furthmüller. *Comp. Mat. Sci.*, 6:15, 1996.

[129] G. Kresse and J. Furthmüller. *Phys. Rev. B*, 54:11169, 1996.

[130] J. P. Perdew, J. A. Chevary, S. H. Vosko, K. A. Jackson, M. R. Pederson, D. J. Singh, and C. Fiolhais. *Phys. Rev. B*, 46:6671, 1992.

[131] D. Murnaghan. *Proc. Natl. Acad. Sci. USA*, 30:244, 1944.

[132] R. C. West. *Handbook of Chemistry and Physics*. CRC, Boca Raton, 1986.

[133] M. Winter. *www.webelements.com*. 2001.

[134] D. W. Stacy, J. K. Johnstone, and D. R. Wilder. *J. Am. Cer. Soc.*, 55:482, 1972.

[135] S. T. Pantelides, D. J. Mickish, and A. B. Kunz. *Phys. Rev. B*, 10:5203, 1974.

[136] M. Houssa, M. Tuominen, M. Naili, V. Afanas'ev, A. Stesmans, S. Haukka, and M. M. Heyns. *J. Appl. Phys.*, 87:8615, 2000.

[137] J. Robertson. *MRS Bull.*, March:217, 2002.

[138] F. Bernardini, V. Fiorentini, and Vanderbilt D. *Phys. Rev. Lett.*, 79:3958–3961, 1997.

[139] R. Resta. *Rev. Mod. Phys.*, 66:899–915, 1994.

[140] R. D. King-Smith and D. Vanderbilt. *Phys. Rev. B*, 47:1651–1654, 1993.

[141] D. Vanderbilt and R. D. King-Smith. *Phys. Rev. B*, 48:4442–4455, 1993.

[142] S. Baroni, S. de Gironcoli, and A. Dal Corso. *Rev. Mod. Phys.*, 73:515–562, 2001.

[143] F. Detraux, P. Ghosez, and Gonze X. *Phys. Rev. Lett.*, 81:3297, 1998.

[144] G. M. Rignanese, X. Gonze, and A. Pasquarello. *Phys. Rev. B*, 63:104305, 2001.

[145] G. M. Rignanese, F. Detraux, X. Gonze, A. Bongiorno, and A. Pasquarello. *Phys. Rev. Lett.*, 89:117601, 2002.

[146] A. P. Mirgorodsky, M. B. Smirnov, and P. E. Quintard. *J. Phys. Chem. Solids*, 60:985–992, 1999.

[147] G. Lukovsky and G.B. Rayner Jr. *Appl. Phys. Lett.*, 77:2912–2914, 2000.

[148] G. D. Wilk, R. M. Wallace, and J. M. Anthony. *J. Appl. Phys.*, 89:5243, 2001.

[149] A. A. Sirenko, C. Bernhard, A. Golnik, A. M. Clark, J. Hao, W. Si, and X. X. Xi. *Nature*, 404:373–376, 2000.

[150] V. Fiorentini and G. Gullery. *Phys. Rev. Lett*, 89:266101, 2002.

[151] A. S. Barker. *Phys. Rev. B*, 12:4071–4084, 1975.

[152] F. J. Feigl, W. B. Fowler, and K. L. Yip. *Solid State Commun.*, 14:225, 1974.

[153] J. K. Rudra and W. B. Fowler. *Phys. Rev. B*, 35:8223, 1987.

[154] B. Henderson and J. E. Wertz. *Adv. Phys.*, 17:749, 1968.

[155] L. E. Halliburton, D. L. Cowan, and L. V. Holroyd. *Phys. Rev. B*, 12(8):3408–3419, 1975.

[156] Y. Chen and J. L. Kolopus W. A. Sibley. *Phys. Rev.*, 186:865, 1969.

[157] G. H. Rosenblatt, M. W. Rowe, G. P. Williams, R. T. Williams, and Y. Chen. *Phys. Rev. B*, 39:10309, 1989.

[158] R. González, M. A. Monge, J. E. Muñoz Santiuste, R. Pareja, Y. Chen, E. Kotomin, M. M. Kuklja, and A. I. Popov. *Phys. Rev. B*, 59(7):4786–4790, 1999.

[159] G. Pacchioni. In P. Woodruff, editor, *The Chemical Physics of Solid Surfaces - Oxide surfaces, Vol. 9*, pages 94–135. Elsevier, Amsterdam, 2000.

[160] R. I. Eglitis, M. M. Kuklja, E. A. Kotomin, A. Stashans, and A. I. Popov. *Comput. Mater. Sci.*, 5:298, 1996.

[161] E. Scorza, U Birkenheuer, and C. Pisani. *J. Chem. Phys.*, 107(22):9645–9658, 1997.

[162] G. Pacchioni, A. M. Ferrari, and G. Ieranò. *Faraday Discuss.*, 106:155, 1997.

[163] A. M. Ferrari and G. Pacchioni. *J. Phys. Chem.*, 99:17010, 1995.

[164] K. Jackson, M. R. Pederson, and B. M. Klein. *Phys. Rev. B*, 43(3):2364–2371, 1991.

[165] B. M. Klein, W. E. Pickett, L. L. Boyer, and R. Zeller. *Phys. Rev. B*, 35(11):5802–5815, 1987.

[166] Q. S. Wang and N. A. W. Holzwarth. *Phys. Rev. B*, 41:3211, 1990.

[167] L. N. Kantorovich, J. M. Holender, and M. J. Gillan. *Surface Sci.*, 343:221–239, 1995.

[168] P. Mori-Sánchez, J. M. Recio, B. Silvi, C. Sousa, A. M. Pendás V. Luaña, and F. Illas. *Phys. Rev. B*, 66:075103, 2002.

[169] R. C. Whited, C. J. Flaten, and W. C. Walker. *Solid St. Commun.*, 13:1903, 1973.

[170] R. Pandey, J. E. Jaffe, and A. B. Kunz. *Phys. Rev. B*, 43:9228–9237, 1991.

[171] M.I. McCarthy and N.M. Harrison. *Phys. Rev. B*, 49:8574–8582, 1994.

[172] A. Zupan and M. Causà. *Int. J. Quantum Chem.*, 56:337–344, 1995.

[173] R. Orlando, R Millini, G Perego, and R. Dovesi. *J. Molec. Catal.*, 119:253, 1997.

[174] B. B. Stefanov and K. Raghavachari. *Phys. Rev. B*, 56:5035, 1997.

[175] G. Pacchioni and G. Ieranò. *Phys. Rev. Lett.*, 79:753, 1997.

[176] G. Pacchioni and A. Basile. *J. Non-Cryst. Solids*, 254:17, 1999.

[177] V. B. Sulimov, C. Pisani, F. Corà, and V. O. Sokolov. *Solid State Commun.*, 90:511, 1994.

[178] V. B. Sulimov and V. O. Sokolov. *J. Non-Cryst. Solids*, 191:260, 1995.

[179] V. B. Sulimov, V. O. Sokolov, and B. Poumellec. *Phys. Status Sol. (b)*, 96:175, 1996.

[180] K. C. Snyder and W. B. Fowler. *Phys. Rev. B*, 48:13238, 1993.

[181] D. Erbetta, D. Ricci, and G. Pacchioni. *J. Chem. Phys.*, 113:10744, 2000.

[182] A. Oshiyama. *Jap. J. Appl. Phys.*, 37(2B):L232, 1998.

[183] M. Boero, A. Pasquarello, J. Sarntheim, and R. Car. *Phys. Rev. Lett.*, 78:887, 1997.

[184] L. Skuja. *J. Non-Cryst. Solids*, 239:16, 1998.

[185] C. M. Carbonaro, V. Fiorentini, and S. Missidda. *J. Non-Cryst. Solids*, 221:89, 1997.

[186] D. C. Allan and M. P. Teter. *J. Amer. Ceramic Soc.*, 73:3247, 1990.

[187] N. Capron, S. Carniato, A. Lagraa, and G. Boureau. *J. Chem. Phys.*, 112:9543, 2000.

[188] N. Capron, S. Carniato, G. Boureau, and A. Pasturel. *J. Non-Cryst. Solids*, 245:146, 1999.

[189] G. Boureau and S. Carniato. *Solid State Commun.*, 98(485), 1996.

[190] D. L. Griscom, E. J. Friebele, and Jr G. H. Sigel. *Sol. State Sommun.*, 15:479, 1974.

[191] D. L. Griscom. In G. Pacchioni, L. Skuja, and D. L. Griscom, editors, *Defects in SiO₂ and Related Dielectrics: Science and Technology, NATO Science Series*, pages 117–161. Kluwer, Dordrecht, 2000.

[192] R. A. Weeks. *J. Appl. Phys.*, 27:1376–1381, 1956.

[193] R. A. Weeks. *J. Non-Cryst. Solids*, 179:1–9, 1994.

[194] M. G. Jani, R. B. Bossoli, and L. E. Halliburton. *Phys. Rev. B*, 27:2285, 1983.

[195] V. A. Radzig. In G. Pacchioni, L. Skuja, and D. L. Griscom, editors, *Defects in SiO₂ and Related Dielectrics: Science and Technology, NATO Science Series*, pages 339–370. Kluwer, Dordrecht, 2000.

[196] D. L. Griscom and E. J. Friebele. *Phys. Rev. B*, 34:7524, 1986.

[197] G. Pacchioni, G. Ierano, and A. M. Marquez. *Phys. Rev. Lett.*, 81:377, 1998.

[198] G. Pacchioni. In G. Pacchioni, L. Skuja, and D. L. Griscom, editors, *Defects in SiO₂ and Related Dielectrics: Science and Technology, NATO Science Series*, pages 161–196. Kluwer, Dordrecht, 2000.

[199] A. Stirling and A. Pasquarello. *Phys. Rev. B*, 66:245201, 2002.

[200] A. Othonos and K. Kalli. *Fiber Bragg Gratings: fundamentals and applications in telecommunication and sensing.* Artech House, Inc., Boston, 1999.

[201] L. Skuja. In G. Pacchioni, L. Skuja, and D. L. Griscom, editors, *Defects in SiO₂ and Related Dielectrics: Science and Technology, NATO Science Series*, pages 73–116. Kluwer, Dordrecht, 2000.

[202] K. Raghavachari, D. Ricci, and G. Pacchioni. *J. Chem. Phys.*, 116(2):825–831, 2002.

[203] W. L. Warren, P. M. Lenahan, B. Robinson, and J. H. Stathis. *Appl. Phys. Lett.*, 53:482, 1988.

[204] V. V. Afanas'ev and A Stesmans. *J. Phys.: Condens. Matter*, 12:2285–2290, 2000.

[205] A. Stesmans, B. Nouwen, and V. V. Afanas'ev. *Phys. Rev. B*, 66:045307, 2002.

[206] Z.-Y. Lu, C. J. Nicklaw, D. M. Fleetwood, R. D. Schrimpf, and S. T. Pantelides. *Phys. Rev. Lett.*, 89:285505, 2002.

[207] D. Donadio, M. Bernasconi, and M. Boero. *Phys. Rev. Lett.*, 87:195504, 2001.

[208] Y. N. Xu, Z. Q. Gu, X. F. Zhong, and W. Y. Ching. *Phys. Rev. B*, 56:7277, 1997.

[209] C. H. Park and D. J. Chadi. *Phys. Rev. B*, 57(22):R13961–R13964, 1998.

[210] N. V. Skorodumova, S. I. Simak, B. I. Lundqvist, I. A. Abrikosov, and B. Johansson. *Phys. Rev. Lett.*, 89:166601, 2002.

[211] A. Eichler. *Phys. Rev. B*, 64:174103, 2001.

[212] J. Aarik, A. Aidla, H. Mändar, V. Sammelsberg, and T. Uuustare. *J. Cryst. Growth*, 220:105, 2000.

[213] D. A. Neumayer and E. Cartier. *J. Appl. Phys*, 90:1801, 2001.

[214] L. N. Kantorovich, J. M. Holender, and M. J. Gillan. *Surf. Sci.*, 343:221, 1997.

[215] K. L. Chopra, S. Major, and D. K. Pandya. *Thin Solid Films*, 102:1–46, 1983.

[216] H. L. Hartnagel, A. L. Dewar, A. K. Jain, and C. Jagadish. *Semiconducting Transparent Thin Films.* IOP Publishing, Bristol, 1995.

[217] J. S. Kim, M. Granstrom, R. H. Friend, N. Johansson, W. R. Salaneck, R. Daik, W. J. Feast, and F. Cacialli. *J. Appl. Phys.*, 84:6859, 1998.

[218] D. S. Ginley and C. Bright(Eds.). *Mater. Res. Bull.*, 25:15(and articles therein), 2000.

[219] C. G. Granqvist and A. Hultaker. *Thin Solid Films*, 411 (and articles in this issue):1–5, 2002.

[220] K. Ellmer. *J. Phys. D: Appl. Phys.*, 34:3097–3108, 2001.

[221] H. Kawazoe, M. Yasukawa, H. Hyodo, M. Kurita, H. Yanagi, and H. Hosono. *Nature*, 389:939, 1997.

[222] J. Robertson, P. W. Peacock, M. D. Towler, and R. Needs. *Thin Solid Films*, 411:96–100, 2002.

[223] K. Hayashi, S. Matsuishi, T. Kamiya, M. Hirano, and H. Hosono. *Nature*, 419:462, 2002.

[224] O. N. Mryasov and A. J. Freeman. *Phys. Rev. B*, 64:233111, 2001.

[225] B. Stierna, C. G. Granqvist, A. Seidel, and L. Häggström. *J. Appl. Phys.*, 68:6241, 1990.

[226] C. Kilic and A. Zunger. *Phys. Rev. Lett.*, 88:095501, 2002.

[227] R. Asahi, A. Wang, J. R. Babcock, N. L. Edleman, A. W. Metz, M. A. Lane, V. P. Dravid, C. R. Kannewurtf, A. J. Freeman, and T. J. Marks. *Thin Solid Films*, 411:101, 2002.

[228] M. Chen, Z. L. Pei, X. Wang, Y. H. Yu, X. H. Liu, and L. S. Wen. *J. Phys. D: Appl. Phys.*, 33:2538–2548, 2000.

[229] P. V. Sushko, A. L. Shluger, K. Hayashi, M. Hirano, and H. Hosono. *Nature Materials*, submitted, 2003.

[230] H. Bartl and T. Scheller. *N. Jb. Miner. Mh.*, 35:547–552, 1970.

[231] A. Stein, G. A. Ozin, and G. D. Stucky. *J. Amer. Chem. Soc.*, 114:8119–8129, 1992.

[232] M. O. Zacate and R. W. Grimes. *J. Phys. Chem. Sol.*, 63:675–683, 2002.

[233] M. J. Frisch, G. W. Trucks, H. B. Schlegel, G. E. Scuseria, M. A. Robb, J. R. Cheeseman, V. G. Zakrzewski, J. A. Montgomery, R. E. Stratmann, J. C. Burant, S. Dapprich, J. M. Millam, A. D. Daniels, K. N. Kudin, M. C. Strain, O. Farkas, J. Tomasi, V. Barone, M. Cossi, R. Cammi, B. Mennucci, C. Pomelli, C. Adamo, S. Clifford, J. Ochterski, G. A. Petersson, P. Y. Ayala, Q. Cui, K. Morokuma, D. K. Malick, A. D. Rabuck, K. Raghavachari, J. B. Foresman, J. Cioslowski, J. V. Ortiz, B. B. Stefanov, G. Lui, A. Liashenko, P. Piskorz, I. Komaromi, R. Gomperts, R. L. Martin, D. J. Fox, T. Keith, M. A. Al-Laham, C. Y. Peng, A. Nanayakkara, C. Gonzalez, M. Challacombe, P. M. W. Gill, B. G. Johnson, W. Chen, M. W. Wong, J. L. Andres, M. Head-Gordon, E. S. Replonge, and J. A. Pople. *GAUSSIAN 98 (Release A1)*. Pittsburgh PA, 1998.

[234] S. J. Zinkle. *Nucl. Instrum. Methods Phys. Res.*, 91:234, 1994.

[235] T. Brudevoll, E. A. Kotomin, and N. E. Christensen. *Phys. Rev. B*, 53:7731, 1996.

[236] A. M. Stoneham and P. W. Tasker. *Phil. Mag. B*, 55:237, 1987.

[237] C. J. Sofield and A. M. Stoneham. *Semicond. Sci. Technol.*, 10:215–244, 1995.

[238] M. A. Szymanski, A. M. Stoneham, and A. L. Shluger. *Microel. Reliab.*, 40:567–570, 2000.

[239] D. Ricci, G. Pacchioni, M. A. Szymanski, A. L. Shluger, and A. M. Stoneham. *Phys. Rev. B*, 64:224101, 2001.

[240] D. R. Hamann. *Phys. Rev. Lett.*, 81:3447, 1998.

[241] J. R. Chelikowsky, D. J. Chadi, and N. Binggeli. *Phys. Rev. B*, 62:R2251, 2000.

[242] G. Pacchioni and G. Ieranò. *Phys. Rev. B*, 56:7304, 1997.

[243] M. I. Heggie, R. Jones, C. D. Latham, and S. C. P. Maynard. *Phil. Mag. B*, 65:463, 1992.

[244] L. Skuja and B. Guttler. *Phys. Rev. Lett.*, 77:2093–2096, 1996.

[245] F. Rochet, S. Rigo, M. Froment, C. D'Anterroches, C. Maillot, H. Roulet, and G. Dufour. *Adv. Phys.*, 35:237, 1986.

[246] Y. Jin and K. J. Chang. *Phys. Rev. Lett.*, 86(9):1793–1796, 2001.

[247] M. A. Szymanski, A. M. Stoneham, and A. L. Shluger. *Sol. State. Elec.*, 45:1233, 2001.

[248] S. Jeon, C. J. Choi, T. Y. Seong, and H. Hwang. *Appl. Phys. Lett.*, 79:245, 2001.

[249] C. S. Kang, H. J. Cho, K. Onishi, R. Nieh, R. Choi, S. Gopalan, S. Krishnan, J. H. Han, and J. C. Lee. *Appl. Phys. Lett.*, 81:2593, 2002.

[250] A. L. P. Rotondaro, M. R. Visokay, A. Shanware, J. J. Chambers, and L. Colombo. *IEEE Elec. Dev. Lett.*, 23:603, 2002.

[251] G. M. Rignanese and A. Pasquarello. *Surf. Sci.*, 490:L614–L618, 2001.

[252] G. M. Rignanese and A. Pasquarello. *Phys. Rev. B*, 63:075307, 2001.

[253] V. V. Afanas'ev and A Stesmans. *Phys. Rev. Lett.*, 80:5176–5179, 1998.

[254] J.H. Stathis. *IBM J. Res. & Dev.*, 46:265–286, 2002.

[255] D. M. Fleetwood. *Microelectronics Reliability*, 42:523–541, 2002.

[256] R. Degreave, B. Kaczer, and G. Groeseneken. *Microelectronics Reliability*, 39:1445–1460, 1999.

[257] R. A. B. Devine. *J. Non-Cryst. Solids*, 107:41, 1988.

[258] R. A. B. Devine. *Nucl. Instrum. Methods Phys. Res., Sect B*, 46:244, 1990.

[259] J.P. Brodholt and Refson K. *J. Geophys. Res.*, 105:18977 –18982, 2000.

[260] N. H. de Leeuw. *J. Phys. Chem. B*, 105:9747–9754, 2001.

[261] G. Pacchioni and M Vitiello. *Phys. Rev. B*, 58:7745–7752, 1998.

[262] P.E. Blöchl and J.H. Stathis. *Phys. Rev. Lett.*, 83:372–375, 1999.

[263] P. E. Bunson, M. Di Ventra, S. T. Pantelides, R. D. Schrimpf, and F. Galloway. *IEEE Trans. Nucl. Sci.*, 46:1568–1573, 1999.

[264] P. E. Bunson, M. Di Ventra, S. T. Pantelides, D. M. Fleetwood, and R. D. Schrimpf. *IEEE Trans. Nucl. Sci.*, 47:2289–2296, 2000.

[265] B. Tuttle. *Phys. Rev. B*, 60:2631–2637, 1999.

[266] T. Bakos, S. N. Rashkeev, and S. T. Pantelides. *Phys. Rev. Lett.*, 88:055508, 2002.

[267] B. Tuttle. *Phys. Rev. B*, 61:4417–4420, 2000.

[268] D.J. Chadi. *Phys. Rev. B*, 64:195403, 2001.

[269] A. Yokozawa and Y Miyamoto. *Phys. Rev. B*, 55:13783–13788, 1997.

222

[270] R. H. Doremus. *J. Non-Cryst. Solids*, 261:101–107, 2000.

[271] K. L. Brower. *Phys. Rev B*, 42:3444, 1990.

[272] T. Sato and M. Shimada. *J. Am. Ceram. Soc.*, 68:356–359, 1985.

[273] X. Guo. *Solid State Ionics*, 112:113–116, 1998.

[274] X. Guo. *J. Phys. Chem. Solids*, 60:539–546, 1999.

[275] T. Nishizaki, M. Okui, K. Kurosaki, M. Uno, S. Yamanaka, K. Takeda, and H. Anada. *J. Alloys Comp.*, 330-332:307–312, 2002.

[276] Ç. Kiliç and A. Zunger. *Appl. Phys. Lett.*, 81:73–75, 2002.

[277] F. Freund, M. M. Freund, and Battlo F. *J. Geophys. Res.*, 98:22209–22229, 1993.

[278] W. B. Fowler, J. K. Rudra, M. E. Zvanut, and F. J. Feigl. *Phys. Rev. B*, 41:8313, 1990.

[279] J. Robertson. *J. Vac. Sci. Technol. B*, 18:1785, 2000.

[280] P. W. Anderson. *Phys. Rev. Lett.*, 34:953, 1975.

[281] A. M. Stoneham and M. J. L. Sangster. *Radiat. Eff.*, 73:267, 1983.

[282] B. E. Deal and A. S. Grove. *J. Appl. Phys.*, 36:3770, 1965.

[283] M. A. Quevedo-Lopez, M. El-Bouanani, M. J. Kim, B. E. Gnade, R. M. Wallace, M. R. Visokay, A. LiFatou, M. J. Bevan, and L. Colombo. IEEE SISC, San Diego, California, 2002.

[284] A. B. Lidiard. *Handbuch der Physik XX*. Springer-Verlag, Berlin, 1957.

[285] A. V. Chadwick and M. Terenzi, editors. *Defects in Solids: Modern Techniques NATO ICI Series B147*. Plenum Press, New York, 1985.

[286] A. Bongiorno and A. Pasquarello. *Phys. Rev. Lett.*, 88:125901, 2002.

[287] A. M. Stoneham. *Phys. Rev. Lett.*, 63:1027, 1989.

[288] G. Mills, H. Jonsson, and G. K. Schenter. *Surf. Sci.*, 324:305, 1995.

[289] B. J. Berne, G. Ciccotti, and D. F. Coker, editors. *Classical and Quantum Dynamics in Condensed Phase Simulations*. World Scientific, London, 1998.

[290] G. Henkelman and H. Jonsson. *J. Chem. Phys.*, 111:7010–7022, 1999.

[291] R. A. Miron and K. A. Fichthom. *J. Chem. Phys.*, 115(19):8742, 2001.

Nano and Giga Challenges in Microelectronics
Greer at al (Editors)

Tunneling Through Single Molecules

John Tomfohr, Jun Li, and Otto F. Sankey

Department of Physics and Astronomy, Arizona State University, Tempe, AZ 85287-1504

Abstract

The theory of electron tunneling through single molecules is introduced with special focus on periodic molecules, where the problem is simply understood in terms of complex wave-vector Bloch states. The molecule length dependence of the current, the Fermi level alignment problem, barrier traversal times, and other important aspects of single molecule transport in general come in to play in this context. The Landauer formalism is a framework for more quantitative treatments of more general structures and a brief introduction to this subject is also given.

Key words: molecular electronics, tunneling transport, complex band, molecular conductance.

1. Introduction

A large amount of theoretical work in molecular electronics is aimed at understanding electron transport through single molecules. An awareness of the basic mechanisms of molecular scale transport and a sense of how they can be harnessed will be important guides for research in this field.

In this chapter we give an overview of some basic aspects of single molecule transport. The system of interest is a small organic molecule (for example, benzene) connected between electrodes. The molecule serves as a bridge for electrons driven from one electrode to the other by an applied bias. While there are a variety of possible transport mechanisms and processes that can influence transport, we focus only on the simplest case of coherent transport. The basic picture is then of a distribution of electrons incident on the molecular barrier and whose wavefunctions in the region of the molecule are essentially fixed; that is, the wavefunctions are not disturbed significantly by fluctuations in geometry or other variables that influence electronic structure.

Coherent transport is expected to play an important role for short (on the order of tens of Angstroms), rigid organic molecules and efforts to calculate transport properties for molecules of this kind have usually proceeded under the assumption of coherence [1–5]. While complete agreement between the predictions of coherent transport models and experimental measurements has not yet been clearly demonstrated, coherent transport is still certainly a significant part of what is seen experimentally [6]. In particular, measured molecule conductances are often much lower (by as much as 3 orders of magnitude [4,7]) than what is expected for pure coherent transport, though they still typically show the exponential dependence on molecule length that is a hallmark of (coherent) tunneling. Further, the rate of decrease of conductance with increasing molecule length, which is a fundamental characteristic of the type of molecule being measured, can be calculated using standard electronic structure methods and convincing agreement with experiment is often found. The agreement in length dependence but not in magnitude of the conductances suggests that there are additional mechanisms or barriers influencing the transport but that coherence remains an important part of the picture.

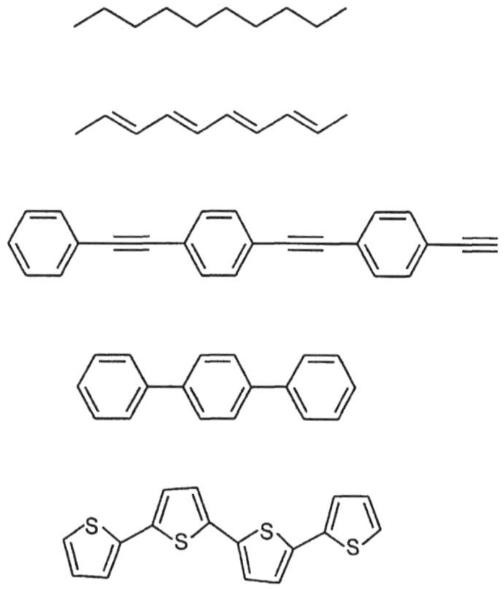

Fig. 1. Some examples of organic periodic molecules. From the top: alkane, polyacetylene (or alkene), phenylene-ethynylene chain, phenylene chain, and thiophene chain.

Our discussion focuses mainly on tunneling through periodic molecules (examples are discussed later; e.g., Fig. 1). This is the simplest example of coherent transport and is easily understood in terms of complex wave-vector Bloch states, the generalization of the decaying and growing solutions ($e^{\pm \kappa x}$) familiar from tunneling through a constant potential barrier. Besides that many molecules of interest in molecular electronics are periodic, the development presented here touches on many of the central issues in single molecule transport generally such as the Fermi level alignment, tunneling times, and the molecule length dependence of the conductance. We present a review of the theory including an illustrative model, a method for computing complex bandstructures (the usual bandstructure extended to the complex wave-vectors), and some applications.

After laying this foundation, we give a brief introduction to the more general and quantitative Landauer formalism. The Landauer approach and equivalent methods based on quantum scattering theory have been in wide use in calculations of molecule transport properties. We derive the main formulas and demonstrate their application in a simple analytic example.

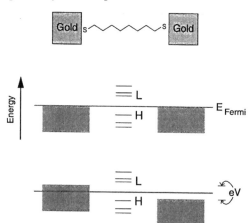

Fig. 2. Illustration of a metal-molecule-metal system. The metals have a continuum of electron states filled up to the Fermi level while the molecule has a discrete spectrum of electron states (molecular orbitals). The application of a voltage V effectively shifts the Fermi levels of the electrodes by the amount eV allowing electrons within the energy window between the Fermi levels of the electrodes to flow from left to right.

2. Overview: the metal-molecule-metal system

The prototype molecular electronic system is shown in Fig. 2. A molecule is connected between metal electrodes, often a dithiolated molecule (sulfur atoms attached at either end) between gold electrodes. The metals have a continuous distribution of electronic states occupied up to a highest energy level (Fermi level) while the molecule contains a discrete spectrum of levels that are thought to act as clear passageways through which electrons can flow from one electrode to the other.

Some examples of realizations of this system are (i) an early experiment in which the attempt was made to bridge a gap (break-junction) between gold electrodes with benzenedithiol molecules [8], (ii) atomic force microscope measurements of transport properties of molecules assembled on a metal substrate (e.g., Refs. [9,10]), and (iii) molecules sandwiched between gold contacts in an etched nanopore [11].

At equilibrium and with no applied voltage, the Fermi levels of the two metals will coincide and normally, if the charge of the molecule is nearly neutral, they will be situated between the molecule HOMO (highest occupied molecular orbital) and LUMO (lowest unoccupied molecular orbital); otherwise one or more electrons would transfer to or from the molecule to a state available at lower energy and the molecule would become charged.

The application of a voltage V effectively displaces the Fermi levels of the electrodes by the amount eV (e is the magnitude of the electron charge). If the molecule and the connections are approximately symmetric, the molecule levels may be imagined to be fixed while the Fermi level of the left electrode μ_L shifts up, and that of the right electrode μ_R down, by the amount $eV/2$. Electrons in the energy range between μ_L and μ_R can transfer from the left electrode to the right and so produce a net current.

At high enough voltages, a molecular level will enter the energy window between μ_L and μ_R and this is expected to cause a rapid increase in current. The simplest ideas for molecular electronic devices are based on manipulating the molecular levels to control the onset of this resonant tunneling.

For example, a molecule designed to have a molecular level that comes into resonance in one bias direction and not the reverse would function as a diode (see, for example, Ref. [12]). This could be accomplished if the molecule is bound more strongly to one electrode than to the other and the HOMO is just below the Fermi level. The situation is depicted in Fig. 3. The molecular level will be "pinned" to the energy levels (e.g., the Fermi level) of the electrode it is bound to (the left electrode in the figure) and the HOMO will enter the energy window between the electrode Fermi levels in one bias direction but not the other.

This pinning of the molecule to one electrode over the other and the resulting asymmetries cre-

ated in the I-V curve are discussed, for example, in Ref. [13].

Another possible way to exploit resonant tunneling would be to arrange to have a third gate electrode that switches the molecule between high and low conductance states by controlling the positions of the molecular levels [14]. Later, we will look at the factors that determine the ratio between the "on" and "off" conductance states for such a molecular transistor.

To get a sense of the amount of current that can be produced, consider the case that the "molecule" is simply a narrow conducting bridge described by a constant potential. The electronic states along the length of the bridge are then planewaves: $\psi(x) = e^{\pm ikx}$. The net current is produced by the electrons within the energy window eV about the equilibrium Fermi level and with k-vectors pointing from left to right [15]:

$$I = e \times (\text{number of electrons in the window}) \\ \times \frac{(\text{speed of the electrons})}{(\text{length of the bridge})} \quad (1)$$

$$I = e \left(eV \frac{dN}{dE} \right) \times \left(\frac{1}{\hbar} \frac{dE}{dk} \right) \times \frac{1}{L} \quad (2)$$

$$I = \frac{e^2}{\hbar} \times \frac{1}{L} \frac{dN}{dk} \times V. \quad (3)$$

Here dN/dk is the k-space density of electrons, and L is the length of the bridge. Using periodic boundary conditions over the bridge, dN/dk is $L/2\pi$ times a factor of 2 for spin. Hence,

$$I = \frac{e^2}{\pi\hbar} V = G_0 V \quad (4)$$

where

$$G_0 \equiv \frac{e^2}{\pi\hbar} = \frac{2e^2}{h} = \frac{1}{12.9 \text{ k}\Omega} = 77 \ \mu\text{S}, \quad (5)$$

is the quantum of conductance.

If there is an obstruction in the bridge (for example, if the bridge is broken and a molecule is inserted in the resulting gap), the net current will be given by the incident current times the transmission probability $T(\mu)$ at the Fermi level μ [16]. This gives the (single channel) Landauer formula

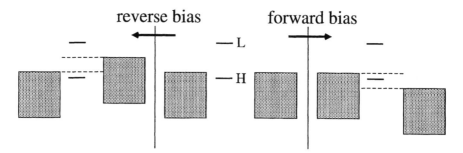

reverse bias   forward bias

Fig. 3. A molecular resonant-tunneling diode. The molecule is bound more strongly to the left electrode so that the molecular levels tend to be "pinned" to the levels of the left electrode (e.g., its Fermi level). Thus, the right electrode Fermi level may be imagined to shift with the applied bias while the molecule levels and left electrode Fermi level stay fixed. If the HOMO (H) is closer than the LUMO (L) to the equilibrium Fermi level, the HOMO will enter the region between the electrode Fermi levels (dashed lines) in the "forward" bias direction but not the reverse.

$$I = \frac{2e^2}{h}T(\mu)V. \qquad (6)$$

More generally, the current at a given voltage is given by integrating a transmission function over the energy window between μ_L and μ_R as will be described in section 4.

3. Tunneling through periodic molecules

In this section we focus on electronic tunneling through periodic molecules or, more specifically, molecules that are composed of a short sequence of repeating fundamental units (see Refs. [2,17,18] for additional discussion and examples). Many molecules that have been considered for molecular electronics applications are of this kind (see Fig. 1) and these are ideal for illustrating the basic physics of coherent electron transport through molecules generally.

The electronic energy eigenstates of a periodic molecule (or any periodic system) are most conveniently taken to be Bloch states,

$$\psi(x) = e^{ikx}u(x) \qquad (7)$$

where k is the Bloch vector and $u(x)$ is a function with the periodicity of the molecule. A given molecular orbital of a (finite) molecule such as, for example, the highest occupied molecular orbital (HOMO) or the lowest unoccupied molecular orbital (LUMO) may be loosely thought of as being formed from linear combinations of Bloch waves

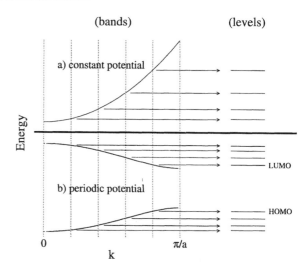

Fig. 4. (a) Particle in a box levels are standing waves formed from states of the free electron (constant potential) parabola ($E \propto k^2$). (b) Molecular levels (HOMO, LUMO) for periodic molecules are standing Bloch waves formed from states of the molecule (periodic potential) bandstructure. (Specifically, the molecule levels shown are for an idealized molecule that consists of 4 repeats of a unit cell and where the cell-to-cell distance is a.)

that satisfy certain boundary conditions; an analogy, Fig. 4, is that particle in a box states (sine waves) are linear combinations of states of an infinite constant potential (plane waves) that vanish at the boundaries of the box. In this sense, the molecular orbitals of periodic molecules are standing Bloch waves.

As described above, electronic tunneling current through a molecule bonded between metals is normally expected to be carried by electrons with energies in the HOMO-LUMO gap region of the molecule. The wavefunctions of these electrons, propagating in the metal, will exhibit an exponential decay as they penetrate from the metal into the molecule region. The electronic states of the molecule that account for this decay are also Bloch states but they are those for which k (Bloch vector) has an imaginary component. These tunneling states are simply a generalization of the exponentially decaying (and growing) solutions ($e^{\pm \kappa x}$) for the square potential barrier of elementary quantum mechanics.

The Bloch k vector is generally written

$$k = k_0 + i\beta/2; \qquad (8)$$

k_0 is a rate of oscillation and β is the spatial rate of decay:

$$e^{ikx} \rightarrow e^{-\beta x/2} e^{ik_0 x}. \qquad (9)$$

In electron tunneling transport, the important quantity is the tunneling probability which is essentially given by $|\psi^2| \approx e^{-\beta x}$. The quantity $\beta(E)$, a function of the energy of the tunneling electron, is fundamental in electronic tunneling. The probability that an electron with energy E will tunnel through a molecule with length L, and therefore the amount of tunneling current, goes roughly like $e^{-\beta(E)L}$. For the textbook example of the square potential barrier of height V_0, $\beta(E)$ has the simple form

$$
\begin{aligned}
\beta(E) &= 2\sqrt{\frac{2m}{\hbar^2}(V_0 - E)} \\
&= 1.0\sqrt{V_0 - E} \text{ eV}^{-1/2} \text{ Å}^{-1}.
\end{aligned}
\qquad (10)
$$

For periodic molecules (and other periodic structures) the form of $\beta(E)$ is generally very different than this. Typically, $\beta(E)$ for a periodic system will have one or more curves connecting valence (HOMO) and conduction (LUMO) band edges; $\beta(E)$ will be small near the band edges and increase as E approaches the middle of the bandgap.

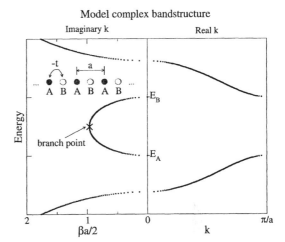

Fig. 5. The complex bandstructure for the poly-AB model. The right half of the figure is the usual (real k) bandstructure while the left region is a plot of $|\text{Im } k|$ for the k solutions with an imaginary part. The valence band branch (maximum at E_A) is connected to the conduction band branch (minimum at E_B) at the branch point of the fundamental β semi-elliptical curve. The k solutions with a non-zero imaginary part decay by $e^{-|\text{Im } k|a}$ from one unit cell to the next. Of particular importance is the reduction in the tunneling probability, $e^{-2|\text{Im } k|a} = e^{-\beta a}$ where β is defined as $\beta(E) = 2\text{Im } k(E)$.

3.1. Model complex bandstructure

These ideas are best illustrated with a simple model. Consider (Fig. 5) a molecule composed of two types of atoms, A and B, each with 1 valence electron, arranged in the sequence $ABABABAB...$ The unit cell is AB and the distance (lattice constant) between cells is a. The wavefunctions are formed from a linear combination of orbitals ϕ_A (one for each A atom) and ϕ_B (for each B) with onsite energies E_A and E_B (we take $E_A < E_B$) and nearest neighbor hopping $-t$:

$$
\begin{aligned}
\int \phi^*_{A(B)}(m) \hat{H} \phi_{A(B)}(m) &= E_{A(B)} \\
\int \phi^*_A(m) \hat{H} \phi_B(m) &= \int \phi^*_A(m) \hat{H} \phi_B(m-1) = -t.
\end{aligned}
\qquad (11)
$$

Here, $\phi_A(m)$ is the orbital on atom A in cell m (similarly for B) and $\hat{H}$ is the Hamiltonian. For a Bloch state, the coefficients $C_A^{(m)}$ and $C_B^{(m)}$ for orbitals $\phi_A(m)$ and $\phi_B(m)$ are related to those in cell $m + 1$ by

$$\mathbf{C}^{(m+1)} = \lambda \mathbf{C}^{(m)} \tag{12}$$

where $\lambda \equiv e^{ika}$ and $\mathbf{C}$ is the (two component) vector containing the coefficients C_A and C_B.

The energy vs. k solutions for this system are easily found to be

$$E_{\pm}(k) = \frac{E_A + E_B}{2}$$
$$\pm \sqrt{\left(\frac{E_A - E_B}{2}\right)^2 + 2t^2(1 + \cos ka)}. \tag{13}$$

These are plotted in Fig. 5. $E_+(k)$ and $E_-(k)$ denote the two bands or branches of the bandstructure and, in the right panel of Fig. 5, they describe conventional valence (E_-) and conduction (E_+) bands. The left panel shows the imaginary part, $\beta a/2 \equiv \text{Im } k$, of the tunneling state Bloch vectors. $\beta(E)$ has a form similar to a semi-ellipse and the two branches E_+ and E_- are connected at its peak, the branch point. The branch point has importance in the Fermi level alignment problem, which will be discussed in the next section. The gap states are increasingly more penetrating (smaller β) as we move from the branch point toward the valence and conduction band edges, so that electrons with energies nearer the band edges have a better chance of tunneling through the molecule. The band edges are each like the top of a square barrier since they mark the crossover point from propagating to tunneling states.

From Eq. 13, we determine the following expression for β:

$$\beta(E)a/2 = \ln\left(\gamma(E) + \sqrt{\gamma(E)^2 - 1}\right) \tag{14}$$

where

$$\gamma(E) \equiv \frac{(E - E_A)(E_B - E)}{2t^2} + 1. \tag{15}$$

Close to the band edges, an expansion of Eq. 14 finds the square-root dependence on the energy distance from the band edges just as for the square barrier, Eq. 10. However, the overall form is obviously very different than a square root.

3.2. Fermi level alignment and conductance estimates

The current through a molecule bonded between metal electrodes is carried by electrons near the Fermi level μ of the metals. The amount of current will depend on how well these electrons penetrate through the molecule from one electrode to the other and this is determined of course by the value of β at the metal Fermi level. For this reason, it is critical to know where the Fermi level of the metal aligns with the molecular orbitals (or the bands from which they are derived). If the Fermi level is near a band edge (HOMO or LUMO), the current will be higher than if it is in the middle of the gap, where β is generally larger.

If the Fermi level μ cuts through one of the conventional (real k) bands, then the conductance will be about equal to the conductance quantum $G_0 = 77$ μS. This follows from the steps leading to Eq. 4 above if k is understood as a Bloch vector. (Note also that if the Fermi level cuts through multiple bands, then the conductance will equal the number of bands crossed times the quantum of conductance). If the Fermi level is in the band (HOMO-LUMO) gap, the conductance G for the molecule, length L, may be expected to be given approximately by G_0 reduced by the factor $e^{-\beta(\mu)L}$:

$$G \approx G_0 e^{-\beta(\mu)L}. \tag{16}$$

The form of this equation has been used, for example, by Magoga and Joachim [2] to summarize the conductance characteristics of a wide variety of periodic molecules.

The Fermi level alignment problem arose in the study of Schottky barriers (metal-semiconductor contacts). For this case, Tersoff proposed [19] that the Fermi level at the metal-semiconductor interface will lie near the branch point of the complex bandstructure of the semiconductor. Physically, this estimate is based on the idea that the character of the gap states (tunneling states) changes from valence to conduction-like as we pass through the gap from the valence to the conduction bands; the branch point is the point of crossover from mostly valence-like to mostly conduction-like [20].

The branch point can simply be identified as the

peak of the fundamental β semi-elliptical curve in the bandgap. The Tersoff branch point estimate has been found to be quite successful for metal-semiconductor contacts and can be taken as a first approximation for the metal-molecule case. From Eq. 16 we then have for the estimated conductance of a molecule,

$$G = G_0 e^{-\beta_{bp}L} \qquad (17)$$

where β_{bp} is the value of β at the branch point of the complex bandstructure of the molecule.

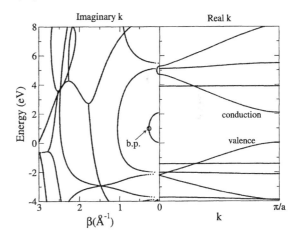

Fig. 6. The "Tour wire", 1,4-bis-phenylethynyl-benzene.

As an example, we estimate the conductance of the "Tour wire", 1,4-bis-phenylethynyl-benzene Fig. 6. In the next section, we will discuss the complex bandstructure of the phenylene-ethynylene chain (the infinite extension of the Tour wire). The branch point β value for the phenylene-ethynylene chain (assuming all phenyl rings are in the same plane) is calculated to be $\beta_{bp} = 0.27$ Å^{-1} and, taking $L \approx 16.3$ Å for 1,4-bis-phenylethynyl-benzene we get

$$G \approx G_0 e^{-\beta_{bp}L} = 0.012\, G_0 \text{ (estimated)}. \qquad (18)$$

Taylor et al. [12] find basically this same result from a fully self-consistent density-functional theory calculation using a realistic model of the entire metal-molecule-metal system. Of course, Eq. 17 is intended just as a useful estimate and not as a substitute for more rigorous calculations. Its advantage is that it depends only on the electronic structure of the molecule (complex bandstructure) and does not need any information about the electrodes and the molecule-electrode coupling.

We mention that we can approximate the on-off conductance ratio (a measure of the strength of the transistor action) of a hypothetical molecular transistor by assuming that the "on" conductance is, ideally, on the order of the conductance quantum while the "off" conductance will just be the value given by Eq. 17. If the Tour wire could somehow be used as a resonant tunneling transistor, we then expect an on-off conductance ratio

of, at best, $e^{\beta_{bp}L} = 75$. Potential mechanisms for transistor action in molecules is another issue and is discussed, for example, in Ref. [14].

3.3. Phenylene-ethynylene molecules

In this section we look at phenylene-ethynylene based molecules using complex bandstructures. Molecules derived from the three ring plus two triple-bond segment, 1,4-bis-phenylethynyl-benzene, of the phenylene-ethynylene chain have exhibited interesting electron transport properties such as negative differential resistance (NDR)–a region in the I-V characteristic where the current decreases for increasing voltage–and the ability to switch between high and low conductance states [11].

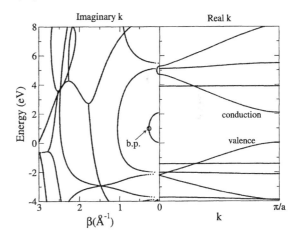

Fig. 7. The complex bandstructure of the phenylene-ethynylene chain (see Fig. 1). The valence band edge defines the zero of energy. The left panel depicts the tunneling states. The small semi-elliptical curve in the energy gap region describes the states that are most penetrating (smallest β). The value of β at the branch point is about 0.27 per Å. The other (higher) β states will have a negligibly small contribution to tunneling current.

The complex bandstructure of the phenylene-ethynylene molecule is shown in Fig. 7. The valence band edge defines the zero of energy and the band gap is about 2 eV. The tunneling bands (left panel) look complicated but the most important band is the small semi-elliptical curve in the bandgap energy region. The states further out to the left (higher β) are much more rapidly decaying and their contribution to current will be negligible.

The value of β at the branch point (labeled "b.p.") is about 0.27 Å^{-1}. This is in rough agreement with an experimentally measured value of $\beta \approx 0.36$ Å^{-1} reported in Ref. [21]. For comparison, alkanes (all single "saturated" bonds) are found to have $\beta_{bp} \approx 0.8$ Å^{-1}. A molecule that is a segment of the phenyl-ethynyl chain will then be expected to have a conductance that is $e^{(0.8-0.27)L}$ times that of an alkane of the same length L. For the Tour wire ($L \approx 16$ Å), this is nearly 5000.

An interesting aspect to consider for phenyl-ethynyl molecules is the effect of rotating a phenyl ring about the molecule axis. Some researchers are interested in the effect of such conformational changes on conductance properties since, conceivably, a molecule whose conductance is changed significantly by a (controllable) change in its geometry could be used, for example, as a switch.

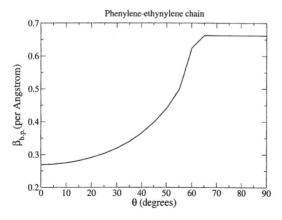

Fig. 8. The effect on β_{bp} of rotating, by the angle θ, every other ring in the phenylene-ethynylene chain.

The rotation of a phenyl ring will destroy the alignment of p_π orbitals with those on neighboring rings and intuitively this should cause a decrease in the conductance. To make this quantitative, we calculate a complex bandstructure for the phenylene-ethynylene chain in which every other phenyl ring is rotated by the angle θ and do this for a range of angles. In this way, we obtain the β_{bp} as a function of θ curve shown in Fig. 8. β_{bp} is seen to increase approximately sinusoidally but then plateau at around 65°. Using $\beta_{bp}(0°) = 0.27$

and $\beta_{bp}(\theta > 65°) = 0.66$ Å^{-1}, we estimate that for the Tour wire, rotation of the middle ring could potentially reduce the conductance by a factor of $e^{(0.66-0.27)\times16} \approx 500$.

A similar calculation can be carried out for the phenylene chain (see Fig. 1). The form of the β vs. θ curve for that structure is found to be similar to that for the phenylene-ethynylene chain but the branch point β values vary from about 0.28 ($\theta = 0°$) to 0.95 ($\theta = 90°$). We see that ring rotation in the phenylene chain has an even larger effect on the conductance. However, it should be noted that steric repulsion between neighboring phenyl rings of the phenylene chain is expected to cause the rings to have a natural rotation around 45°, and $\beta_{bp}(45°)$ is found to be about 0.47 Å^{-1}. For comparison, experimental measurements of current for phenylene oligomers find β values around 0.4 to 0.6 per Å[6].

3.4. Tunneling transport times

An interesting question in tunneling generally is how long each electron spends in the barrier region (see, for example, the review Ref. [22]). Though probably more of theoretical interest, an estimate of the tunneling time may help give some indication of the importance of inelastic electron-molecule interactions (for example, molecular vibrations induced by tunneling electrons).

One of the most interesting approaches to this problem uses the concept of a spin clock [23–27]. The basic idea is to introduce an imaginary perturbing magnetic field, in the barrier region only, that causes the spin of the tunneling electron to rotate. Given the spin orientation of the incident electron, an expression for the probability of a spin flip can easily be worked out. An estimate of the tunneling time is then given by relating this rotation in space (over the length of the barrier) to the temporal spin rotation rate expected for an electron, with the given initial spin orientation, in a magnetic field. An analysis using these concepts has been carried out and it is found that at the energy of the branch point the transit time reaches a minimum [28]. Thus electrons at this energy are expected to cause minimal disturbance of the molecule.

A controversial point arises when one defines a transit speed. In a crystal, the group velocity of an electron wave-packet is of course given by $v_g = \frac{1}{\hbar}\frac{dE(k)}{dk}$ so the time elapsed in passing through a region of length L is L/v_g. It is not clear that the expression for v_g can be applied to gap regions where k is complex, but the spin clock can be used to make this extension seem reasonable [28]. This has the interesting implication that electrons with energies near the branch point, where $dE/dk \to \infty$, will have infinite group velocities (see also Ref. [29]). This behavior may be difficult to observe for electrons, but superluminal tunneling speeds have been reported in experiments with light tunneling through the evanescent regions of photonic crystals (e.g., Ref. [30]). The usual interpretation of this is that the wave packet has been reshaped producing an apparent superluminal speed, so that there is no violation of relativity or causality. A recent contribution interprets the apparent superluminal effect as coming from the input wave modulating interference between forward and backward scattered waves [31].

3.5. Method for obtaining a complex bandstructure

To generate a conventional (real k) bandstructure, one solves a standard energy eigenvalue equation for a range of chosen k points. For a complex bandstructure, it is impractical and inefficient to do an exhaustive search over the complex k plane for the solutions of interest and, instead, the problem is inverted: determine the complex k points (output) associated with a range of energies E (input).

In this section we describe a method for accomplishing this. We assume a local orbital picture: wave-functions are expanded in terms of a set of local orbitals $\{\phi_\mu\}$ and the Hamiltonian H and overlap S matrices are defined by

$$H_{\mu\nu} \equiv \int \phi_\mu^* \hat{H} \phi_\nu \text{ and } S_{\mu\nu} \equiv \int \phi_\mu^* \phi_\nu \quad (19)$$

where $\hat{H}$ is the Hamiltonian operator.

We consider a general infinite sequence of identical cells with nearest neighbor cell distance a and denote the Hamiltonian and overlap interactions

between mth neighbor cells by $H^{(m)}$ and $S^{(m)}$, i.e., $H^{(0)}$ contains the Hamiltonain matrix elements between orbitals within a cell, $H^{(1)}$ contains the interactions between orbitals in one cell and those of the right nearest neighbor cell ($H^{(-1)} = H^{(1)\dagger}$ has those for the left nearest neighbor), etc. Assuming there are no interactions beyond Nth neighbor cells, the standard energy eigenvalue equation ($HC = ESC$) can be written

$$\sum_{m=-N}^{N} H^{(m)} \mathbf{C}^{(m)} = \sum_{m=-N}^{N} ES^{(m)} \mathbf{C}^{(m)} \quad (20)$$

where $\mathbf{C}^{(m)}$ contains the coefficients for orbitals in the mth cell and, assuming the Bloch form for the eigenstates,

$$\mathbf{C}^{(m+1)} = \lambda \mathbf{C}^{(m)} \quad (21)$$

where

$$\lambda \equiv e^{i(k_0+i\beta/2)a}. \quad (22)$$

We rewrite Eq. (20) as

$$\sum_{m=-N}^{N} Z^{(m)} \mathbf{C}^{(m)} = 0 \quad (23)$$

where $Z^{(m)} = H^{(m)} - ES^{(m)}$. This is transformed into an eigenvalue equation for λ by using $\mathbf{C}^{(N)} = \lambda \mathbf{C}^{(N-1)}$ to eliminate $\mathbf{C}^{(N)}$:

$$-\sum_{m=-N}^{N-1} Z^{(m)} \mathbf{C}^{(m)} = \lambda Z^{(N)} \mathbf{C}^{(N-1)}. \quad (24)$$

Together with Eq. 21 for $-N \leq m \leq N-2$ we then have the generalized eigenvalue problem

$$A\mathbf{c} = \lambda B\mathbf{c} \quad (25)$$

where

$$A \equiv \begin{pmatrix} -Z^{(N-1)} & -Z^{(N-2)} & \dots & -Z^{(-N+1)} & -Z^{(-N)} \\ 1 & 0 & \dots & 0 & 0 \\ 0 & 1 & \dots & 0 & 0 \\ \vdots & \vdots & & \vdots & \vdots \\ 0 & 0 & \dots & 1 & 0 \end{pmatrix}, \quad (26)$$

$$B \equiv \begin{pmatrix} Z^{(N-1)} & 0 & \dots & 0 & 0 \\ 0 & 1 & \dots & 0 & 0 \\ \vdots & \vdots & & \vdots & \vdots \\ 0 & 0 & \dots & 1 & 0 \\ 0 & 0 & \dots & 0 & 1 \end{pmatrix}, \quad (27)$$

and

$$\mathbf{c} \equiv \begin{pmatrix} \vec{C}^{N-1} \\ \vec{C}^{N-2} \\ \vdots \\ \vec{C}^{-N+1} \\ \vec{C}^{-N} \end{pmatrix}. \qquad (28)$$

For example, for $N = 1$, we have

$$\begin{pmatrix} -Z^{(0)} & -Z^{(-1)} \\ 1 & 0 \end{pmatrix} \begin{pmatrix} \vec{C}^{(0)} \\ \vec{C}^{(-1)} \end{pmatrix}$$

$$= \lambda \begin{pmatrix} Z^{(1)} & 0 \\ 0 & 1 \end{pmatrix} \begin{pmatrix} \vec{C}^{(0)} \\ \vec{C}^{(-1)} \end{pmatrix}. \qquad (29)$$

This form was first obtained by Boykin [32,33]. There are subroutines available for solving the generalized eigenproblem Eq. 25. Equation 25 will generate a spectrum of λ values for each input energy E and we get corresponding k_0 and β values from Eq. 22. A computer program for computing the complex bandstructure, given local orbital Hamiltonian and overlap matrices, is available for download [34].

4. Landauer theory

Until now, we have focused entirely on the electronic structure of the molecule. Details about the electrodes and the electrode-molecule connection have barely entered the discussion. To obtain more quantitative results and to deal with more general structures requires considering the entire metal-molecule-metal system. In this section we introduce the widely used Landauer approach to this problem (see the book by Datta [15] for a more complete development).

In the following, we assume the Hamiltonian of the metal-molecule-metal system has the form

$$H = H_L + H_R + H_M + V \qquad (30)$$

where $H_{L(R)}$ are the Hamiltonian matrices of the isolated left and right electrodes, H_M is the Hamiltonian of the molecule, and $V \equiv H_{LM} + H_{RM}$ is the coupling between the electrodes and the molecule. It is assumed that there is no direct coupling between the left and right electrodes ($H_{LR} = 0$).

In the Landauer view, the current is produced by electrons occupying a distribution of scattering states (a scattering state in one dimension is just, collectively, an incident wave, its reflected and transmitted parts, and its wavefunction in the barrier region). The scattering states originating in the left (L) electrode are filled up to the Fermi level μ_L and those of the right (R) electrode are filled up to μ_R, with $|\mu_L - \mu_R| = eV$. The current I is determined by summing up the currents contributed by all of these states. This leads to the Landauer or Landauer-Büttiker expression [15]

$$I = \frac{2e}{h} \int T(E)(f(E - \mu_L) - f(E - \mu_R))dE \qquad (31)$$

where $T(E)$ is the transmission function.

In one dimension, $T(E)$ is just the squared modulus of the transmission amplitude familiar from elementary barrier tunneling problems in quantum mechanics. In a moment, we will give a general expression for $T(E)$, but first we look at a simple example.

Consider the model shown in Fig. 9 of a sequence of atoms, a wire, with an s-like orbital on each atom. The orbital onsite energies are ϵ_0 and there is nearest neighbor hopping $-\tau_0$. Replace one atom in the middle of the sequence with a "barrier" atom which is identical to the others except that it has an extra onsite energy b (barrier) in addition to ϵ_0.

The wire is assumed to be connected between electrodes that have potentials separated by the amount eV where V is the applied bias. In the wire atoms leading up to the barrier atom, the "leads", the electron energy eigenstates are just planewaves,

$$|k\rangle = \sum_k e^{ikna}|n\rangle \qquad (32)$$

with energies

$$E(k) = \epsilon_0 - 2\tau_0 \cos ka. \qquad (33)$$

Here a is the distance between nearest neighbors and $|n\rangle$ is the orbital on the nth atom.

The transmission and reflection probabilities for an electron with energy E incident on the barrier is determined simply by constructing solutions of the form

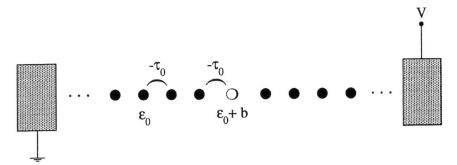

Fig. 9. Model molecular wire: a sequence of atoms (onsite energies ϵ_0) connected to a single "barrier" atom (onsite energy $\epsilon_0 + b$). The interaction matrix element between all sites is $-\tau_0$. The wire is connected between bulk electrodes.

$$|\psi_l\rangle = |k\rangle + r|-k\rangle,$$
$$|\psi_{\text{barrier}}\rangle = c|b\rangle,$$
$$|\psi_r\rangle = t|k\rangle. \tag{34}$$

$|\psi_l\rangle$ is the sum of incident and reflected waves in the left lead, $|\psi_r\rangle$ is the transmitted wave in the right lead, and $|\psi_{\text{barrier}}\rangle$, the wavefunction on the barrier, is just a constant, c, times an orbital on the barrier $|b\rangle$. The transmission amplitude t is easily found from the three equations determined by projecting $H|\psi\rangle = E|\psi\rangle$ onto the barrier orbital $|b\rangle$, the rightmost orbital of the left lead (just to the left of $|b\rangle$), and the leftmost orbital of the right lead. This yields the expressions

$$e^{ika} + e^{-ika}r = c$$
$$e^{-ika}t = c$$
$$-\tau_0(1 + r + t) = (E(k) - \epsilon_0 - b)c. \tag{35}$$

Solving for t we find

$$t = \frac{2ie^{ika}\tau_0 \sin ka}{2\tau_0 + (E(k) - \epsilon_0 - b)e^{-ika}} \tag{36}$$

so the transmission function ($|t|^2$) is

$$T(E) = \frac{(2\tau_0)^2 - (E - \epsilon_0)^2}{b^2 + (2\tau_0)^2 - (E - \epsilon_0)^2}. \tag{37}$$

Assuming the voltage is low, the current is given by $I = \frac{2e^2}{\hbar}T(\mu)V = G_0 T(\mu)V$ where μ is the equilibrium Fermi level.

Most often, it is more convenient to work with a more general expression for the transmission function which we now derive (see Ref. [35] for a similar development). The transition rate associated

with an electron scattering from a state $|l\rangle$ of the left electrode (satisfying $H_L|l\rangle = E_l|l\rangle$) to a state $|r\rangle$ of the right electrode is given by a well-known expression

$$\Gamma_{l \to r} = \frac{2\pi}{\hbar}|T_{rl}|^2\delta(E_l - E_r) \tag{38}$$

where $T = V + VGV$ is the T-matrix of scattering theory [36] (different from the transmission function) and $G = (z - H)^{-1}$ ($z \equiv E + i\eta$) is the full (retarded) Green's function. The limit $\eta \to 0+$ is always implied. The current carried from the left electrode to the right electrode is given by

$$I = 2e\sum_{lr} f(E_l - \mu_L)\Gamma_{l \to r} \tag{39}$$

where the factor of 2 is for spin and $f(E) = (e^{E/kT} + 1)^{-1}$ is the Fermi function. Equation 39 is simplified as follows

$$I = \frac{4\pi e}{\hbar}\sum_{lr}\langle l|T|r\rangle\langle r|T^\dagger|l\rangle f(E_l - \mu_L)\delta(E_l - E_r)$$

$$= \frac{4\pi e}{\hbar}\sum_{lr}\int dE \,\langle l|T|r\rangle\langle r|T^\dagger|l\rangle f(E - \mu_L)\delta(E - E_l)\delta(E - E_r)$$

$$= \frac{4\pi e}{\hbar}\sum_{lr}\int dE \,\langle l|\rho_L(E)T|r\rangle\langle r|\rho_R(E)T^\dagger|l\rangle f(E - \mu_L) \tag{40}$$

where $\rho_{L(R)}(E) \equiv \delta(E - H_{L(R)})$ is the density of states operator for the left and right electrodes.

Summing over r, we can replace $|r\rangle\langle r|$ by the identity to get

$$I = \frac{4\pi e}{\hbar} \int dE \; \text{tr}(\rho_L(E)T\rho_R(E)T^\dagger)f(E-\mu_L) \quad (41)$$

where tr is the trace. Equation 41 is the current carried from the left electrode to the right electrode and the net current is simply given by subtracting off the right to left current:

$$I = \frac{4\pi e}{\hbar} \int dE \; \text{tr}(\rho_L(E)T\rho_R(E)T^\dagger)(f(E-\mu_L)-f(E-\mu_R)). \quad (42)$$

Defining the transmission function as

$$T(E) = 4\pi^2 \text{tr}(\rho_L T \rho_R T^\dagger). \quad (43)$$

we get Eq. 31.

Equation 43 for the transmission function is usually written in a different form (see, for example, Ref. [15]). The density of states operator $\rho(E) = \delta(E - H)$ is related to the Green's function $G = (z - H)^{-1}$ through

$$\rho(E) = \delta(E - H) = \frac{i}{2\pi}(G(E) - G^\dagger(E)). \quad (44)$$

Using this and the cyclic property of the trace we can write

$$T(E) = \text{tr}(\Gamma_L G_M \Gamma_R G_M^\dagger) \quad (45)$$

where the quantities $\Gamma_{L(R)}$ are

$$\Gamma_{L(R)} = i(\Sigma_{L(R)} - \Sigma_{L(R)}^\dagger), \quad (46)$$

the self energies are

$$\Sigma_{L(R)} \equiv V G_{L(R)}^0 V, \quad (47)$$

and G^0 is the Green's function for the *isolated* electrode,

$$G_{L(R)}^0 = (z - H_{L(R)})^{-1}. \quad (48)$$

The superscript 0 on the electrode Green's function is to emphasize that these are for the *isolated* (disconnected from the molecule) electrodes. G_M contains the Green's function matrix elements (of the coupled system) between orbitals of the molecule. This "dressed" Green's function of the molecule is determined by taking the inverse of

$$\begin{pmatrix} z - H_L & H_{LM} & 0 \\ H_{LM}^\dagger & z - H_M & H_{RM} \\ 0 & H_{RM}^\dagger & z - H_R \end{pmatrix}. \quad (49)$$

G_M is the submatrix of the inverse that contains matrix elements between molecule basis orbitals (i.e., the block matrix in the center of the inverse) and is easily found by inverting Eq. 49 and is

$$G_M = (z - H_M - \Sigma_L - \Sigma_R)^{-1}. \quad (50)$$

Generally, when a system s described by a Hamiltonian H_s is coupled to an environment with Hamiltonian H_e through a coupling V, the dressed Green's function G_s for system s is given by

$$G_s = (z - H_s - V G_e V)^{-1}. \quad (51)$$

For illustration, we consider again the problem of a wire, with a single barrier atom, connected between electrodes (Fig. 9). To be clear, H_L and H_R in this case are the Hamiltonians of the leads connected to the electrodes but disconnected from the barrier atom, H_M is just the number $\epsilon_0 + b$, and V is the hopping interaction between the barrier atom and the neighboring lead atoms.

We require first the Green's function for the isolated leads (disconnected from the barrier atom). Ignoring the possibility of electrons reflecting from the lead-electrode interface back toward the barrier atom (as was implicitly done above), the lead Green's function will be that of a semi-infinite sequence of atoms. We need only the Green's function matrix element g^0 for the endpoint or "surface" atom. This can be obtained from recognizing that extending the surface by one atom (coupling the surface atom to another atom which becomes the new surface) gives the same system again (a semi-infinite sequence of atoms). From Eq. 51 we then have the relation

$$g^0 = (z - \epsilon_0 - \tau_0 g^0 \tau_0)^{-1}. \quad (52)$$

This gives two roots, one of which is discarded on the grounds that its imaginary part is positive (the imaginary part is proportional to the negative of the density of states and so should be negative). This gives, for the endpoints of the left and right leads,

$$g_{L(R)}^0(E) = g^0(E) =$$
$$\frac{E - \epsilon_0 - i\sqrt{(2\tau_0)^2 - (E - \epsilon_0)^2}}{2\tau_0^2}. \quad (53)$$

The superscript 0 indicates that this is a Green's function matrix element for the leads disconnected from the barrier atom. For the barrier, we require the full Green's function which, in this example, is just a number. Using equation 50 with $H_M = \epsilon_0 + b$ we find

$$G_M(E) = (E - (\epsilon_0 + b) - \tau_0 g_L^0 \tau_0 - \tau_0 g_R^0 \tau_0)^{-1}$$
$$= \left(-b + i\sqrt{(2\tau_0)^2 - (E - \epsilon_0)^2} \right)^{-1}. \quad (54)$$

Finally,

$$\Gamma_{L,R}(E) = \Gamma \equiv i(\tau_0 g^0(E)\tau_0 - \tau_0 g^0(E)^* \tau_0)$$
$$= \sqrt{(2\tau_0)^2 - (E - \epsilon_0)^2}, \quad (55)$$

so that the transmission function ($\Gamma G_M \Gamma G_M^*$) is

$$T(E) = \frac{(2\tau_0)^2 - (E - \epsilon_0)^2}{b^2 + (2\tau_0)^2 - (E - \epsilon_0)^2}. \quad (56)$$

This is just the expression Eq. (37) obtained previously.

The evaluation of the transmission function for more complicated (and realistic) systems generally follows along the same lines as the example only there are more basis orbitals and interactions so that the Green's functions, molecule-electrode interactions, etc. are represented by matrices. Equation 52 for the Green's function of semi-infinite electrodes, for example, is then a matrix equation that must be solved self-consistently. The most important obstacles to implementation are including bias-induced changes in charge distribution and geometry. Both of these topics have been the focus of much recent theoretical work (see, for example, Refs. [7,5,4,37–40]) A computer program that calculates I-V curves following the basic procedure just outlined is available for download [34].

5. Conclusion

We have described some of the basic features of single molecule coherent transport. For periodic molecules, the physics is a natural generalization of square barrier transport: molecule orbitals are derived from the bands of an infinite extension of the finite molecule, and the simple dispersion relation k or $\kappa \propto \sqrt{E - V_0}$ is replaced by a complex bandstructure. We have also introduced the Landauer approach, which has been the foundation for much of the theoretical work on single molecule transport.

As mentioned in the introduction, coherent transport theory alone is often not adequate for describing what is observed in experiments on molecular scale transport. For example, the experiments described in Ref. [41] measured current through a complex of organic molecules surrounding a single cobalt atom. In this case, the current is dominated by sequential transport: the electrons are moving one at a time through the molecular complex or, in other words, the complex is being repeatedly reduced by one electrode and oxidized by the other. Sequential transport, which includes Coulomb blockade (see, e.g., Refs. [42,43]) and essentially most of redox chemistry [44], is fundamentally different from that described by the Landauer formalism and is an important direction for study in molecular electronics. A comparison between sequential and coherent transport models for a metal-molecule-metal system is given in Ref. [45]. Coherent transport is just one component of a more complete, and still emerging, theory of single molecule electron transport.

6. Acknowledgment

We thank the National Science Foundation (NIRT ECS-0103175) and (DMR-99-86706) for support. This work was stimulated by the molecular electronics group at Arizona State University and Motorola. We thank Ganesh Ramachandran, Jin He, Stuart Lindsay, Alex Primak, Xristo Zarate, Devins Gust, Anna Moore, Tom Moore, Salah Boussaad, Huixin He, Nongjian Tao, Larry Nagahara, Alex Demkov, Gil Speyer, and Xiodong Zhang for discussions.

References

[1] M. P. Samanta, W. Tian, S. Datta, J. I. Henderson, and C. P. Kubiak, Phys. Rev. B **53**, R7626 (1996).

236

[2] M. Magoga and C. Joachim, Phys. Rev. B **56**, 4722 (1997).

[3] E. G. Emberly and G. Kirczenow, Phys. Rev. B **58**, 10911 (1998).

[4] M. Di Ventra, S.T. Pantelides, and N.D. Lang, Phys. Rev. Lett. **84**, 979 (2000).

[5] J. Taylor, H. Guo, and J. Wang, Phys. Rev. B **63**, 245407 (2001).

[6] M. A. Rampi and G. M. Whitesides, Chem. Phys. **281**, 373 (2002).

[7] P. S. Damle, A. W. Ghosh, and S. Datta, Phys. Rev. B **64**, 201403 (2001).

[8] M. A. Reed, C.Zhou, C.J. Muller, T.P. Burgin, and J.M. Tour, Science **278**, 252 (1997).

[9] D. J. Wold and C. D. Frisbie, J. Am. Chem. Soc. **123**, 5549 (2001).

[10] X. D. Cui, A. Primak, X. Zarate, J. Tomfohr, O.F. Sankey, A.L. Moore, T.A. Moore, D. Gust, G. Harris, and S.M. Lindsay, Science **294**, 571 (2001).

[11] J. Chen, M. A. Reed, A. M. Rawlett, and J. M. Tour, Science **286**, 1550 (1999).

[12] J. Taylor, M. Brandbyge, and K. Stokbro, Phys. Rev. Lett. **89**, 138301-1 (2002).

[13] W. Tian, S. Datta, S. Hong, R. Reifenberger, J. I. Henderson, and C. P. Kubiak, J. Chem. Phys. **109**, 2874 (1998).

[14] Di Ventra, M., S. T. Pantelides, and N. D. Lang, Appl. Phys. Lett. **76**, 3448 (2000).

[15] S. Datta, *Electronic Transport in Mesoscopic Systems*. (Cambridge University Press, 1995) 377 pp.

[16] Y. Imry and R. Landauer, Rev. Mod. Phys. **71** s306 (1999).

[17] J. K. Tomfohr and O. F. Sankey, Phys. Rev. B **65**, 245105 (2002).

[18] C. Joachim and M. Magoga, Chem. Phys. **281**, 347 (2002).

[19] J. Tersoff, Phys. Rev. Lett. **52**, 465 (1984).

[20] J. J. Rehr and W. Kohn, Phys. Rev. B **9**, 1981 (1974).

[21] S. Creager, C. J. Yu, C. Bamdad, S. O'Connor, T. Maclean, E. Lam, Y. Chong, G. T. Olsen, J. Luo, M. Gozin, and J. F. Kayyem, J. Am. Chem. Soc. **121**, 1059 (1999).

[22] E. H. Hauge, J. A. Støvneng, Rev. Mod. Phys. **61**, 917 (1989).

[23] A. I. Baz', Sov. J. Nucl. Phys. **4**, 182 (1967).

[24] V. F. Rybachenko, Sov. J. Nucl. Phys. **5**, 635 (1967).

[25] M. Büttiker, Phys. Rev. B **27**, 6178 (1983).

[26] A. Nitzan, J. Jortner, J. Wilkie, A.L. Burin, and M. A. Ratner, J. Phys. Chem. B **104**, 5661 (2000).

[27] M. Büttiker and R. Landauer, Phys. Rev. Lett. **49**, 1739 (1982).

[28] J. K. Tomfohr, O.F. Sankey, and S. Wang, Phys. Rev. B **66**, 235105-1 (2002).

[29] R. Landauer and Th. Martin, Rev. Mod. Phys. **66**, 217 (1994).

[30] A. Haché and Louis Poirier, Appl. Phys. Lett. **80**, 518 (2002).

[31] H.G. Winful, Phys. Rev. Lett. **90**, 23901 (2003).

[32] T. B. Boykin, Phys. Rev. B **54**, 7670 (1996).

[33] T. B. Boykin, Phys Rev. B **54**, 8107 (1996).

[34] http://physics.asu.edu/sankey/moltronics

[35] V. Mujica, M. Kemp, and M. A. Ratner, J. Chem. Phys. **101**, 6849 (1994).

[36] B. A. Lippmann and J. Schwinger, Phys. Rev. **79**, 469 (1950).

[37] V. A. Sablikov, S. V. Polyakov, and M. Büttiker, Phys. Rev. B **61**, 13763 (2000).

[38] M. Brandbyge, J. Mozos, P. Ordejón, J. Taylor, and K. Stokbro, Phys. Rev. B **65** 165401 (2002).

[39] M. Paulsson and S. Stafström, Phys. Rev. B **64**, 035416 (2001).

[40] M. Di Ventra, S. T. Pantelides, and N. D. Lang, Phys. Rev. Lett. **88**, 046801 (2002).

[41] J. Park, A. N. Pasupathy, J. I. Goldsmith, C. Chang, Y. Yaish, J. R. Petta, M. Rinkoski, J. P. Sethna, H. D. Abruña, P. L. McEuen, and D. C. Ralph, Nature **417**, 722 (2002).

[42] D. V. Averin and K. K. Likharev, *Mesoscopic Phenomena in Solids*, edited by B. L. Altshuler, P. A. Lee, and R. A. Webb (Elsevier, 1991) p. 169.

[43] A. E. Hanna and M. Tinkham, Phys. Rev. B **44**, 5919 (1991).

[44] Bard, A. J., and L. R. Faulkner, *Electrochemical Methods: Fundamentals and Applications.* (John Wiley and Sons 2001), 856 pp.

[45] C. Kergueris, J. P. Bourgoin, S. Palacin, D. Esteve, C. Urbina, M. Magoga, and C. Joachim, Phys. Rev. B **59**, 12505 (1999).

Nano and Giga Challenges in Microelectronics
Greer at al (Editors)
© *2003 Elsevier Science B.V. All rights reserved*

Practical Quantum Computing

P.M. Lenahan [1]

The Pennsylvania State University, University Park, PA 16802

Abstract

Recent publications suggest that quantum computing may have enormous advantages over the "classical" computation of the present day. For example, Shor has demonstrated that a quantum computer could, at least in principle, provide a remarkable advantage in the factoring of very large numbers. This paper reviews the fundamental requirements for a quantum computation system as well as some of the most important ideas involved in potential applications of such systems. Pioneering "practical" applications of quantum computation, primarily involving magnetic resonance will also be reviewed; possible advances in resonance based quantum computation will also be discussed.

Key words: spin, quantum computing

1. Introduction

Interest in the possibilities of quantum computing has been growing since 1982, when Richard Feynmann[1] proposed that computers based upon quantum mechanical principles might be quite useful in simulating quantum mechanical systems. In 1985, David Deutsch[2] showed that a quantum computer could have capabilities greater than that of classical computers in the solution of certain problems.

Interest in the potential of quantum computation increased dramatically in 1994 when Peter Shor[3,4] demonstrated that, at least in principle, a quantum computer could exhibit an enormous advantage in the factoring of very large numbers. This factoring problem is of great significance to cryptography. Several other potential important applications of quantum computation have also been identified recently, the list sorting algorithm developed by Grover[5] is, next to the factoring algorithm of Shor, likely the most significant application.

Although the potential applications of quantum computation are great, so are the challenges in the path towards practical quantum computation. As yet, no "practical" quantum computer has been built. However, many ingenious ideas appear in the recent literature and some very promising, though very small scale, practical demonstrations have recently taken place. Arguably, the most promising results have come through magnetic resonance.

This paper reviews the basic principles and the most important potential applications of quantum computation so far identified. With regard to potential practical applications, the focus is primarily upon magnetic resonance. The paper is written from the viewpoint of an experimentalist in semiconductor device physics and materials physics.

---

[1] pmlesm@engr.psu.edu

2. What is Quantum Computation?

Quantum computation is computation using quantum mechanical systems[6]. The bit of classical computation becomes the quantum bit or qubit of quantum mechanical computation. The qubit is a physical object. A classical bit is either a 0 or a 1. Two possible states for a qubit are |0> or |1>; however, these are possible states of a quantum mechanical system which might also exist in a superposition of those states

$$|\Psi\rangle = a\,|0\rangle + b\,|1\rangle\,, \tag{1}$$

where a and b are, most generally, complex numbers.

In classical computation, we can examine a bit to determine whether it is in the 1 or 0 state. We can't do quite the same thing with the qubit. We measure the qubit and obtain |0> with a probability $|a|^2$ and obtain |1> with a probability of $|b|^2$, where $|a|^2 + |b|^2 = 1$. Note that this measurement *changes* the state of the qubit, no longer a superposition but the specific state consistent with the measurement. Yet, in some sense, the information conveyed in a and b is present in the qubit prior to measurement. The exploitation of this *quantum information* is key to the development of quantum computation.

Suppose we have a system of three classical bits. The eight possible states, would be 000, 001, 010, 011, 100, 101, 110, and 111. Three qubit systems would also have eight possible *basis states* which we could write as |000>, |001>, |010>, |011>, |100>, |101>, |110>, and |111>. However, the three qubits could also exist in a superposition of these eight states.

$$
\begin{aligned}
|\Psi\rangle = \quad & a_{000}\,|000\rangle + a_{001}\,|001\rangle \\
+ \; & a_{010}\,|010\rangle + a_{011}\,|011\rangle \\
+ \; & a_{100}\,|100\rangle + a_{101}\,|101\rangle \\
+ \; & a_{110}\,|110\rangle + a_{111}\,|111\rangle
\end{aligned} \tag{2}
$$

The result of a measurement would occur with a probability $|a_{\alpha\beta\gamma}|^2$ for the state $|\alpha\beta\gamma>$. The sum of the probabilities of all eight outcomes, of course, must be one.

In general, a system of n qubits would have a quantum states specified by 2^n complex amplitudes. For fairly modest numbers of qubits, the number of amplitudes is enormous. For example, if n = 100, the number of amplitudes exceeds 10^{30}! In effect, Mother Nature "computes" with this amount of information for a hundred particles. The challenge for quantum computation is exploiting this computational power.

3. The Quantum Fourier Transform and Shor's Algorithm

The most powerful potential application of quantum computation yet discovered is a capability for efficient factoring of very large numbers. This process, discovered by Peter Shor, is now called Shor's algorithm[3,4]. The algorithm utilizes the quantum Fourier transform and modular arithmetic.

The quantum Fourier transform is almost identical to the conventional discrete ("classical") Fourier transform. The input for the discrete Fourier transform is a series of N complex numbers: x_0, x_1, x_2,...x_N. It produces an output of a second series of N complex numbers (y_0, y_1, y_2,...y_N) according to the expression

$$y_K = \frac{1}{\sqrt{N}} \sum_{j=0}^{N-1} x_j \exp\left[\frac{2\pi ijk}{N}\right] \tag{3}$$

It's easy to see that if the input series x_0, x_1, x_2,...x_N, repeats with a period P, then the discrete Fourier transform output will repeat with a period N/P. Table 1 illustrates this process for some particularly simple series of eight numbers.

Table 1
The outputs have been multiplied by constants to simplify the presentation.

Period (P)	Input	Output	N/P	
8	10000000	11111111	1	(a)
4	10001000	10101010	2	(b)
2	10101010	10001000	4	(c)
1	11111111	10000000	8	(d)

The transformation carried out by the quantum Fourier transform is the same as that of the discrete Fourier transform. However, the quantum transform operates on the complex numbers which represent the phase and the amplitude contributions of the various contributions of a superposition of states. Thus, the series of eight numbers illustrated in Table 1 might represent a superposition of various states of the three qubit system of equation (2). (In this case however, for simplicity of presentation, the states have not been normalized.) We might imagine the three qubit systems to be three spin 1/2 nuclei in a magnetic field. We simplify the notation further by indicating $|000>$ as $|0>$, $|001>$ by $|1>$, $|011>$ by $|2>$, etc.

We could envision a superposition of the eight possible states (again neglecting normalizing constants):

$$|0\rangle + |1\rangle + |2\rangle + |3\rangle + |4\rangle + |5\rangle + |6\rangle + |7\rangle \qquad (4)$$

This series corresponds to the input series (d) of Table 1. The quantum Fourier transform would thus transform this state to the state $|0>$. Another example, the input superposition state $|0> +|4>$ would quantum Fourier transform to $|0> +|2> +|3> +|6>$.

In Shor's algorithm, a number M is factored in several steps. A group of quantum bits is partitioned into two registers. Register I has $2\log_2 M$ qubits and register II has $\log_2 M$ qubits. Register I is prepared in an equally weighed superposition of all M possible states. Following our (somewhat oversimplified) notation, this would be a series of M^2 ones; 11111... Register II is prepared in the ground state, all zeros; 00000...

The state of the two register system would then be

(Register I) (Register II)

$$(|0\rangle + |1\rangle + |2\rangle + |3\rangle + |4\rangle + |5\rangle \ldots) \quad |0\rangle \qquad (5)$$

With both registers prepared in this way, the function $f(x) = a^x$ mod M is evaluated, where x represents the values of the first register and the output value is stored in the second register. The function $f(x)$ is simply the remainder after division of a^x by M. The (small) number "a" can be chosen at random but may not have a factor in common with M. It can be shown that this function is periodic and that the periodicity is simply related to factors of M. It can also be shown from number theory that if the period of a^x mod M is r, then the product $(a^{r/2}-1)(a^{r/2}+1)$ is an integer multiple of M, the number we wish to factor. Therefore, we expect that at least one of these two terms, $(a^{r/2}-1)$ or $(a^{r/2}+1)$ must have a (non-trivial) factor in common with M.

As a simple example, suppose M = 15 and a = 2. The results of Table 2 show that the periodicity r of 2^x mod 15 is 4.

Table 2
Results for $f(x) = 2^x \mathrm{mod} 15$ showing that periodicity $r = 4$.

x	0	1	2	3	4	5	6	7	8	9
f(x)	1	2	4	8	1	2	4	8	1	2

Continuing with this example, we evaluate the system described in the superposition state of equation (5) using $f(x) = 2^x$ mod 15:

$$|0\rangle |1\rangle + |1\rangle |2\rangle + |2\rangle |4\rangle + |3\rangle |8\rangle + |4\rangle |1\rangle +$$
$$|5\rangle |2\rangle + |6\rangle |4\rangle + |7\rangle |8\rangle + |8\rangle |1\rangle + |9\rangle |2\rangle$$
$$= (|0\rangle + |4\rangle + |8\rangle + \ldots) |1\rangle + \qquad (6)$$
$$(|1\rangle + |5\rangle + |9\rangle + \ldots) |2\rangle +$$
$$(|2\rangle + |6\rangle + \ldots) |4\rangle +$$
$$(|3\rangle + |7\rangle + \ldots) |8\rangle$$

Since the first register had been prepared in an equal superposition of all $|x>$, the function is evaluated for all values, in parallel.

If we now evaluate the second register, the state of the first register will collapse into one of only four possibilities:

$$(|0\rangle + |4\rangle + |8\rangle + \ldots) \quad \text{or}$$
$$(|1\rangle + |5\rangle + |9\rangle + \ldots) \quad \text{or}$$
$$(|2\rangle + |6\rangle + \ldots) \quad \text{or} \qquad (7)$$
$$(|3\rangle + |7\rangle + \ldots)$$

At this point, a quantum Fourier transform is performed on the first register qubits. Remember, the quantum Fourier transform will have a period of the number of bits divided by the period r. With r known, the factor of M can easily be determined by finding the greatest common divisor (gcd) of $(a^{r/2}+1, M)$ and $(a^{r/2}-1, M)$. This last step can be carried out by a classical computer.

Continuing with the example, a = 2 and M = 15, we would obtain a periodicity (100010001→ 1010101) of <u>two</u> from the quantum Fourier transform and gcd (2+1, 15) and (2-1, 15): <u>three</u>. Of course, three times five is, in fact, fifteen.

4. The Fundamental Requirements of a Quantum Computing System

The fundamental requirements for a quantum computing system are quite challenging. The quantum computing system would be made of qubits which interact in, a quantum mechanical sense, are *entangled*. A great challenge is that the qubits must maintain the quantum mechanical nature of their interactions during the manipulations involved in quantum computation (this implies weak coupling with the rest of the universe) but nevertheless allow rapid manipulation and measurement (this implies strong coupling with at least some aspect of the rest of the universe).

DiVincenzo has summarized five fundamental requirements for a quantum computing system[7].

(i) There must be a scalable physical system with well characterized entangled qubits. A simple example could be a system of coupled (entangled) spin $1/2$ nuclei in a magnetic field. The states of the two spins could be written as a $|00>$ +b $|01>$ +c $|10>$ +d $|11>$. Consider the state

$$\frac{|00\rangle + |11\rangle}{\sqrt{2}}, \qquad (8)$$

often called the Bell state or the EPR (Einstein, Podolsky, Rosen) pair.

The Bell state has a very interesting property. In measuring the first qubit, two results are possible: 0 or 1, each with a probability of $1/2$. If the measurement yields a 0 for the first qubit, the post-measurement state can only be $|00>$. If the measurement yields a 1 for the first qubit, the post-measurement state can only be $|11>$. So, the measurement of the second qubit always yields the same result as the measurement of the first qubit. (John Bell proved that these measurement correlations are stronger than any which could exist in a classical system[8].) This coupling is called *entanglement*. If the system is to be useful, we must be able to scale up this process to many entangled qubits. A three register qubit could have a state $|000>$ + $|001>$ +$|010>$ +$|011>$ +$|100>$ +$|101>$ +$|110>$ +$|111>$ (neglecting normalizing factors) which could simultaneously represent numbers from 0 to 7.

The qubits must also be well characterized: its internal Hamiltonian, its couplings with the rest of the universe (including other qubits) must be known.

(ii) It must be possible to "initialize" the state of the qubits to a simple state such as $|000...>$. This is, in part, a common sense requirement: in any sort of computation registers should be in known values prior to the start of computation. A second, more subtle reason for this is that quantum error correction will almost certainly be an important part of any quantum computing scheme and a supply of $|0>$ state qubits play an important role in quantum error correction[6].

(iii) The qubits must have decoherence times much longer than the times required to operate the computer's quantum gates. Decoherence is simply quantum noise. In our somewhat idealized discussion so far, we have implicitly assumed that the quantum mechanical systems under discussion (our system of qubits) is *closed*, that is, aside from the well controlled operations we perform on the system in computation, it is perfectly isolated from the rest of the universe. Unfortunately, no such system exists in reality. In fact arguably, the greatest problem in developing quantum computers is in the apparently contradicting requirements that the qubits of the system must be almost perfectly isolated from their environment while simultaneously interacting strongly with each other and also interacting strongly with some measurement device.

Decoherence times are typically described in terms of a phase coherence time T_2 and an energy relaxation time T_1. Here, T_2 would, for a one qubit system, represent a time corresponding to a phase shift of π; whereas T_1 would represent a time corresponding to a shift for a $|1>$ to a $|0>$.

(iv) There must be a "universal" set of quantum logic gates. A quantum logic gate is an input-output device. The inputs and the outputs of this device are quantum mechanical variables such as spin. The action of a quantum gate on an input is described by a unitary operator, U. Quantum gates can be connected together to form a quantum circuit. That is, the output variable from one gate can be moved to the input of another gate.

A quantum gate is "universal" if copies of it can be connected together to evaluate any classical logical function and to enact any unitary transformation on a set of quantum variables.

Quite simple quantum logic gates are universal; Lloyd[9] has shown that almost any quantum logic gate with two or more inputs is a universal gate. If qubits are envisioned as electron spins or spin-1/2 nuclei, we could consider a one qubit gate in terms of the Bloch sphere representation. In the Bloch sphere representation of a two level system, for example a spin-1/2 nucleus, the state of the spin is represented by a point on the surface of the sphere. One qubit gates cause rotations on the sphere as a result of radiofrequency pulses corresponding at the magnetic resonance frequency of the spin.

A two qubit gate involves interaction between two particles, for example two spins. A particularly simple example is given by two coupled nuclear spins in a liquid. The resonance frequency of spin A is affected by the spin quantum number of spin B and vice versa. The most obvious way to utilize a two spin gate would be to apply an rf pulse at a frequency which will excite spin A only if spin B is in one of its two spin eigenstates[10].

(v) It must be possible to measure the state of the specific qubits. This is another common sense requirement. Having carried out the computation, we gain nothing if we can't measure the result. Although this requirement is obvious, the process by which the requirement may be achieved may not be straightforward.

5. Quantum Error Correction

One of the greatest obstacles in the path of quantum computation is decoherence or quantum noise. The accumulation of errors in computation could easily render the outputs of computation meaningless. However, effective schemes to correct errors have been developed, at least in principle. These schemes must work very well. Estimates of the required accuracy range from 10^{-4} to 10^{-6} per operation[12].

In order to protect "quantum information" from decoherence or quantum noise, the information is encoded with additional information. Even if some of the information is lost, there will be enough redundancy to extract the encoded information.

An extremely simple example of this redundancy would be to replace one bit with three copies. That is, replace a 0 with 000; replace a 1 with 111.

We process all three bits, and measure them. Suppose that we input 000 and that whatever operation performed would always transform a 0→1. if the probability of error is precisely zero, we will inevitably measure output of 111. However, suppose thtat there is a small but finite possibility of error. In this case it is possible that the output would be 110 or 101 or 011 (one error among the three digits). It is far less likely that a second error would occur. Thus, taking the majority vote, 1, would with very high probability, provide the correct output.

Far more complex error correcting codes have been developed; however, most or all are based upon this redundancy principle.

6. A Leading Candidate for Quantum Computation: Magnetic Resonance

As of yet, no practical quantum computer has been built. However, numerous proposals, many quite ingenious, have appeared in the recent literature. Among the most promising ideas are those involving ion traps[14], quantum dots[15], quantum waveguide arrays[16], liquid nuclear magnetic resonance (NMR)[17,18], and solid state magnetic resonance involving both electron spin resonance (ESR) and NMR[19,20].

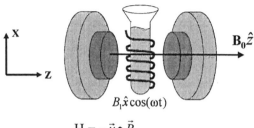

$$H = -\vec{\mu} \bullet \vec{B}$$
$$= B_0\hat{z} + B_1\left(\hat{x}\cos(\omega t)\right)$$

Fig. 1. A simplified schematic of an NMR system; NMR involves a static magnetic field and an RF field.

Although it is not yet clear which, if any, of these candidates will be the winner, or even if there will be a winner, various magnetic resonance schemes appear to be leading contenders. This is so because both NMR and ESR based systems offer relatively long decoherence times, well characterized qubits, and well established means for manipulating spins (quantum gates) and for qubit measurement.

6.1. *Liquid State NMR*

Liquid state NMR had already been utilized to build very small quantum computers[17,18]. It has been the first quantum computer approach to actually implement quantum algorithms. Most notably, Vandersypen *et al.*[17] have successfully demonstrated Shor's algorithm with a seven qubit liquid NMR computer, demonstrating the simplest

possible case for the algorithm: 15=3x5. Although a detailed description of this benchmark study is beyond the space limits of this paper, the basic principles are not.

Recall that any quantum computation may be carried out if one may couple any type of "universal" quantum gates. A universal quantum gate readily accessible to NMR is the CNOT gate[9], which can be constructed from the coupling between two spin-1/2 nuclei. A little background in NMR is required to understand how this works[21]. A rudimentary NMR spectrometer is schematically illustrated in figure 1. Note the presence of a very large static field $B_0\hat{z}$ and a small rf field $B_1\hat{x}\cos(\omega t)$. It is easy to understand the operation of the spectrometer for the case of a single uncoupled spin.

The Hamiltonian for an isolated spin in a magnetic field $\vec{B}$ is

$$H = -\vec{\mu} \bullet \vec{B}, \tag{9}$$

where $\vec{\mu}$ is the magnetic moment of the spin. The magnetic moment is parallel and proportional to the particles angular momentum $\vec{J}$, so

$$\vec{\mu} = \gamma \vec{J}, \tag{10}$$

where γ si a scalar called the gyromagnetic ratio. The angular momentum is generally expressed in terms of a dimensionless angular momentum operator $\vec{I}$ such that

$$\vec{J} = \hbar \vec{I}. \tag{11}$$

With only a static magnetic field B_o applied along the z-direction, we obtain a very simple Hamiltonian from (10):

$$H = -\vec{\mu} \bullet \vec{B} = -\gamma \hbar B_o I_Z. \tag{12}$$

The energy eigenvalues of this Hamiltonian are multiples of the eigenvalues of I_Z: m = I, I-1, I-2... -I. For the relevant case of a spin 1/2 nucleus, m = +1/2, -1/2. This results in two energy eigenvalues $-\gamma \hbar \frac{B_o}{2}$ and $+\gamma \hbar \frac{B_o}{2}$.

Transitions between these two states may be induced by the perturbing Hamiltonian due to the small rf magnetic field $B_1\hat{x}\cos(\omega t)$:

$$H_{pert} = -\gamma \hbar B_1 \cos(\omega t) I_x. \tag{13}$$

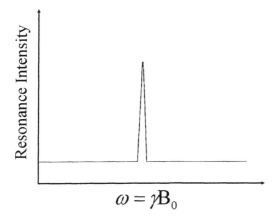

Fig. 2. A schematic illustration of an NMR spectrum of an isolated nuclear spin, for example, an isolated proton.

This perturbing Hamiltonian induced transitions between the two states when energy conservation is satisfied, that is when $\hbar\omega = \Delta E = \gamma\hbar B_o = \hbar\omega$. Thus, the resonance condition for an isolated spin is simply

$$\omega = \hbar B_o \tag{14}$$

If we were to plot resonance versus frequency, for a single spin, figure 2 would result.

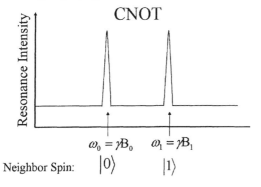

Fig. 3. The effect of a second nuclear spin on the NMR spectrum. (Two coupled spin $\frac{1}{2}$ particles can implement a CNOT gate.)

Now consider the effect of a second nuclear spin nearby. The details of the coupling between the two spins depend upon the details of the experiment. However, for the purposes of this discussion, the important point to realize is simply that the spin quantum number of the second spin alters the resonance condition; that is, it shifts the resonance

frequency at constant field or resonance field at a constant frequency of the first spin. We might envision, for example a spin-1/2 hydrogen nucleus coupled to a spin-1/2 ^{13}C nucleus. The spin state of the carbon nucleus determines the resonance frequency of the hydrogen, as illustrated in figure 3. We can think of this as a controlled NOT gate or a CNOT gate[22].

Suppose the rf generator is set to one of the two hydrogen resonance frequencies. The CNOT gate performs a NOT operation on one bit, corresponding to the hydrogen, flipping it form a 0 to a 1 or a 1 to a 0 if and only if the value of the second bit, corresponding to the ^{13}C is, for example, a 1. A truth table for this gate is illustrated in Table 3.

Table 3
CNOT Truth Table. A and B represent the two inputs.

	A=0	A=1
B=0	0	1
B=1	1	0

As discussed previously, *any* computation can be carried out with a sufficiently large array of universal quantum gates, in this case, pairs of spin-1/2 nuclei; with the appropriate application of rf pulses, to "flip" the spins can carry out any quantum calculation. The above discussion is slightly oversimplified. (A more detailed discussion is provided by J.A. Jones[10] and by Vandersypen et al.[17].)

Although much has been accomplished with liquid state NMR quantum computing, it has serious, probably insurmountable problems. One very big problem is temperature. Liquid state NMR can only be carried out on *liquids*. The relatively high temperatures mean relatively small differences in the thermodynamic equilibrium population of the various states. This leads to quite low sensitivity for a large number of qubits (nuclear spins). Another problem is that the difference in excitation frequency of spins involved, decreases sharply with increasing numbers of spins in the computing molecule[10]. (The problem is, fundamentally measuring the spins in a system sufficiently large to allow for practical calculations.) These and other problems are variously estimated to limit the ultimate size of liquid state NMR computers to somewhere between 10 and 30 qubits, numbers small

243

enough to preclude practical quantum computation on liquid state NMR.

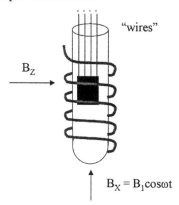

Fig. 4. A block diagram of the Kane computer: a silicon based integrated circuit at very low temperature ($T \approx 100$ milliKelvin) in an NMR system.

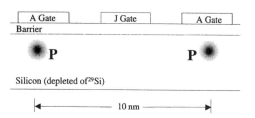

Fig. 5. The physical basis of the Kane computer: phosphorous donor atoms precisely placed close to one another near a gated silicon surface.

Magnetic resonance, in solids, both electron spin resonance (ESR) and solid state NMR may overcome these obstacles. Although solid state magnetic resonance quantum computing is in its infancy, it holds considerable promise. No quantum calculations have as yet been carried out in solid state resonance based systems, but several ideas under investigation may eventually allow for far more complex computation.

6.2. Solid State Magnetic Resonance

Kane[19] recently proposed an ingenious solid state magnetic resonance "quantum computer" based upon closely spaced (≈ 100Å) phosphorus donor atoms in silicon. The system utilizes solid state NMR at quite low $\cong 100$ milliKelvin temperature. His proposal is schematically illustrated in

the block diagram of figure 4 and the more detailed schematic of figure 5. Since the phosphorus donor electron is in a rather highly delocalized wave-function, the shape of the wave-function can be altered by application of a gate voltage. A positive gate voltage, for example, would pull the donor electron towards the gate away from the spin-1/2 ^{31}P nucleus. The NMR resonance condition of the ^{31}P depends upon both the spin state and wave-function of the donor electron. Without the unpaired donor electron nearby, the ^{31}P NMR spectrum would be essentially a single sharp line. The donor electron spin, in effect, supplies a local field through the electron nuclear-hyperfine interaction. This local field would either add to or subtract from the applied field density on the electron spin quantum number: $|1>$ an increased field and increased resonance frequency; $|0>$ a decreased field and decreased resonance frequency. To first order, the size of this interaction, called the hyperfine interaction, may be expressed in the Fermi –contact interaction[21]

$$A = \frac{8}{3}\pi\mu_B g_n \mu_n \left|\Psi(0)\right|^2 \tag{15}$$

where μ_B is the electron Bohr magneton, g_n and μ_n are the appropriate (phosphorus) g factor and nuclear Bohr magneton, and $\Psi(0)$ represents the electron wave function of the ^{31}P nucleus. Since $\Psi(0)$ can be varied with gate voltage, the NMR resonance frequency may be controlled by the gate.

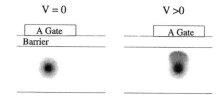

Fig. 6. The effect of the Kane computer A gate: the donor electron wave function is shifted by the gate bias – altering the electron nuclear hyperfine interaction and thus the resonance condition of the phosphorous nucleus. The NMR frequency can be controlled by the gate, allowing specific gates to be accessed with specific RF frequencies.

This process is schematically illustrated in figure 6. The application of a positive voltage to the A-gate distorts the ^{31}P electron wave-function and the electron is attracted to the gate reducing the electron nuclear hyperfine interaction, shifting the ^{31}P NMR frequency towards that of the isolated ^{31}P nucleus.

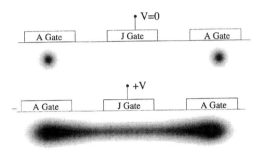

Fig. 7. In the Kane computer, coupling between the phosphorous nuclei can be controlled through indirect electron-nuclear coupling – via application of voltages to the J gates.

Coupling between two adjacent ^{31}P nuclei can be varied from very weak to quite large by application of a voltage to the J-gate between them. As schematically illustrated in figure 7, a positive J-gate voltage attracts the donor electron wavefunction into the J-gate region, substantially increasing the overlap between the electron wavefunctions. As shown, by Ruderman and Kittel[23] and Bloembergen and Rowland[24] many years ago, an indirect nuclear spin-nuclear spin coupling can be established through electrons. In Kane's proposal, the J-gate bias effectively turns the coupling on and off.

Since two coupled spins provide a CNOT gate, a universal gate, the array of phosphorus atoms under A- and J-gates can provide the basis for quantum computing, if the spin states can be measured.

Although virtually every aspect of the Kane computer involves extremely challenging engineering problems, the spin measurement scheme is probably the most daunting. If the donor electron or adjacent phosphorus atoms have opposite spin orientations ($\uparrow\downarrow$) both electron can occupy the donor orbital of either phosphorus. If, on the other hand both electrons have the same spin quantum number, this can't occur because of the Pauli exclusion principle. The electron and nuclear spins are strongly correlated under some biasing conditions; therefore, the nuclear spin quantum numbers could be discerned, if one could distinguish between the parallel/anti-parallel or triplet/singlet states of the two electrons. Kane points out that the *capacitance* of the A-gate P system is larger for the singlet ($\downarrow\uparrow$) electron state. An exceedingly sensitive (but in principle possible) capacitance measurement could provide this data.

Thus, in principle, the Kane proposal constitutes a reasonably complete plan for a quantum computer. Although the plan is reasonably complete, the actual implementation of the plan would be extremely difficult on any scale. The plan requires single phosphorus atoms precisely placed about 100Å from one another and gate lengths of much less than 50Å are required. It is difficult to see how one might avoid quite large capacitive coupling between adjacent gates. Perhaps the most challenging aspect in the implementation of the Kane computer is the measurement of quite small capacitance changes resulting from the response of the phosphorus donor electron. Although not a problem in principle, the requirement of very low temperatures, T≈100milliKelvin would make the operation of such a computer quite complex.

Fig. 8. The Vrijen *et al.* computer is based on a Si/Ge heterostructure with phosphorous donors on the boundary between regions of composition, yielding different g tensors for the donor electrons.

Vrijen *et al.*[20] have suggested a solid state quantum computer which is in some respects quite similar to that of Kane but which appears to overcome some of the greatest challenges to its practical implementation. The Vrijen computer also utilizes phosphorous donors and nearby biasing gates, but in their system the P donor is in a silicon germanium superlattice, on the boundary between SiGe layers of quite different composition. The devices is schematically illustrated in figure 8.

The SiGe heterostructure offers several advantages. The electron effective mass can be quite small in these heterostructures; the dielectric constant, κ, can also be considerably larger than that of silicon. These two factors considerably increase the Bohr radius, r_B,

$$r_B \cong \left(\frac{h^2 \varepsilon_0}{\pi q^2}\right) \frac{\kappa}{m} \qquad (16)$$

Since the dimensions of the device scale approximately with the Bohr radius, the approximately one order of magnitude increase in r_B makes practical implementation significantly more feasible.

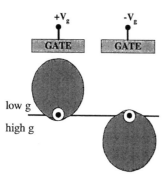

Fig. 9. The application of a positive or negative gate voltage shifts the resonance condition, making it possible to access specific sites with a specific ESR microwave frequency.

In the Vrijen *et al.* device, only one type of gate is required. The application of a positive voltage does two things. First, as illustrated in figure 9, it pulls the donor electron wave function toward the low q side of the heterostructure, altering the field or frequency at which electron spin resonance (ESR) may be detected. In ESR, $\omega = g\beta H/\hbar$, where β is the Bohr magneton and H is the magnetic field at resonance. In general, g is a tensor of second rank. For the purposes of this discussion, however, it may be regarded as a scalar in this case. The ability to arbitrarily adjust the resonance condition at a specific site allows one to address a specific gate with a specific combination of applied field and microwave frequencies.

The gates play a second role illustrated in figure 10. With increasing positive gate voltage, increasing overlap in the donor electron wave functions can occur. This overlap provides for an exchange coupling between the two electrons. Again, two coupled spin 1/2 particles, here two electrons, provide the basis for a CNOT gate, a "universal" gate. From a system of universal gates one may, in principle, construct a quantum computer.

A particularly appealing aspect of the Vrigen model is its measurement technique, illustrated in

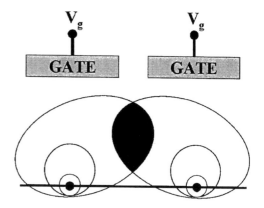

Fig. 10. Coupling between the adjacent donor electrons can be controlled by the gate voltages. The two coupled electrons can implement a CNOT gate.

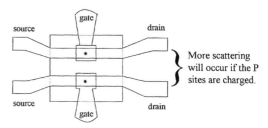

Fig. 11. Detection of the electron spin states in the Vrijen *et al.* computer. The singlet state electrons can both occupy one of the two phosphorous donor orbitals the triplet state cannot. If both electrons occupy one orbital, both the donor sites will be charged, one positively, one negatively. The charged sites will lead to more scattering and less source drain current. Thus one may detect the electron spin states non-destructively via current measurement.

figure 11. (Conduction below the P atoms carry current paths.)

As in the Kane proposal, measurement involves differentiating between the P electron singlet ($\uparrow\downarrow$) and triplet ($\uparrow\uparrow + \downarrow\downarrow$) states. In the singlet state, both electrons can occupy one of the P sites. This renders one site negative, the other site positive. The charged sites will increase scattering of charge carriers in both source drain paths. In the triplet state, ($\uparrow\uparrow + \downarrow\downarrow$), both P sites remain neutral, leading to less scattering through the source drain channels. Thus *many* electrons can be utilized to sample the spin state of the P sites.

The Vrigen *et al.* proposal is, like the Kane proposal, quite clever and reasonably complete. Nevertheless, practical implementation would also be extremely challenging. As in Kane's proposal, extremely low temperatures are required for operation. Although the device dimensions are not as challenging as those of the Kane proposal, they still involve complex heterostructures with dimensions of tens to of order one hundred nanometers. Another problem (not as acute as in the Kane proposal) is the accurate measurement of the electron spin quantum numbers.

As pointed out by Kane[19], "The primary challenge in using nuclear spin in quantum computing lies in measuring the spins. There are several well established solid state resonance ideas and techniques which might help address this primary challenge. The techniques are spin dependent recombination (SDR)[25-29] and Pound-Overhauser[30-34] double resonance. Combining the advantages of SDR and solid state double resonance with recent advances in ultra high frequency ESR[35], it could be possible to achieve single electron and simple nuclear spin detection using essentially conventional MOS technology.

By combining this very high ESR sensitivity SDR with Pound-Overhauser double resonance techniques[30-34] and a novel but quite simple approach to the measurement of nuclear spin quantum numbers, it should be possible to: (1) program an initial set of spin quantum numbers, (2) carry out an almost arbitrarily complex sequence of nuclear spin manipulations and (3) then evaluate nuclear spin quantum numbers via robust changes in a current or voltage. (This could provide a basis for practical quantum computing in the relatively near future, since essentially all the building blocks are already available.)

SDR can be observed in semiconductor devices in which the current is dominated by electron-hole recombination[25-29]. Electron/hole recombination in indirect semiconductors like silicon proceeds via electron and hole capture at deep levels[36,37]. Elegant work published long ago by Grove and coworkers[38,39] demonstrated how MOS devices could be configured so as to yield currents dominated by recombination processes at and very near the Si/dielectric boundaries of these devices. This can be done by configuring the devices as gate controlled diodes and applying appropriate bias to the diode junction and gate contacts.

At very high magnetic field strength, the simple SDR model proposed by Lepine[28] should provide a lower limit for the magnitude of the effect. This is so because the Lepine model is based upon very basic physical principles. In short, when a conduction electron is captured by a paramagnetic deep level defect, say a dangling bond, the capture event is spin dependent. This is so because the Pauli exclusion principle does not allow occupation of the dangling bond orbital by two electrons with the same spin quantum number. If both the conduction electron and dangling bond electrons "line up" with an applied magnetic field, the electron capture event will be forbidden. (Essentially the same argument would hold for hole capture at the site.)

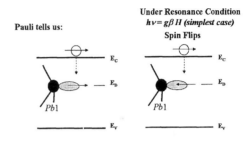

Fig. 12. The Lepine model for spin dependent recombination is schematically illustrated. Two spin parallel electrons cannot occupy the same deep level orbital whereas two anti-parallel spins can.

The application of a large magnetic field produces a decrease in the recombination dominated current. In the Lepine model, this decrease is equal the product of the electron (or hole) polarization and the polarization of the deep level centers. If we satisfy the ESR condition of the deep level "dangling bond" site, the electrons at the site are "flipped" and the recombination rate is thus increased. If the spin temperature of the deep level system is greatly increased, the recombination current approaches the larger value observed with no applied magnetic field. This process is illustrated in figure 12. (A *much larger* effect is actually observed at modest field strength, for reasons which have been discussed in the literature[31].)

For a system of N weakly interacting electrons, the ratio of the electron population, N_L, in the

248

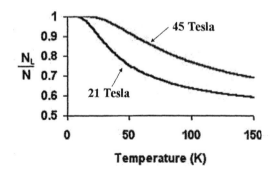

Fig. 13. A plot of electronic polarization versus absolute temperature. Utilizing the Overhauser effect, nuclear polarizations can equal these values.

lower spin energy level to the total N is given by[40]

$$\frac{N_L}{N} = \frac{\exp\left(2\beta H/kT\right)}{\exp\left(2\beta H/kT\right) + \exp\left(-2\beta H/kT\right)} \quad (17)$$

where both g values are taken to be 2, β is the Bohr magneton, H is the applied magnetic field, k is Boltzmann's constant, and T is the absolute temperature. If $2\beta H/kT = 1$, this ratio is 0.88. Commercially available superconducting magnets generate up to about 21 Tesla. (Much higher fields are available (45 Tesla) at certain specialized facilities, for example, at the U.S. National High Magnetic Fields Laboratory in Tallahassee, Florida). At 21 Tesla $2\beta H/kT = 1$ at 28 K; at 45 Tesla, $2\beta H/kT = 1$ at 60 K. Expression 18 is evaluated as a function of temperature in figure 13. Recent advances in very high frequency ESR allows application of the time varying magnetic fields appropriate for resonance at these fields[35]. (Again if we simply take g = 2, the resonance frequency νgs $2\beta H/h$, where h is Planck's constant.) The large electron polarization results in a giant SDR effect. At sufficiently high fields and low temperatures, that is when $2\beta H/kT \geq 1$, at an appropriately biased MOS gate controlled diode could be turned essentially "on" or "off" by, respectively, applying or removing microwave radiation satisfying the resonance condition of the dominating deep level defect. At the highest possible fields, quite large fractional changes ($\approx$50%) in current could be achieved at temperatures of about 100 K. At the highest fields, significant current changes would be observed even at room temperature.

The large electronic polarization also provides the opportunity to generate an enormous nuclear spin polarization via the Overhauser effect. As originally proposed by Overhauser[30,31] and later demonstrated by Carver and Slichter[32], through straightforward double resonance, it is possible to effectively achieve nuclear polarizations about equal to that of electrons. As figure 13 illustrates, it should thus be possible to achieve very high nuclear spin polarization at moderately low temperatures, if the ESR conditions can be met. Recent developments in high field ESR demonstrate this possibility[35].

So it should be possible to simultaneously achieve a giant SDR effect and a giant nuclear polarization. This combination allows for a novel means to evaluate nuclear spin quantum numbers. Consider an MOS gate controlled diode in which the Si/SiO_2 interface region deep levels are dominated by sites in which ^{29}Si sites are bonded to one hydrogen and two oxygen atoms. The ^{28}Si/hydrogen site has already been observed in an MOS device with SDR[41].

Four possible sets of nuclear spin quantum numbers exist for this site. With near complete nuclear polarizations possible one could initialize the two spin system with the ^{29}Si and ^{1}H nuclei energy spin quantum numbers, say $|->$, and with an appropriate rf pulse sequence generate any of the four possible combinations $|->$, $|-+>$, $|+->$, $|++>$, including the initial state. Each of the four combinations would yield quite different ESR resonance conditions.

The four possible sets of nuclear spin quantum numbers could be determined by sweeping the magnetic field over the tens of milliTesla range over which the electron-nuclear hyperfine interactions would shift the ESR resonance condition. At quite high fields, one would observe an extremely large increase in the current of the appropriately biased gate controlled diode when the ESR resonance condition is satisfied. This observation yields the nuclear spin quantum numbers. One could then envision implementing pairs of the nuclear spins as CNOT gates. The output of computation via these gates could be measured via an SDR electron current.

More complex nuclear spin/defect systems can be imagined. It is almost certain that near interface paramagnetic centers involving three nitrogen atoms bonded to a silicon can dominate deep levels near the Si/dielectric interface under appropriate circumstances[42]. Such a site could involve four "entangled" nuclear spins; if the silicon atom nucleus were to be ^{29}Si and the nitrogens ^{15}N, the system would have four spin $- 1/2$ nuclei. If the nitrogen sites were ^{14}N, with spin of one, more computational capability (and complexity) would be available.

The enormous efforts involved in the development of a near perfect Si/SiO$_2$ boundary region allow for an extremely well controlled environment suitable for well-isolated systems of several entangled nuclear spins. For example, consider a deep level defect in the oxide very near the Si/SiO$_2$ boundary. Essentially all oxygen atoms possess spin-zero nuclei. The natural abundance of spin-zero ^{28}Si is about 95%. Although some (nuclei spin-1/2) hydrogen will inevitably be present, to a great extent specifically introduced deep level defects will be only weakly coupled to their surroundings by nuclear spin-spin interactions. The electron paramagnetism of the sites can be controlled by biases applied to the gate and junction. The density of paramagnetic charge carriers near these sites can also be controlled by applied voltages and by the presence or absence of light (hc/$\lambda \geq E_g$ (silicon) $\cong 1.1$eV). The defect sites involved in this scheme therefore have nuclear spin systems which can be well isolated from their surroundings during "computation," that is during application of rf pulses to manipulate the nuclear spin quantum numbers.

7. Conclusions

Quantum computation offers great potential advantages in the solution of certain problems, most notably in the factoring of very large numbers. At the present time, quantum computation is primarily a "theoretical" topic as no practical quantum computer has yet been built. Although enormous challenges must be overcome, it appears that there is some possibility of achieving quantum computation.

References

[1] R.P. Feynmann, *Intr. J. Theor. Phys.*, **21**, 467 (1982)

[2] D. Deutsch, *Proc. R. Soc. Lond.* **A400**, 97-117 (1985)

[3] P.W. Shor, in *Proc. 35th Annual Symposium on Foundations of Computer Science.* (ed. by Goldwasser, S.) 124-134 (IEEE Computer Society, Los Alamitos, CA 1994).

[4] A. Ekert and R. Jozsa, *Reviews of Modern Phys.* **68**, 733 (1996)

[5] L.K. Grover, Phys. Rev. Lett., **79**, 4709 (1997)

[6] An excellent review, particularly of the theoretical aspects of the field, is provided by M.A. Nielsen and I.L. Chuang, *Quantum Computation and Quantum Information*, Cambridge University Press, Cambridge (2000)

[7] D.P. DiVincenzo, *Fortschr. Phys.* **48**, 771 (2000)

[8] J. Baggot, *The meaning of Quantum Theory*, Oxford Science Publication, Oxford, U.K. (1992)

[9] S. Lloyd, *Phys. Rev. Lett.* **75**, 346 (1995)

[10] J.A. Jones, *Fortschr. Phys.* **48**, 909 (2000)

[11] A.M. Steane, *Nature*, **399**, 124 (1999)

[12] J. Preskill, *Proc. R. Soc. Lond. A*, **454**, 469 (1998)

[13] D.P. DiVincenzo and D. Loss, *J. Mag. and Magn. Matls.*, **200**, 202 (1999)

[14] D.J. Wineland, C. Monroe, D.M. Meekhof, B.E. King, D. Leibfried, W.M. Itano, J.C. Bergquist, D. Berkeland, J.J. Bollinger and J. Miller, *Proc. R. Soc. Lond. A*, **454**, 411 (1998)

[15] L.B. Ioffe, V.B. Geshkenbein, M.V. Fiegel'man, A.L. Fauchere, and G. Blatter, *Nature*, **398**, 679 (1999)

[16] R. Akis and D.K. Ferry, *Appl. Phys. Lett.*, **79**, 2823 (2001)

[17] L.M.K. Vandersypen, M. Steffen, G. Breyta, C.S. Yannoni, M.H. Sherwood, and I.L. Chuang, *Nature*, **414**, 883 (2001)

[18] I.L. Chuang, N. Gershenfeld, and M. Kubinec, *Phys. Rev. Lett.*, **80**, 3408 (1998)

[19] B.E. Kane, *Nature* **393**, 133 (1998)

[20] R. Vrijen, E. Yablonovitch, K. Wang, H.W. Jiang, A. Balandin, V. Roychowdhury, T. Mor, and D.P.. DiVincenzo, *Phys. Rev. A* **62**, 12306.1-12306.10 (2000)

[21] C.P. Slichter, *Principles of Magnetic Resonance*, Second Edition, Springer-Verlag. Berlin (1978)

[22] D.P. DiVincenzo, *Phys. Rev.* **A51**, 1015 (1995)

[23] N. Bloembergen and T.J. Rowland, *Phys. Rev.*, **97**, 1679 (1955)

[24] M.A. Rudermann and C. Kittel, *Phys. Rev.* **96**, 99 (1954)

[25] P.M. Lenahan, and M.A. Jupina, *Colloids and Surfaces* **45**, 211 (1990)

[26] J.W. Gabrys, P.M. Lenahan, and W. Weber, *Microelectronic Eng.* **22**, 273 (1993)

[27] T.D. Mishima, and P.M. Lenahan, *IEEE Trans Nucl. Sci.* **47**, 2249 (2000).

[28] D.J. Lepine, *Phys. Rev.* **B6**, 436 (1972)

[29] D. Kaplan, I. Solomon, and N.F. Mott, *J. Phys. Lett* (Paris) **39**, L51 (1978)

[30] A.W. Overhauser, *Phys. Rev.* **89**, 689 (1953)

[31] A.W. Overhauser, *Phys. Rev.* **92**, 411 (1953)

[32] T.R. Carver, and C.P. Slichter, *Phys. Rev.* **92**, 212 (1953)

[33] C.D. Jeffries, *Phys. Rev.* **106**, 164 (1957)

[34] C.D. Jeffries, *Phys. Rev.* **117**, 1056 (1960)

[35] A.K. Hassan, L.A. Pardi, J. Krzystek, A. Sienkiewicz, P. Goy, M. Roher, and L.C. Brunel, *Journal of Magnetic Resonance* **142**, 300 (2000)

[36] W. Shockley, and T.W. Read, *Phys. Rev.* **87**, 835 (1952)

[37] R.N. Hall, *Phys. Rev.* **87**, 387 (1952)

[38] A.S. Grove, D.J. and Fitzgerald, *Solid State Electronics* **9**, 783-806 (1966)

[39] E.H. Snow, A.S. Grove, and D.J. Fitzgerald, *Proc. IEEE* **55**, 1168 (1967).

[40] C. Kittel, *Introduction to Solid State Physics,* Fifth Edition. John Wiley and Sons, New York (1976) Chapter 14

[41] T. Winbauer, K. Ito, Y. Mochizuki, M. Horikawa, T. Kitano, M.S. Brandt, M. Stutzmann, *Appl. Phys. Lett.* **76**, 2280 (2000).

[42] N.L. Cohen, R.E. Paulsen, M. White, *IEEE Trans Electron Dev.* **42**, 2004 (1995).

[43] Electronic aspects of deep level defects in Si/SiO_2 systems are reviewed by E.H. Nicollian, J.R. Brews, in *Metal Oxide Semiconductor (MOS) Physics and Technology.* John Wiley and Sons, New York (1982).

INDEX

Printed and bound by CPI Group (UK) Ltd, Croydon, CR0 4YY

14/05/2025

01871478-0001